COMPLEX ANALYSIS AND APPLICATIONS

SECOND EDITION

THE WADSWORTH MATHEMATICS SERIES

Series Editors
Raoul H. Bott, Harvard University
David Eisenbud, Brandeis University
Hugh L. Montgomery, University of Michigan
Paul J. Sally, Jr., University of Chicago
Barry Simon, California Institute of Technology
Richard P. Stanley, Massachusetts Institute of Technology

W. Beckner, A. Calderón, R. Fefferman, P. Jones, *Conference on Harmonic Analysis in Honor of Antoni Zygmund*
M. Behzad, G. Chartrand, L. Lesniak-Foster, *Graphs and Digraphs*
J. Cochran, *Applied Mathematics: Principles, Techniques, and Applications*
W. Derrick, *Complex Analysis and Applications*, Second Edition
A. Garsia, *Topics in Almost Everywhere Convergence*
K. Stromberg, *An Introduction to Classical Real Analysis*
R. Salem, *Algebraic Numbers and Fourier Analysis*, and L. Carleson, *Selected Problems on Exceptional Sets*

COMPLEX ANALYSIS AND APPLICATIONS

SECOND EDITION

WILLIAM R. DERRICK
University of Montana

Wadsworth International Group
Belmont, California
A Division of Wadsworth, Inc.

Acquisitions Editor: John Kimmel

Production Editor: Andrea Cava

Copy Editor: Janet Greenblatt

Technical Illustrator: Blakeley Graphics

Cover and Interior Design: Lois Stanfield

Typesetter: RDL Artset Limited, Sutton, Surrey, England

ISBN 0-534-02853-5

Printed in the United States of America

1 2 3 4 5 6 7 8 9 10——88 87 86 85 84

Library of Congress Cataloging in Publication Data

Derrick, William R.
   Complex analysis and applications.

   (Wadsworth mathematics series)
   Rev. ed. of: Introductory complex analysis and applications. 1972.
   Bibliography: p.
   Includes index.
   1. Mathematical analysis. 2. Functions of complex variables.
   I. Title. II. Series.
QA300.D455   1983   515.9   83-6865
ISBN 0-534-02853-5

CONTENTS

PREFACE

Complex analysis is one of the most fascinating and successful areas of mathematics. Its results help prove important theorems and provide the inspiration for many useful concepts in other areas of mathematics. Many of the most powerful techniques used in the application of mathematics to other sciences and engineering are based on complex function theory. Because of its wide applicability, its blend of geometric and analytic concepts, and the simplicity of many of its results, complex analysis provides an excellent introduction to modern mathematics. Recent developments in complex function theory and in the theory of several complex variables promise to provide useful applications in many areas of engineering.

One of my goals in writing this book has been to reach the topic of complex integration as quickly as possible. This requires delaying the treatment of the geometric properties of elementary functions. The advantages of reaching the heart of the subject rapidly are obvious: The subsequent development is richer in applications and broader in its treatment of series, singularities, and contour integrals. Furthermore, the cost of delaying the geometric properties is greatly reduced because they can now be viewed as conformal mappings. Thus, greater cohesion of the topics is achieved, and the time gained can be used in other directions.

Another goal in writing this book has been to provide a wider choice of applications and a more comprehensive range of techniques than those found in the traditional texts. I have included applications in optics, jet flows, and wakes, as well as the traditional examples in fluid and heat flow and electrostatics. The contour integral techniques given provide ample background for calculating Regge poles and inverse Laplace transforms. The text includes a development of integral transforms, a subject that is usually studied in a real-variable setting but assumes far greater significance when viewed as complex variables. It is not intended that all these topics be covered; rather, the material is included for instructors to tailor the course to their interests.

ORGANIZATION AND COVERAGE

This book is intended as a text for a one-term junior-level introductory course in complex analysis. Considerably more material has been included than can be covered at leisure in one term, thereby giving instructors the opportunity to select the topics they deem most important.

Chapters 1 through 5 include most of the material that is covered in a one-term introductory course. Instructors who want to minimize the theoretical aspects of the course should omit the optional sections 2.5 and 3.5. Those wishing to omit lengthy applications should avoid the optional sections 1.10, 4.5, 5.7, and 5.8. A brief treatment of harmonic functions is included in Section 6.1 and can be studied anytime after the material in Section 2.3 has been presented.

Chapter 6 is an introduction to integral transforms from a complex variable viewpoint. It is hoped that some instructors will elect to introduce some of these topics into their course syllabus. Integral transforms are extremely powerful techniques in the sciences and engineering. This introduction should prepare students for advanced courses in applied mathematics.

A syllabus for a one-term course could include:

For Engineering Students
Chapters 1 through 5 including
Section 1.10 or 4.5 or 5.7–5.8
plus Sections 6.1 and 6.3 or
6.6–6.7

For Mathematics Students
Chapters 1 through 5 including
Sections 2.5 and 3.5 plus
Sections 6.2 and 6.4–6.5

LEVEL

The text has been written for the "average" engineering student. Special care has been taken to explain each concept as clearly as possible: Each concept is preceded by a motivating example or discussion. Each section includes a number of fully worked out examples. The more challenging proofs are identified by a dagger (†) or have been placed in the optional sections. The book can be used at several different levels, depending on the sections and examples that are chosen.

ACCURACY

All examples and answers in this text have been carefully checked by the author and several reviewers to prevent errors. I would appreciate your bringing to my attention any errors that remain, and I pledge to incorporate all corrections in the next printing of the book.

IN-TEXT PEDAGOGY

EXAMPLES
Numerous worked examples, ranging from the most immediate to fairly sophisticated applications, are included in each section. Spacing is provided to separate examples from other material in the text.

EXERCISES
Special care has been taken in preparing the exercise sets to guarantee that each exercise provides a valuable learning experience. The more challenging exercises are identified by an asterisk (*). Each set contains a full range of exercises presented in increasing order of difficulty. Many of the exercises consist of useful results and theorems; the instructor is urged to carefully select those that will most benefit the class.

Solutions are provided for the odd numbered exercises. These are not simply answers, nor are they completely worked out solutions;

instead, they are hints on a direction that can be taken to achieve the given answer. Students are urged to try their own ideas *before* using the hints provided. An Instructor's Manual containing solutions for the even numbered exercises is available from the publisher.

CHAPTER NOTES

Each chapter concludes with a brief indication of other results and sources of collateral and supplementary material. Interested students are urged to examine these leads for a deeper understanding of the material.

SYMBOL TABLE AND APPENDIXES

A table of symbols used in this book follows this preface. The appendixes at the end of the book contain tables of conformal mappings, Laplace transforms, and a brief review of line integrals and Green's theorem.

ACKNOWLEDGMENTS

I wish to thank the following individuals for their many helpful suggestions in reviewing various drafts of this manuscript: Janos Aczel (U. of Waterloo), Eric Bedford (Princeton U.), Douglas Campbell (Brigham Young U.), Michael O'Flynn (San Jose State U.), Donald Hartig (Cal Poly San Luis Obispo), Harry Hochstadt (Polytechnic Institute of New York), William Jones (U. of Colorado), John Kogut (U. of Illinois), Steven Krantz (Pennsylvania State U.), E. Macskasy (U. British Columbia), James Morrow (U. of Washington), David Sanchez (U. of New Mexico), Franklin Schroeck, Jr. (Florida Atlantic U.), Jerry Schuur (Michigan State U.), Brian Seymour (U. British Columbia), R. O. Wells, Jr. (Rice U.), Lawrence Zalcman (U. of Maryland).

William R. Derrick

TABLE OF SYMBOLS

The number indicates the page on which the symbol is defined.

$\mathfrak{C}$	2	$\sinh z$	53	PV	167
i	5	$\log z$	55	V	206
z	5	$\text{Log } z$	57	V_n	206
$\text{Re } z$	6	z^a	58	V_s	206
$\text{Im } z$	6	pws	71, 267	Δ	243
$\bar{z}$	7	$\int_\gamma f(z)\, dz$	73	$\Gamma + iQ$	255
$|z|$	11	$-\gamma$	89	$U(\phi + 0)$	266
$\arg z$	12	$|dz|$	89	$U(\phi - 0)$	266
$\text{Arg } z$	12	$P_n(z)$	98	$U'(\phi + 0)$	266
$\text{Int } S$	24	∂R	103	$U'(\phi - 0)$	266
$\mathfrak{M}$	26	$L_n(z)$	112	$\hat{U}$	272
∞	26	S_n	115	$H(\phi - a)$	276
$f(z)$	28	lub	126	$\mathcal{L}_2\{U\}$	277
e^z	47	$J_n(z)$	138	$\mathcal{L}\{U\}$	277
$\mathfrak{R}$	49	(f, G)	143	$U * V$	283
$\cos z$	51	$\Gamma(z)$	150	$Z(s)$	284
$\sin z$	51	a_{-1}	152	$\text{Si}(\phi)$	287
$\cosh z$	53	$\text{Res}_z f(z)$	152		

1 ANALYTIC FUNCTIONS

Complex numbers were first proposed by Girolamo Cardano, in a monumental treatise on the solution of cubic and quartic equations entitled *Ars Magna* in 1545. To appreciate the audacity of this proposal, one must realize that the concept of negative numbers had just gained acceptance and there was still controversy about their properties. Cardano's "fictitious" quantities were ignored by most mathematicians until the mathematical genius Carl Friedrich Gauss gave them their present name and used them in proving the Fundamental Theorem of Algebra, which states that every nonconstant polynomial has at least one zero. In this book we shall explore the properties of complex numbers and of complex-valued functions of a complex variable. We shall see that the theory of functions of a complex variable extends the concepts of calculus to the complex plane. In so doing, differentiation and integration acquire new depth and elegance, and the two-dimensional nature of the complex plane yields many results useful in applied mathematics.

1.1 COMPLEX NUMBERS AND THEIR ALGEBRA

The numbers used in elementary algebra and calculus are called the **real numbers**. They consist of all numbers that may be geometrically represented by the points on an infinitely long straight line (see

1

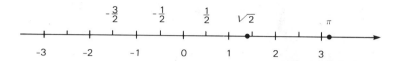

FIGURE 1.1. Model of real number system

Figure 1.1). The line is marked off in equally spaced intervals representing each of the integers, with the positive real numbers to the right of zero and the negative real numbers to the left. Every real number is represented by a single point on the line. The real numbers satisfy the following five rules of algebra, called the **field axioms.**

---

1. Commutative laws
$$a + b = b + a \quad \text{and} \quad ab = ba.$$

2. Associative laws
$$(a + b) + c = a + (b + c) \quad \text{and} \quad (ab) c = a(bc).$$

3. Distributive laws
$$a(b + c) = ab + ac \quad \text{and} \quad (a + b)c = ac + bc.$$

4. Identities. The *additive identity* 0 and *multiplicative identity* 1 satisfy $0 \neq 1$ and
$$a + 0 = a = 0 + a \quad \text{and} \quad a \cdot 1 = a = 1 \cdot a.$$

5. Inverses. Each real number a has an *additive inverse* $(-a)$ and, if $a \neq 0$, a *multiplicative inverse* a^{-1} satisfying
$$a + (-a) = 0 = (-a) + a \quad \text{and} \quad aa^{-1} = 1 = a^{-1}a.$$

The real numbers have one basic flaw: They do not provide all possible solutions to polynomial equations. For example, the equation $x^2 + 1 = 0$ cannot be solved using real numbers, since the square of any real number is nonnegative. To remedy this defect, we define the set of **complex numbers** $\mathbb{C}$ consisting of all ordered pairs

$$z = (x, y)$$

of real numbers x and y, with the following operations of addition and multiplication:

$$(x, y) + (a, b) = (x + a, y + b),$$
$$(x, y) \cdot (a, b) = (xa - yb, xb + ya).$$

Since complex numbers are represented by ordered pairs of real numbers, we can assign the point in the Cartesian plane with coordinates x and y to the complex number $z = (x, y)$. However, it is more useful to let z be the **vector** (directed line segment) from the origin to the point (x, y). Using this model for each complex number, we see that the sum of two complex numbers

$$(x, y) + (a, b) = (x + a, y + b)$$

corresponds to the **parallelogram law of vector addition** illustrated in Figure 1.2.

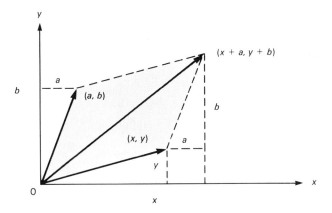

FIGURE 1.2. Parallelogram law of vector addition

Vectors can also be used to interpret the multiplication

$$(x, y) (a, b) = (xa - yb, xb + ya)$$

of two complex numbers. Note that

$$(x, 0) (a, b) = (xa, xb),$$

so that the vector (a, b) is stretched or shrunk by a factor of x if $x > 0$ and is also reflected with respect to the origin if $x < 0$ (see Figure 1.3). Observe further that

$$(0, 1) (a, b) = (-b, a).$$

Using similar triangles, we see that multiplying any complex number by $(0, 1)$ rotates the vector counterclockwise by $\pi/2$ radians (see

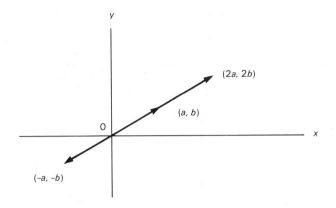

FIGURE 1.3.   Stretching and reflecting a vector

Figure 1.4). Since $(x, y) = (x, 0) + (0, 1) (y, 0)$, we can express the product of (x, y) and (a, b) in the form

$$(x, y) (a, b) = [(x, 0) + (0, 1) (y, 0)] (a, b)$$
$$= (x, 0) (a, b) + (0, 1) (y, 0) (a, b)$$
$$= (xa, xb) + (0, 1) (ya, yb).$$

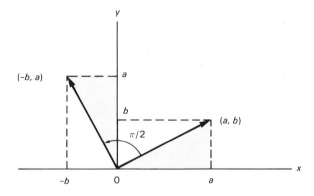

FIGURE 1.4.   Vector rotation: $(0, 1) (a, b) = (-b, a)$

Thus, complex multiplication involves the sum of two stretchings of (a, b) with the second one rotated $\pi/2$ radians counterclockwise. For example, the product

$$(1, 2)\,(2, 3) = (2, 3) + (0, 1)\,(4, 6)$$
$$= (2, 3) + (-6, 4)$$
$$= (-4, 7)$$

is illustrated in Figure 1.5.

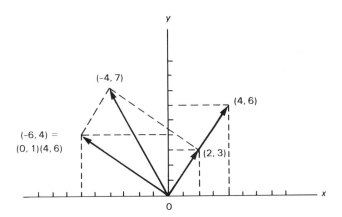

FIGURE 1.5. Complex multiplication

If we identify the ordered pair $(x, 0)$ with the real number x, we note that the addition and multiplication of such pairs satisfy the usual operations of addition and multiplication of real numbers:

$$(x, 0) + (a, 0) = (x + a, 0) \quad \text{and} \quad (x, 0)\,(a, 0) = (xa, 0).$$

Thus, the set of complex numbers includes the real numbers. Since

$$(x, y) = (x, 0) + (0, 1)\,(y, 0),$$

if we represent $(x, 0)$ by x and denote $(0, 1)$ by the symbol i, we can rewrite $z = (x, y)$ in the form

$$z = x + iy.$$

This is the standard notation for complex numbers. The symbol i is called the **imaginary unit***, and it satisfies the property

$$i^2 = (0, 1)\,(0, 1) = (-1, 0)$$

or

$$i^2 = -1.$$

---

*Engineering books often denote the imaginary unit by the symbol j.

The **origin** of the coordinate system is denoted by the complex number 0. The Cartesian plane model of the complex numbers is called the **complex plane**.

When we refer to the complex number $z = x + iy$, we call the number x the **real part** of z and denote it by Re z. The number y is called the **imaginary part** of z and is denoted by Im z. If $x = 0$, we have $z = iy$ and we say that z is **pure imaginary**.

EXAMPLE 1.

Find the real and imaginary parts of $z = 2 + 3i$.

SOLUTION: We have Re $z = 2$ and Im $z = 3$.

Using the notation $z = x + iy$ for complex numbers, the definitions of addition and multiplication allow us to add and multiply complex numbers in the same way that we add and multiply polynomials, except that $i^2 = -1$:

$$(1 + 2i) + (2 + 3i) = 3 + 5i,$$
$$(1 + 2i) \cdot (2 + 3i) = 2 + (4i + 3i) + 6i^2 = -4 + 7i.$$

It is easy to check that the operations of addition and multiplication of complex numbers are commutative, associative, and distributive. The numbers 0 and 1 serve as additive and multiplicative identities for the complex numbers.

We can subtract complex numbers by noting that

$$z_1 - z_2 = z_1 + (-z_2) = z_1 + (-1)z_2.$$

For example,

$$(7 + 2i) - (3 - 4i) = (7 + 2i) + (-3 + 4i)$$
$$= 4 + 6i.$$

Thus $-z$ is the additive inverse of z. To check that the complex numbers form a field (see Exercise 33), we must verify the existence of a multiplicative inverse for any number $a + bi \neq 0$. If we multiply $a + bi$ by its *complex conjugate* $a - bi$, we have

$$(a + bi)(a - bi) = a^2 + (abi - abi) - b^2i^2 = a^2 + b^2.$$

Hence, the multiplicative inverse of a + bi is

$$(a + bi)^{-1} = \frac{a - bi}{a^2 + b^2}.$$

Division of complex numbers is accomplished by multiplying by the multiplicative inverse of the divisor. For example, suppose we wish to divide x + yi by a + bi ≠ 0. Then

$$\frac{x + yi}{a + bi} = (x + yi) \left(\frac{a - bi}{a^2 + b^2} \right) = \left(\frac{ax + by}{a^2 + b^2} \right) + \left(\frac{ay - bx}{a^2 + b^2} \right) i.$$

Alternatively, we can multiply the numerator and denominator of the quotient by the complex conjugate of the divisor:

$$\frac{x + yi}{a + bi} = \frac{x + yi}{a + bi} \frac{a - bi}{a - bi} = \left(\frac{ax + by}{a^2 + b^2} \right) + \left(\frac{ay - bx}{a^2 + b^2} \right) i.$$

EXAMPLE 2.

Express the quotient $\dfrac{1 - 2i}{3 - 4i}$ as a complex number.

SOLUTION: Multiplying numerator and denominator by the complex conjugate $3 - (-4i) = 3 + 4i$ of the denominator, we have

$$\frac{1 - 2i}{3 - 4i} = \frac{(1 - 2i)(3 + 4i)}{(3 - 4i)(3 + 4i)} = \frac{3 - 6i + 4i - 8i^2}{9 + 12i - 12i - 16i^2}$$

$$= \frac{11}{25} - \frac{2}{25} i.$$

We denote the complex conjugate of the complex number z by the symbol $\bar{z}$. Note that if z = x + iy, then

$$z + \bar{z} = (x + iy) + (x - iy) = 2x = 2 \, \text{Re} \, z,$$
$$z - \bar{z} = (x + iy) - (x - iy) = 2iy = 2i \, \text{Im} \, z,$$

and

$$z\bar{z} = (x + iy)(x - iy) = x^2 + y^2.$$

Thus, we have the identities

$$\operatorname{Re} z = \frac{z + \bar{z}}{2}, \quad \operatorname{Im} z = \frac{z - \bar{z}}{2i},$$

and the Pythagorean theorem tells us that $(\text{length } z)^2 = z\bar{z}$.

EXAMPLE 3.

   Find the length of the vector $z = 5 + 7i$.

   SOLUTION: Multiplying z by $\bar{z}$, we have

$$(\text{length } z)^2 = z\bar{z} = (5 + 7i)(5 - 7i) = 25 + 49,$$

   so that length $z = \sqrt{74}$.

If $z_1 = x_1 + iy_1$ and $z_2 = x_2 + iy_2$, then

$$\overline{z_1 + z_2} = \overline{(x_1 + x_2) + i(y_1 + y_2)} = (x_1 + x_2) - i(y_1 + y_2)$$
$$= (x_1 - iy_1) + (x_2 - iy_2) = \bar{z}_1 + \bar{z}_2.$$

Thus, the complex conjugate of a sum of complex numbers is the sum of their conjugates:

$$\overline{z_1 + z_2} = \bar{z}_1 + \bar{z}_2.$$

Similarly, we can show (see Exercises 27-29) that

$$\overline{z_1 - z_2} = \bar{z}_1 - \bar{z}_2,$$
$$\overline{z_1 z_2} = \bar{z}_1 \bar{z}_2$$

and

$$\overline{(z_1/z_2)} = \bar{z}_1/\bar{z}_2, \quad z_2 \neq 0.$$

EXERCISES

In Exercises 1-12, find the sum, difference, product, and quotient of each pair of complex numbers.

   1. $i, 2$ 2. $i, -i$
   3. $1 + i, i$ 4. $2 - i, 3 + i$
   5. $1 + i, 1 - i$ 6. $2 + i, 3 - 4i$

7. $5, 2 + i$ 8. $5i, 2 + i$
9. $3 - 2i, 4 + i$ 10. $2 + i, 2 - i$
11. $4 + 5i, 1 - i$ 12. $2 + i, 2i$

In Exercises 13–20, write the given number in the form $x + iy$.

13. $(1 - i)^2$ 14. $(1 - i)^3$
15. $(1 - 2i)^2$ 16. $i^2 (1 + i)^3$

17. $\dfrac{2 + i}{3 - i} - \dfrac{4 + i}{1 + 2i}$ 18. $\dfrac{3 + 2i}{1 + i} + \dfrac{5 - 2i}{-1 + i}$

19. $(1 + i)(1 + 2i)(1 + 3i)$ 20. $(1 - i)(1 - 2i)(1 - 3i)$
21. Prove that Re $(iz) = -$Im z.
22. Prove that Re $(z) = Im(iz)$.
23. Prove that if $z_1 z_2 = 0$, then $z_1 = 0$ or $z_2 = 0$.
24. Show that if Im $z > 0$, then Im $(1/z) < 0$.
*25. Suppose $z_1 + z_2$ and $z_1 z_2$ are both negative real numbers. Prove that z_1 and z_2 must be real.
*26. Prove the binomial theorem for complex numbers,

$$(z_1 + z_2)^n = z_1^n + \binom{n}{1} z_1^{n-1} z_2 + \binom{n}{2} z_1^{n-2} z_2^2 + \ldots + z_2^n,$$

where n is a positive integer and $\binom{n}{k} = \dfrac{n!}{k!(n - k)!}$.

(*Hint:* Use induction.)

In Exercises 27–29, let $z_1 = x_1 + iy_1$ and $z_2 = x_2 + iy_2$.

27. Show that $\overline{z_1 - z_2} = \bar{z}_1 - \bar{z}_2$.

28. Show that $\overline{z_1 z_2} = \bar{z}_1 \bar{z}_2$.

29. Show that $\overline{(z_1/z_2)} = \bar{z}_1/\bar{z}_2$, where $z_2 \neq 0$.

In his book *Ars Magna*, Girolamo Cardano included a method for finding the roots of the general cubic equation

$$z^3 + pz^2 + qz + r = 0$$

that had been discovered by Niccolò Tartaglia.

*30. Show that the substitution $w = z + p/3$ reduces the general cubic to an equation of the form

$$w^3 + aw + b = 0.$$

---

*An asterisk (*) indicates more challenging exercises.

*31. Show that the roots of the equation in Exercise 30 are

$$w = A + B, \quad -\frac{A + B}{2} + \frac{A - B}{2}\sqrt{3}i, \quad -\frac{A + B}{2} - \frac{A - B}{2}\sqrt{3}i,$$

where $A = \sqrt[3]{-\dfrac{b}{2} + D}$, $B = \sqrt[3]{-\dfrac{b}{2} - D}$, and $D = \sqrt{\dfrac{b^2}{4} + \dfrac{a^3}{27}}$.

*32. Show that complex numbers are needed even for finding the *real* roots of

$$w^3 - 19w + 30 = 0$$

by Tartaglia's method.

33. Prove that complex numbers satisfy the field axioms.
34. Prove that the additive identity of $\mathbb{C}$ is unique.
35. Prove that the multiplicative identity of $\mathbb{C}$ is unique.

1.2 POLAR REPRESENTATION

We have seen that complex numbers can be represented as vectors in the complex plane. In this section we will use the concept of a directed line segment to determine the properties of the length and the angle of inclination of a vector in the complex plane.

Consider the nonzero vector $z = x + iy$ shown in Figure 1.6. The

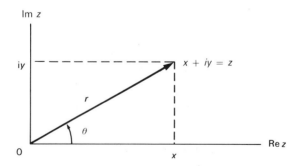

FIGURE 1.6. Polar representation

length of the vector z can be determined by the Pythagorean theorem:

$$r = \sqrt{x^2 + y^2}.$$

We call this length the **absolute value** (or **modulus** or **magnitude**) of the complex number z and denote it by

$$|z| = \sqrt{x^2 + y^2}.$$

Note that $|z| \geqslant \mathrm{Re}\ z$, $|z| \geqslant \mathrm{Im}\ z$, and $|\bar{z}| = |z|$. Further, recall that in Section 1.1 we proved that $z\bar{z} = x^2 + y^2$. Hence, $z\bar{z} = |z|^2$.

The interpretation of complex addition as vector addition is very useful in proving the following important result.

THE TRIANGLE INEQUALITY

$$|z_1 + z_2| \leqslant |z_1| + |z_2|.$$

PROOF: Remember that the length of one side of a triangle is less than or equal to the sum of the lengths of the other two sides. Thus, the triangle inequality follows immediately from the shaded triangle

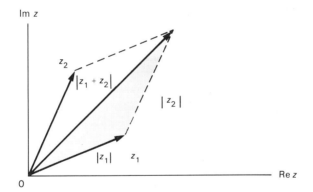

FIGURE 1.7. Triangle inequality: $|z_1 + z_2| \leqslant |z_1| + |z_2|$

in Figure 1.7. The triangle inequality may also be proved algebraically (see Exercise 38). ∎

Returning to Figure 1.6, we see that the angle that the vector $z = x + iy$ makes with the positive real axis is given by the expression

$$\theta = \arctan \frac{y}{x}.$$

This expression, however, will not be valid in the second or third quadrant, since the values of the arctangent lie in the interval

$(-\pi/2, \pi/2)$. Furthermore, the angle of inclination of the vector is determined up to *a multiple of* 2π, since the angles

$$\theta + 2\pi k, \qquad k = 0, \pm 1, \pm 2, \ldots,$$

all give the same direction in the complex plane. The angle of inclination of the vector z, determined except for a multiple of 2π, is called the **argument** of z and is denoted by arg z. That value of arg z satisfying

$$-\pi \leqslant \arg z < \pi$$

is called the **principal value** of the argument and is designated by Arg z. When working with the argument, it is convenient to adopt the convention that the notation arg z ignores multiples of 2π and to use the expression

$$\text{Arg } z + 2\pi k, \qquad k \text{ a fixed integer,}$$

to indicate a particular angle.

Returning to the original vector $z = x + iy$, $z \neq 0$, note that

$$x = r \cos \theta = |z| \cos(\arg z)$$

and

$$y = r \sin \theta = |z| \sin(\arg z).$$

Thus,

$$z = x + iy = r (\cos \theta + i \sin \theta)$$

can be rewritten in the form

$$z = |z| [\cos(\arg z) + i \sin(\arg z)], \quad z \neq 0.$$

This is called the **polar representation** of the complex number z.*

EXAMPLE 1.

Find the polar representation of $1 - i$.

SOLUTION: Consider Figure 1.8. The absolute value of $1 - i$ is

$$|1 - i| = \sqrt{1^2 + (-1)^2} = \sqrt{2},$$

while the principal value of the argument $1 - i$ is

$$\text{Arg}(1 - i) = -\frac{\pi}{4}.$$

---

*Engineering books often use the notations $r\,\underline{/\theta}$ and r cis θ for the polar representation of z.

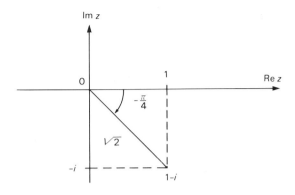

FIGURE 1.8. Polar representation of $1 - i$

Since polar angles are not uniquely determined, the argument is

$$\arg(1-i) = \frac{-\pi}{4} + 2\pi k,$$

where k is any integer. Thus, the polar representation of $1 - i$ is

$$1 - i = \sqrt{2}\left[\cos\left(\frac{-\pi}{4} + 2\pi k\right) + i\sin\left(\frac{-\pi}{4} + 2\pi k\right)\right].$$

The multiplication of two complex numbers z and w has interesting geometric interpretations when we write both numbers in their polar representations. Let $\theta = \arg z$ and $\phi = \arg w$. Writing z and w in their polar representations, we have

$$z = |z|(\cos\theta + i\sin\theta) \quad \text{and} \quad w = |w|(\cos\phi + i\sin\phi).$$

Then

$$zw = |z||w|(\cos\theta + i\sin\theta)(\cos\phi + i\sin\phi)$$
$$= |z||w|[(\cos\theta\cos\phi - \sin\theta\sin\phi) + i(\sin\theta\cos\phi + \cos\theta\sin\phi)]$$

and by the addition formulas of trigonometry,

$$zw = |z||w|[\cos(\theta + \phi) + i\sin(\theta + \phi)]. \tag{1}$$

Since

$$|\cos(\theta + \phi) + i\sin(\theta + \phi)| = 1,$$

equation (1) yields

$$|zw| = |z||w| \tag{2}$$

and

$$\arg zw = \arg z + \arg w. \tag{3}$$

Hence, the length of the vector zw is the *product* of the lengths of the vectors z and w, whereas the polar angle of the vector zw is the *sum* of the polar angles of the vectors z and w. Since the argument is determined up to a multiple of 2π, equation (3) is interpreted to mean that if particular values are assigned to any two of the terms, then there is a value of the third term for which equality holds. The geometric construction of the product zw is shown in Figure 1.9. For multiplication the angle between w and zw must be identical to the angle between 1 and z in Figure 1.9. It follows that the triangles of $0 \quad 1 \quad z$ and $0 \quad w \quad zw$ are similar.

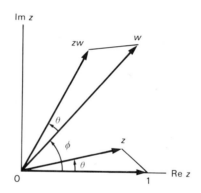

FIGURE 1.9. Complex multiplication

Division of complex numbers leads to the following equation:

$$\frac{z}{w} = \frac{z\bar{w}}{w\bar{w}} = \frac{|z|(\cos\theta + i\sin\theta)|\bar{w}|(\cos\phi - i\sin\phi)}{|w|^2}, \qquad w \neq 0.$$

Since $|\bar{w}| = |w|$, we obtain by the addition formulas of trigonometry

$$\frac{z}{w} = \frac{|z|}{|w|}[\cos(\theta - \phi) + i\sin(\theta - \phi)].$$

Hence,

$$\frac{z}{w} = \frac{|z|}{|w|} \qquad\qquad (4)$$

and

$$\arg(z/w) = \arg z - \arg w, \qquad\qquad (5)$$

with equation (5) subject to a similar interpretation as equation (3).
The product

$$zw = |z||w|[\cos(\theta + \phi) + i \sin(\theta + \phi)],$$

where $\theta = \arg z$ and $\phi = \arg w$, leads to a very interesting result when $z = w$. Since $\theta = \phi$, we have

$$z^2 = |z|^2 [\cos(2\theta) + i \sin(2\theta)].$$

Setting $w = z^2$, we obtain

$$z(z^2) = |z||z|^2 [\cos(\theta + 2\theta) + i \sin(\theta + 2\theta)]$$

or

$$z^3 = |z|^3 [\cos(3\theta) + i \sin(3\theta)].$$

Since $z = |z|(\cos \theta + i \sin \theta)$, we have shown that

$$(\cos \theta + i \sin \theta)^2 = \cos(2\theta) + i \sin(2\theta)$$

and

$$(\cos \theta + i \sin \theta)^3 = \cos(3\theta) + i \sin(3\theta).$$

Continuing this process, we obtain **De Moivre's theorem,** named in honor of the French mathematician Abraham De Moivre (1667–1754),

$$(\cos \theta + i \sin \theta)^n = \cos n\theta + i \sin n\theta,$$

where n is a positive integer. De Moivre's theorem has many useful applications.

EXAMPLE 2.

Calculate $(1 - i)^{23}$.

SOLUTION: We could multiply $1 - i$ by itself 23 times to obtain the answer, but using De Moivre's theorem, we can find the answer quite easily. We saw in Example 1 that

$$1 - i = \sqrt{2} \left[\cos \left(\frac{-\pi}{4} + 2\pi k \right) + i \sin \left(\frac{-\pi}{4} + 2\pi k \right) \right].$$

Using the principal value of the argument we have

$$(1 - i) = \sqrt{2} \left[\cos \left(\frac{-\pi}{4} \right) + i \sin \left(\frac{-\pi}{4} \right) \right].$$

Then, by De Moivre's theorem,

$$(1 - i)^{23} = (\sqrt{2})^{23} \left[\cos \left(\frac{-\pi}{4} \right) + i \sin \left(\frac{-\pi}{4} \right) \right]^{23}$$

$$= 2^{23/2} \left[\cos \left(\frac{-23\pi}{4} \right) + i \sin \left(\frac{-23\pi}{4} \right) \right].$$

But $-23\pi/4 = \pi/4 - 6\pi$, and $\sin(\pi/4) = \cos(\pi/4) = 1/\sqrt{2}$, so we get

$$(1 - i)^{23} = 2^{23/2} \frac{(1 + i)}{\sqrt{2}} = 2048(1 + i).$$

De Moivre's theorem can also be used to find the roots of a complex number. If z is an nth root of the complex number w, then

$$z^n = w.$$

To find z, set

$$z = |z|(\cos \theta + i \sin \theta) \quad \text{and} \quad w = |w|(\cos \phi + i \sin \phi),$$

where $\theta = \arg z$ and $\phi = \arg w$. Then, from De Moivre's theorem, we have

$$|z|^n (\cos n\theta + i \sin n\theta) = |w|(\cos \phi + i \sin \phi).$$

Thus, we may take

$$|z| = |w|^{1/n}$$

and

$$\theta = \frac{1}{n} \arg w = \frac{1}{n} (\text{Arg } w + 2\pi k), \qquad k = 0, \pm 1, \pm 2, \ldots . \qquad (6)$$

Although equation (6) provides infinitely many values for θ, only n different polar angles are obtained because

$$\frac{2\pi(k+n)}{n} = \frac{2\pi k}{n} + 2\pi,$$

so the polar angles repeat every n integers. Therefore, we restrict our attention to the n polar angles

$$\theta = \frac{1}{n}(\text{Arg } w + 2\pi k), \qquad k = 0, 1, \ldots, n - 1.$$

EXAMPLE 3.

Find the three cube roots of $w = 1 - i$.

SOLUTION: Let z be a cube root of $1 - i$. Then

$$z^3 = 1 - i,$$

and by De Moivre's theorem,

$$|z|^3 (\cos 3\theta + i \sin 3\theta) = \sqrt{2}\left[\cos\left(\frac{-\pi}{4} + 2\pi k\right) + i \sin\left(\frac{-\pi}{4} + 2\pi k\right)\right].$$

Thus,

$$|z| = 2^{1/6} \qquad \text{and} \qquad \theta = \frac{-\pi}{12} + \frac{2\pi k}{3}, \qquad k = 0, 1, 2.$$

Hence, the three cube roots of $1 - i$ are

$$z_0 = \sqrt[6]{2}\left[\cos\left(\frac{-\pi}{12}\right) + i \sin\left(\frac{-\pi}{12}\right)\right]$$

$$= \sqrt[6]{2}\left[\cos\left(\frac{\pi}{12}\right) - i \sin\left(\frac{\pi}{12}\right)\right],$$

$$z_1 = \sqrt[6]{2}\left[\cos\left(\frac{7\pi}{12}\right) + i \sin\left(\frac{7\pi}{12}\right)\right].$$

$$z_1 = \sqrt[6]{2}\left[\cos\left(\frac{5\pi}{4}\right) + i \sin\left(\frac{5\pi}{4}\right)\right].$$

Conic sections provide further examples of the concepts in this section. Although the usual formulas of analytic geometry can be used (with $x = \text{Re } z$ and $y = \text{Im } z$), it is easy to define the conic sections in terms of distance.

EXAMPLE 4.

An **ellipse** is defined as the set of all points in $\mathbb{C}$ the sum of whose distances from two fixed points, called the **foci**, is a constant. What is the equation of the ellipse passing through i having ± 1 as its foci?

SOLUTION: Since $z - z_0$ is the vector from z_0 to z, the definition of an ellipse yields

$$|z - 1| + |z + 1| = c,$$

where c is a real constant. Since $z = i$ must satisfy this equation, we have (see Figure 1.10)

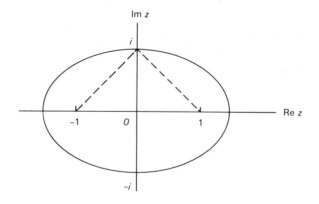

FIGURE 1.10. Ellipse $|z - 1| + |z + 1| = 2\sqrt{2}$

$$c = |i - 1| + |i + 1| = 2\sqrt{2}.$$

Thus, the ellipse is given by the equation

$$|z - 1| + |z + 1| = 2\sqrt{2}.$$

EXAMPLE 5.

A **parabola** is defined as the set of all points in $\mathbb{C}$ whose distance from a given point, called the **focus**, equals their distance from a straight line called the **directrix**. Find the equation of the parabola that has i as its focus and the line $\operatorname{Im} z = -1$ as its directrix.

SOLUTION: By definition, we get

$$|z - i| = \text{Im } z + 1,$$

since the closest point on the directrix to z lies vertically below z (see Figure 1.11). If we want to find the corresponding formula from analytic geometry, we square both sides, obtaining

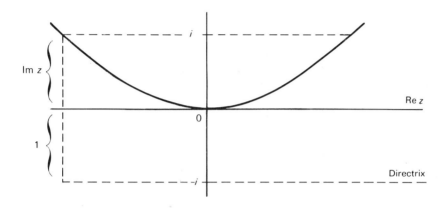

FIGURE 1.11. Parabola $|z - i| = \text{Im } z + 1$

$$|z|^2 + 1 + 2 \text{ Re } z i = (\text{Im } z + 1)^2$$

or

$$|z|^2 - 2 \text{ Im } z = (\text{Im } z)^2 + 2 \text{ Im } z.$$

Letting $y = \text{Im } z$ and $|z|^2 = x^2 + y^2$, we get the parabola

$$y = x^2/4.$$

EXAMPLE 6.

A **hyperbola** consists of all points z in $\mathbb{C}$ such that the absolute value of the difference between the distances from z to two fixed points, called the *foci*, is a constant. What is the equation of the hyperbola with foci ± 1 that passes through the point $1 + i$?

SOLUTION: By definition, we have
$$||z - 1| - |z + 1|| = c,$$
where c is a constant. Since the point $z = 1 + i$ satisfies the equation, we find that $c = \sqrt{5} - 1$.

EXERCISES

In Exercises 1-9, find the absolute value, argument, and polar representation of the given complex numbers.

1. i
2. $-i$
3. $1 + i$
4. $-3 + 4i$
5. $4 + 3i$
6. $5 - 12i$
7. $2 + 7i$
8. $2 - i$
9. $5 + 2i$

In Exercises 10-15, use De Moivre's theorem to express each number in the form $x + iy$, where x and y are real.

10. $(1 + i)^{29}$
11. $(-1 + i)^{17}$
12. $(-1 - i)^{36}$
13. $(2 + 2i)^{12}$
14. $(\sqrt{3} + i)^{15}$
15. $(-\sqrt{3} + i)^{13}$

Find all the solutions of the following equations in Exercise 16-23.

16. $z^2 = i$
17. $z^2 = 1 + i$
18. $z^2 = 2 - i$
19. $-z^2 = \sqrt{3} + i$
20. $z^3 = 2 + i$
21. $z^3 = 1 + \sqrt{3}i$
22. $z^4 = i$
23. $z^4 = -1$

24. Find the equation of the ellipse with foci $\pm i$ that passes through the point $1 + i$. What is the corresponding formula from analytic geometry?

25. Find the equation of the ellipse with foci 1 and i that passes through the origin. What is the corresponding formula from analytic geometry?

26. Find the parabola with focus $1 + i$ and the line $\text{Re } z + \text{Im } z = 0$ as directrix.

27. Write the general equation in complex form of a hyperbola with foci a and b.

28. Prove that $|z| \leqslant |\text{Re } z| + |\text{Im } z| \leqslant \sqrt{2}|z|$.

*29. Prove that if $|z_1| = |z_2| = |z_3|$ and $z_1 + z_2 + z_3 = 0$, then z_1, z_2, z_3 are the vertices of an equilateral triangle.
(*Hint:* Show that $|z_1 - z_2|^2 = |z_2 - z_3|^2 = |z_3 - z_1|^2$.)

*30. Prove that the triangle with vertices z_1, z_2, z_3 is equilateral if and only if

$$z_1^2 + z_2^2 + z_3^2 = z_1 z_2 + z_2 z_3 + z_3 z_1.$$

31. Prove that $|z_1 - z_2| \geq ||z_1| - |z_2||$.

32. Prove that $\left| \sum_{k=1}^{n} z_k \right| \leq \sum_{k=1}^{n} |z_k|$.

33. Prove that $|z_1 \pm z_2|^2 = |z_1|^2 + |z_2|^2 \pm 2 \operatorname{Re} z_1 \bar{z}_2$.

34. Prove that $|z_1 + z_2|^2 + |z_1 - z_2|^2 = 2(|z_1|^2 + |z_2|^2)$.

35. Prove that

$$\left| \frac{z - a}{1 - \bar{a}z} \right| < 1$$

if $|z| < 1$ and $|a| < 1$.

36. Show that the triangle inequality is an equality for nonzero numbers z_1 and z_2 if and only if $\arg z_1 = \arg z_2$.

37. Show that if z_0 is a root of a polynomial $P(z)$ with real co-efficients, then $\bar{z}_0$ is also a root of $P(z)$.

38. Expand $|z_1 + z_2|^2$ to prove the triangle inequality. (*Hint:* $\operatorname{Re} z_1 \bar{z}_2 \leq |z_1 \bar{z}_2| = |z_1| |z_2|$.)

39. The n roots of the equation $z^n = 1$ are called **nth roots of unity**. Show that the nth roots of unity are given by

$$z_k = \cos\left(\frac{2\pi k}{n}\right) + i \sin\left(\frac{2\pi k}{n}\right), \qquad k = 0, 1, \ldots, n - 1.$$

40. Let z_k be any nth root of unity. Prove that

$$1 + z_k + z_k^2 + \ldots + z_k^{n-1} = 0, \qquad \text{if } z_k \neq 1.$$

41. If $1, z_1, z_2, \ldots, z_{n-1}$ are the nth roots of unity, show that

$$(z - z_1)(z - z_2) \cdot \ldots \cdot (z - z_{n-1}) = 1 + z + z^2 + \ldots + z^{n-1}.$$

*42. Find all possible times that the hands of a clock can be inter-changed obtaining another position that actually does occur on an ordinary clock.

*43. By minimizing the expression $\sum_{k=1}^{n} (|a_k| - \lambda |z_k|)^2$, where $a_1, \ldots, a_n, z_1, \ldots, z_n$ are complex numbers, for arbitrary real λ, show that

$$\left(\sum_{k=1}^{n} |a_k z_k| \right)^2 \leq \left(\sum_{k=1}^{n} |a_k|^2 \right) \left(\sum_{k=1}^{n} |z_k|^2 \right).$$

*44. Prove *Lagrange's identity:*

$$\left| \sum_{k=1}^{n} a_k z_k \right|^2 = \left(\sum_{k=1}^{n} |a_k|^2 \right) \left(\sum_{k=1}^{n} |z_k|^2 \right) - \sum_{1 \leq j < k \leq n} |a_j \bar{z}_k - a_k \bar{z}_j|^2.$$

*45. *Enestrom–Kakeya theorem.* Let $P(z)$ be a polynomial with real coefficients,

$$P(z) = a_n z^n + a_{n-1} z^{n-1} + \ldots + a_1 z + a_0,$$

satisfying $a_0 > a_1 > \ldots > a_n > 0$. Prove that all roots of $P(z)$ satisfy $|z| > 1$. (*Hint:* Apply the triangle inequality to $(1 - z) P(z) = a_0 - [(a_0 - a_1)z + (a_1 - a_2)z^2 + \ldots + (a_{n-1} - a_n)z^n + a_n z^{n+1}]$.)

1.3 SETS IN THE COMPLEX PLANE

Let z_0 be a complex number. Then an **ϵ-neighborhood** of z_0 is the set of all points z whose distance from z_0 is less than ϵ, that is, all z satisfying $|z - z_0| < \epsilon$ (see Figure 1.12). Pictorially, this is the interior of a disk centered at z_0 of radius ϵ.

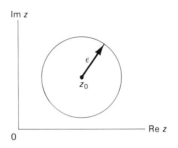

FIGURE 1.12. An ϵ-neighborhood of z_0

Let S be a set of points in the complex plane $\mathcal{C}$. The point z_0 is said to be an **interior point** of S if some ϵ-neighborhood of z_0 is contained entirely in S; the set of all interior points of S is called the **interior** of S and is denoted by Int S. The **complement** of S is the set $\mathcal{C} - S$ of all points not in S. The set Int $(\mathcal{C} - S)$ is referred to as the

exterior of S. A point z_0 is a **boundary point** of S if every ϵ-neighborhood of z_0 contains points in S and points not in S. Note that every boundary point of S is not in the interior or exterior of S. The set of all boundary points of S is called the **boundary** of S (see Figure 1.13).

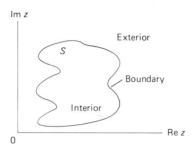

FIGURE 1.13. Interior, exterior, and boundary of a set

A point z_0 is called an **accumulation point** of a set S if each neighborhood of z_0 contains at least one point of S distinct from z_0.

EXAMPLE.

Let S_0 be the set of all points z such that $|z| < 1$. Find the interior, boundary, and exterior of the set S_0.

SOLUTION: Let z_0 be any point in S_0. Note that the disk $|z - z_0| < \epsilon$ lies entirely in S_0 whenever $\epsilon < 1 - |z_0|$. Thus, every point of S_0 is an interior point. Similarly, every point z_0

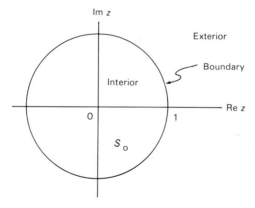

FIGURE 1.14. Interior, boundary, and exterior of the set $|z| < 1$

satisfying $|z_0| > 1$ is exterior to S_0. If $|z_0| = 1$, then every ϵ-neighborhood of z_0 will contain points in S_0 and points not in S_0. Hence, the boundary of S_0 consists of all points on the circle $|z| = 1$, the interior of S_0 is the set $|z| < 1$, and the exterior of S_0 is the set of all points satisfying $|z| > 1$ (see Figure 1.14).

A set is **open** if all its points are interior points; that is, $S = \text{Int } S$ when S is open. Thus, the set S_0 in our example above is an open set. The complement of an open set is said to be **closed**. For example, the set T of all points z such that $|z| \geq 1$ is closed. Similarly, the set $|z| \leq 1$ is closed.

A set S is said to be **bounded** if there is a positive real number R such that all z in S satisfy $|z| < R$. If this condition does not hold, we say that S is **unbounded**. For example, the set S_0 in our earlier example is bounded, but $T = \{z : |z| \geq 1\}$ is unbounded.

A set S is **connected** if it cannot be represented as the union of two nonempty disjoint sets A and B, neither containing a boundary point of the other. Intuitively, what this says is that S consists of a single piece. For example S_0 is connected, but the set of all z for which $|z - 2| < 1$ or $|z + 2| < 1$ is not connected, as we can let A be the set of all z such that $|z - 2| < 1$, and B be the set of all z for which $|z + 2| < 1$ (see Figure 1.15). Then A and B are disjoint open sets, and such sets cannot contain a boundary point of the other (why?).

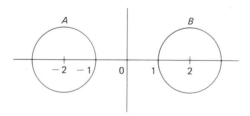

FIGURE 1.15. $A \cup B$ is not connected

A **region*** is an open connected set. It is intuitively clear that any two points in a region can be joined by a polygon contained in that

---

*Many books call an open connected set a **domain.** We will avoid this usage to prevent possible confusion when describing the domain of definition of a complex function.

region, but this fact requires verification. The proof is slightly complicated, but should be studied, as the technique will be used again.

THEOREM

Any two points of a region can be joined by a polygon that lies in the region.

PROOF† Call the region S, and suppose z_0 lies in S. Denote by S_1 all those points in S that can be joined to z_0 by a polygon; denote by S_2 those points that cannot be so joined. If z_1 is in S_1, and hence in S, it is an interior point of S. Thus, there is an ϵ-neighborhood of z_1 lying in S: $|z - z_1| < \epsilon$. All these points are in S_1, as each can be joined to z_1 by a straight line lying in S and hence can be joined to z_0 by a polygon in S. Thus, every point in S_1 is an interior point of S_1, and so S_1 is open. If z_2 is in S_2, let $|z - z_2| < \epsilon$ be a neighborhood contained in S. No point in this neighborhood can be in S_1, for otherwise z_2 is in S_1. Thus, every point of S_2 is an interior point of S_2, so S_2 is open. Neither set can contain a boundary point of the other, as both are open and the two sets are disjoint. Since S is connected, one of these sets must be empty. But z_0 is in S_1, so S_2 is

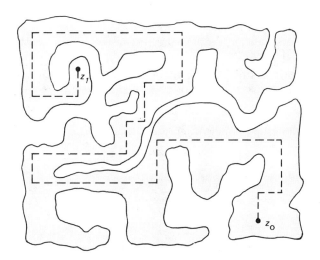

FIGURE 1.16. Polygon joining two points z_0, z_1

---

†A dagger (†) indicates more challenging proofs or proofs that are optional.

empty. Thus, any two points can be joined to z_0 by a polygonal path in S and thus to each other by a polygonal path by way of z_0. The proof is now complete. ∎

Furthermore, it is even possible to insist that all lines in the polygon be parallel to the coordinate axes. The proof using this additional requirement is identical, since we can always join the center of an open disk to one of its points by at most two line segments parallel to the axes. (See Figure 1.16.)

A region is **simply connected** if its complement is connected. This implies that a simply connected region has no "holes" in it. For example, the set S_0 in our earlier example is simply connected, but the set of all z satisfying $0 < |z| < 1$ is not, since the origin forms a "hole" for this set.

For many purposes it is useful to extend the system $\mathbb{C}$ of complex numbers by including a **point at infinity** denoted by the symbol ∞. This new set is called the **extended complex plane** $\mathfrak{M}$ and the point ∞ satisfies the following algebraic rules:

$$a + \infty = \infty + a = \infty, \qquad \frac{a}{\infty} = 0, \qquad a \neq \infty,$$

$$b \cdot \infty = \infty \cdot b = \infty, \qquad \frac{b}{0} = \infty, \qquad b \neq 0.$$

As a geometric model for $\mathfrak{M}$ we use $x_1^2 + x_2^2 + x_3^2 = 1$ the unit sphere in three-dimensional space. We associate to each point z in the plane that point Z in the sphere where the ray originating from the north pole N and passing through z intersects the sphere. Thus, N corresponds to ∞ (see Figure 1.17) and ϵ-neighborhoods of N on

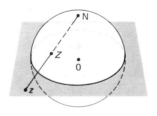

FIGURE 1.17. The Riemann sphere

the unit sphere correspond to neighborhoods of the point at infinity. This model is called the **Riemann sphere** and the point correspondence is referred to as **stereographic projection**. It can be shown that all straight lines in $\mathbb{C}$ correspond to circles passing through ∞ in $\mathfrak{M}$.* We shall prove this assertion in Chapter 5.

EXERCISES

In Exercises 1-10, classify the sets according to the terms *open, closed, bounded, connected,* and *simply connected.*

1. $|z + 3| < 2$ 2. $|\operatorname{Re} z| < 1$
3. $|\operatorname{Im} z| > 1$ 4. $0 < |z - 1| \leqslant 1$
5. $|z| \leqslant \operatorname{Re} z + 2$ 6. $|z - 1| - |z + 1| > 2$
7. $|z + 1| + |z + i| \geqslant 2$
8. $|z - 1| < \operatorname{Im} z$
9. $2\sqrt{2} < |z - 1| + |z + 1| < 3$
10. $||z - i| - |z + i|| < 1$

11. What are the boundaries of the sets in Exercises 1-10?

In Exercises 12-15, prove the indicated properties of open and closed sets.

12. The intersection of finitely many open sets is open.
13. The union of finitely many closed sets is closed.
14. The intersection of any collection of closed sets is closed.
15. The union of any collection of open sets is open.
16. Prove that if any two points in an open set can be joined by a polygon lying in the set, then the set is a region.
17. The **closure** of a set S is the intersection of all closed sets containing S. Prove that the closure of a connected set is connected.
18. Prove that S is closed if and only if it contains all its accumulation points.
19. What is the accumulation point of the set consisting of all points $z = 1/n$, n a positive integer? (This exercise shows that an accumulation point need not belong to the set.)
20. Let S be the set of all z satisfying $|z| \geqslant 1$ or $z = 0$. Show that $z = 0$ is not an accumulation point of S.
21. What is the accumulation point of the set of all points $z = -in$, n a positive integer, in the extended plane $\mathfrak{M}$? Does this set have an accumulation point in $\mathbb{C}$?
22. Show that any point in a region is an accumulation point of that region.

---

*Other common notations for $\mathfrak{M}$ are $\bar{S}$, Σ, $\mathbb{C} \cup \{\infty\}$.

1.4 CONTINUOUS FUNCTIONS OF A COMPLEX VARIABLE

A complex-valued function of a complex variable is a rule that assigns a complex number w to each complex number z in a set S. We write $w = f(z)$ and say that w is the value of the function at the point z in the domain of definition S of the function. Writing $w = f(z)$ in terms of the real and imaginary part decompositions $z = x + iy$ and $w = u + iv$ of each complex variable,

$$w = u(z) + iv(z) = u(x, y) + iv(x, y),$$

we note that a complex-valued function of a complex variable consists of a *pair* of real-valued functions of two real variables.

EXAMPLE 1.

Express $w = z^2$ as a pair of real-valued functions of two real variables.

SOLUTION: Setting $z = x + iy$, we obtain

$$w = z^2 = (x + iy)^2 = (x^2 - y^2) + i(2xy).$$

Thus, $u(x, y) = x^2 - y^2$ and $v(x, y) = 2xy$.

Real-valued functions of a real variable $y = f(x)$ can be described geometrically by a graph in the xy-plane. No such convenient representation is possible for $w = f(z)$, as it would require four dimensions, two for each complex variable. Instead, information about the function is displayed by drawing separate complex planes for the

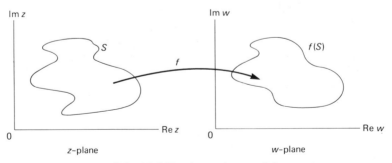

FIGURE 1.18. A mapping $w = f(z)$

variables z and w and indicating correspondences between points or sets of points in the two planes (see Figure 1.18). The function f is said to be a **mapping** of the set S in the z-plane into the w-plane. A function f mapping a set S into a set S', $f : S \rightarrow S'$, is said to be **one-to-one** if $f(z_1) = f(z_2)$ only for $z_1 = z_2$; it is said to be **onto** if $S' = f(S)$ where $f(S)$ is the set of all values assumed by f on the set S.* We call $f(S)$ the **image** set of S under the mapping f.

EXAMPLE 2.

Analyze the function $w = 3z$.

SOLUTION: Setting $z = x + iy$, we obtain

$$w = u + iv = 3x + i(3y).$$

Thus, $u = 3x$, $v = 3y$, and each nonzero vector in the z-plane is stretched into a vector, with the same argument but three times its length, in the w-plane. Since any point $a + ib$ in the w-plane is the image of the point $(a/3) + i(b/3)$ in the z-plane, the function $w = 3z$ is onto. The function is also one-to-one, since $3z_1 = 3z_2$ only when $z_1 = z_2$.

EXAMPLE 3.

Describe the image set of the function $w = z^2$ defined on the disk $|z| < 2$, and state if this mapping is one-to-one.

SOLUTION: Writing each point of the disk in its polar representation

$$z = r(\cos \theta + i \sin \theta),$$

where $0 \leqslant r = |z| < 2$ and $0 \leqslant \theta < 2\pi$, we obtain

$$w = z^2 = r^2 (\cos 2\theta + i \sin 2\theta).$$

Hence, each argument is doubled, indicating that the disk $|z| < 2$ is mapped onto $|w| < 4$ with each point of $0 < |w| < 4$ the

---

*One-to-one functions are often called **injections** and onto functions are referred to as **surjections**. A function that is both an injection and a surjection is called a **bijection**.

image of *two* points of $0 < |z| < 2$. For example, $z = \pm i$ both map to $w = -1$. Hence, the mapping is not one-to-one. (See Figure 1.19.)

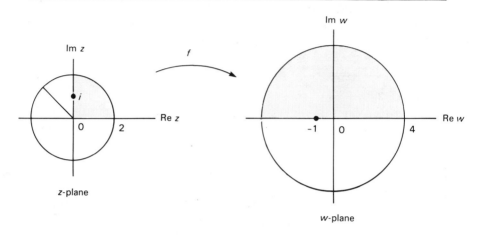

FIGURE 1.19.   The mapping $w = z^2$

EXAMPLE 4.

Determine if the function

$$w = \frac{z - 1}{z - 2}$$

is one-to-one, and state where the function can be defined.

SOLUTION: Suppose z_1 and z_2 yield the same value of w:

$$\frac{z_1 - 1}{z_1 - 2} = \frac{z_2 - 1}{z_2 - 2}.$$

Cross-multiplying, we have

$$z_1 z_2 - 2z_1 - z_2 + 2 = z_1 z_2 - z_1 - 2z_2 + 2.$$

Cancelling like terms and gathering all the terms involving z_1 on one side and z_2 on the other, we obtain $z_1 = z_2$, implying that the function is one-one-one.

The answer to the second question depends on what values of w we wish to allow. If we restrict w to the complex plane $\mathbb{C}$, then the function is not defined at $z = 2$, since the denominator vanishes. However, if we allow w to assume all values in the extended plane $\mathfrak{M}$, then the function can be defined on $\mathfrak{M}$, with $z = 2$ mapped to $w = \infty$. The image of the point at ∞ is obtained by evaluating

$$w = \frac{z - 1}{z - 2} = \frac{1 - \dfrac{1}{z}}{1 - \dfrac{2}{z}}$$

as z tends to ∞. Thus, $z = \infty$ is mapped to $w = 1$ (see Exercise 26).

Suppose f is defined on a region G and a is a point in G. Then limits and continuity are defined in the same way as in the real variable case.

DEFINITION

The function $f(z)$ is said to have **limit** A as z approaches a,

$$\lim_{z \to a} f(z) = A,$$

provided that for every $\epsilon > 0$ there exists a number $\delta > 0$ such that

$$|f(z) - A| < \epsilon$$

whenever $0 < |z - a| < \delta$. The function $f(z)$ is said to be **continuous** at a if and only if

$$\lim_{z \to a} f(z) = f(a)$$

(see Figure 1.20). A continuous function is one that is continuous at all points where it is defined.

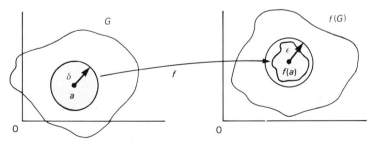

FIGURE 1.20. Continuity of f at a

Geometrically, the definition of a limit states that *any* ϵ-neighborhood of A contains all the values of f in some δ-neighborhood of a, except possibly the value $f(a)$. The next example will illustrate the usual procedure for determining δ for a given $\epsilon > 0$.

EXAMPLE 5.

Prove that $\lim\limits_{z \to 3} \dfrac{z-1}{z-2} = 2.$

SOLUTION: Simplifying the expression $|f(z) - A|$, we obtain

$$\left| \frac{z-1}{z-2} - 2 \right| = \left| \frac{3-z}{z-2} \right| < \frac{\delta}{|z-2|},$$

since we are assuming that $0 < |z - 3| < \delta$, with δ yet to be determined in terms of ϵ. If $\delta < \frac{1}{2}$, using the triangle inequality we have

$$|z - 2| = |1 - (3 - z)| \geqslant 1 - |3 - z| > 1 - \delta > \tfrac{1}{2}$$

so that

$$\left| \frac{z-1}{z-2} - 2 \right| < 2\,\delta.$$

Thus, given any small number $\epsilon > 0$, if we select $\delta < \min\left(\frac{1}{2}, \frac{1}{2}\,\epsilon\right)$, we obtain

$$\left| \frac{z-1}{z-2} - 2 \right| < \epsilon.$$

Since the definition of a limit of a complex-valued function of a complex variable is identical to that of a real-valued function of a real variable and absolute values behave as in the real case, precisely the same rules of limits apply. The verification of the following properties are exact analogs of the usual proofs of elementary calculus.

RULES OF LIMITS

Let $\lim_{z \to a} f(z) = A$ and $\lim_{z \to a} g(z) = B$.

Then

(i) $\lim_{z \to a} [f(z) \pm g(z)] = A \pm B$,

(ii) $\lim_{z \to a} f(z)g(z) = AB$,

(iii) $\lim_{z \to a} \dfrac{f(z)}{g(z)} = \dfrac{A}{B}$, for $B \ne 0$.

PROOF: Given $\epsilon > 0$, there is a number $\delta_1 > 0$ such that $|f(z) - A| < \epsilon$, whenever $|z - a| < \delta_1$, and a number $\delta_2 > 0$ such that $|g(z) - B| < \epsilon$, whenever $|z - a| < \delta_2$. Let $|z - a| < \delta$, where $\delta = \min(\delta_1, \delta_2)$. Then, by the triangle inequality,

$$|[f(z) + g(z)] - (A + B)| = |[f(z) - A] + [g(z) - B]|$$
$$\leqslant |f(z) - A| + |g(z) - B| < \epsilon + \epsilon = 2\epsilon$$

and

$$|[f(z) - g(z)] - (A - B)| = |[f(z) - A] + [B - g(z)]|$$
$$\leqslant |f(z) - A)| + |B - g(z)| < \epsilon + \epsilon = 2\epsilon$$

Since $\epsilon > 0$ is arbitrary, this shows that $f(z) \pm g(z)$ can be made to be arbitrarily close to $A \pm B$ by selecting z close enough to a. Hence, rule (i) holds. Furthermore,

$$|f(z)g(z) - AB| = |f(z)g(z) - f(z)B + f(z)B - AB|$$
$$= |f(z)[g(z) - B] + B[f(z) - A]|$$
$$\leqslant |f(z)| |g(z) - B| + |B| |f(z) - A|$$

and

$$\left| \frac{f(z)}{g(z)} - \frac{A}{B} \right| = \left| \frac{f(z)}{g(z)} - \frac{f(z)}{B} + \frac{f(z)}{B} - \frac{A}{B} \right|$$

$$= \left| \frac{f(z)[B - g(z)]}{Bg(z)} + \frac{f(z) - A}{B} \right|$$

$$\leqslant \frac{|f(z)|}{|B| |g(z)|} |B - g(z)| + \frac{|f(z) - A|}{|B|} .$$

If $0 < \epsilon < \tfrac{1}{2} |B|$, we have

$$|B| = |B - g(z) + g(z)| \leqslant \epsilon + |g(z)|,$$

so that

$$|g(z)| \geqslant |B| - \epsilon > \tfrac{1}{2} |B|$$

and

$$|f(z)| = |f(z) - A + A| \leqslant |A| + \epsilon.$$

Hence,

$$|f(z)g(z) - AB| \leqslant \epsilon (|A| + |B| + \epsilon)$$

and

$$\left| \frac{f(z)}{g(z)} - \frac{A}{B} \right| < \frac{\epsilon}{|B|} \left(\frac{|A| + \epsilon}{\tfrac{1}{2}|B|} + 1 \right),$$

so we can make $f(z)g(z)$ and $f(z)/g(z)$ arbitrarily close to AB and A/B, respectively, by selecting z sufficiently close to a. This proves rules (ii) and (iii). ∎

The rules of limits can be used to prove that every polynomial function in z

$$f(z) = a_n z^n + a_{n-1} z^{n-1} + \ldots + a_1 z + a_0$$

is continuous in $\mathbb{C}$. Note that the identity function

$$f(z) = z$$

is clearly continuous at any point by letting $\delta = \epsilon$. Applying the second rule of limits repeatedly, we see that $f(z) = z^n$ is continuous for every positive integer n. Every constant function $f(z) = c$ is trivially continuous, since all δ-neighborhoods of any point are mapped into any ϵ-neighborhood of c. Again, applying the second rule of limits, we see that $f(z) = a_n z^n$ is continuous. Finally, using the first rule of limits repeatedly, we see that all polynomials are continuous. Indeed, using the third rule of limits, we find that all quotients of polynomials

$$\frac{a_n z^n + \ldots + a_1 z + a_0}{b_m z^m + \ldots + b_1 z + b_0}$$

are continuous at those points where the denominator does not vanish. It also follows from the rules of limits that the sum $f(z) + g(z)$ and the product $f(z)g(z)$ of two continuous functions are continuous, and the quotient $f(z)/g(z)$ is defined and continuous at all points where $g(z)$ does not vanish.

EXAMPLE 6.

Determine if the function

$$f(z) = \begin{cases} \dfrac{z^2 - 1}{z - 1}, & z \neq 1 \\ 3, & z = 1 \end{cases}$$

is continuous.

SOLUTION: Clearly, f is continuous in the set $z \neq 1$, since the denominator is nonzero. Thus, the only point where we still need to check the continuity is $z = 1$. However,

$$\lim_{z \to 1} \frac{z^2 - 1}{z - 1} = 2$$

because

$$\frac{z^2 - 1}{z - 1} = z + 1,$$

if $z \neq 1$. But by definition, $f(1) = 3 \neq 2$. Hence, f is not continuous.

EXERCISES

Use the ϵ-δ definition of a limit to verify Exercises 1–10.

1. $\lim\limits_{z \to 1} 2z = 2$

2. $\lim\limits_{z \to i} iz = -1$

3. $\lim\limits_{z \to -i} z + i = 0$

4. $\lim\limits_{z \to i} z^2 + 1 = 0$

5. $\lim\limits_{z \to 1+i} 2z - 3 = -1 + 2i$

6. $\lim\limits_{z \to 1+i} z^2 = 2i$

7. $\lim\limits_{z \to 2} \dfrac{z^2 - 4}{z - 2} = 4$

8. $\lim\limits_{z \to -i} \dfrac{z^2 + 1}{z + i} = -2i$

9. $\lim\limits_{z \to 1} \dfrac{z^3 - 1}{z - 1} = 3$

10. $\lim\limits_{z \to 2} \dfrac{z^2 - 3z + 2}{z - 2} = 1$

Prove that the functions in Exercises 11–14 are continuous in $\mathbb{C}$:

11. $w = \text{Re } z$

12. $w = \text{Im } z$

13. $w = \bar{z}$

14. $w = |z|$

Suppose $f(z)$ is a continuous function on a domain G. Prove that the functions in Exercises 15–18 are continuous on G.

15. $\operatorname{Re} f(z)$ 16. $\operatorname{Im} f(z)$

17. $|f(z)|$ 18. $f(\bar{z})$

19. At what points is the function

$$f(z) = \begin{cases} \dfrac{z^3 - 1}{z^2 - 1}, & z \neq \pm 1 \\[3mm] \dfrac{3}{2}, & z = \pm 1 \end{cases}$$

continuous?

Prove that the functions in Exercises 20–23 are continuous for $z \neq 0$. Can the function be defined so as to make it continuous at $z = 0$?

20. $f(z) = \dfrac{z \operatorname{Re} z}{|z|^2}$

21. $f(z) = \dfrac{|z|^2}{z}$

22. $f(z) = \dfrac{(\operatorname{Re} z)(\operatorname{Im} z)}{|z|^2}$

23. $f(z) = \dfrac{(\operatorname{Re} z)^2 - (\operatorname{Im} z)^2}{|z|^2}$

24. Prove that every function of the form

$$w = \frac{z - a}{z - b}, \qquad a \neq b$$

is a one-to-one mapping of the extended plane $\mathfrak{M}$ onto itself.

25. Prove that every function of the form

$$w = \frac{az + b}{cz + d}, \qquad ad \neq bc$$

is a one-to-one mapping of $\mathfrak{M}$ onto itself.

26. The function $f(z)$ has the limit A as z approaches ∞,

$$\lim_{z \to \infty} f(z) = A,$$

if for every $\epsilon > 0$ there exists a number $\delta > 0$ such that

$$|f(z) - A| < \epsilon \qquad \text{whenever} \qquad |z| > \delta.$$

Use this definition to prove that

$$\lim_{z \to \infty} \frac{z - 1}{z - 2} = 1.$$

27. Suppose the coefficients of the polynomial

$$P(z) = a_n z^n + \ldots + a_1 z + a_0$$

satisfy $|a_0| \geqslant |a_1| + |a_2| + \ldots + |a_n|$. Prove that $P(z)$ has no roots in the unit disk $|z| < 1$. (*Hint:* Note that

$$|P(z)| \geqslant |a_0| - [|a_1| |z| + \ldots + |a_n| |z|^n].)$$

1.5 NECESSARY CONDITIONS FOR ANALYTICITY

The derivative of a complex-valued function of a complex variable is defined in precisely the same way as the real-valued case of elementary calculus.

DEFINITION
The **derivative** f' of f at a is given by

$$f'(a) = \lim_{h \to 0} \frac{f(a + h) - f(a)}{h},$$

provided the limit exists. The function f is said to be **analytic** (or **holomorphic**) on the region G if it has a derivative at each point of G, and f is said to be **entire** if it is analytic on all of $\mathcal{C}$.

Observe that in the definition above, h is a complex number, as is the quotient $[f(a + h) - f(a)]/h$. Thus, for the derivative to exist, it is necessary that the quotient above tend to a unique complex number $f'(a)$ independent of the manner in which h approaches zero.

LEMMA: If f has a derivative at a, then f is continuous at a.

PROOF: $\displaystyle \lim_{h \to 0} f(a + h) = \lim_{h \to 0} \left\{ \left[\frac{f(a + h) - f(a)}{h} \right] \cdot h + f(a) \right\}$

$$= f(a). \blacksquare$$

Manipulations of the definition of derivative lead to the usual rules of differentiation:

$$(f \pm g)' = f' \pm g',$$
$$(fg)' = fg' + gf',$$

$$\left(\frac{f}{g}\right)' = \frac{gf' - fg'}{g^2}, \quad g \neq 0,$$

$$(f(g(z)))' = f'(g(z))g'(z), \quad \text{Chain rule.}$$

The proofs are identical to those in any elementary calculus text. For example,

$$(fg)'(a) = \lim_{h \to 0} \frac{f(a + h)g(a + h) - f(a)g(a)}{h}$$

$$= \lim_{h \to 0} \frac{f(a + h)g(a + h) - f(a + h)g(a) + f(a + h)g(a) - f(a)g(a)}{h}$$

$$= \lim_{h \to 0} \left[f(a + h) \cdot \frac{g(a + h) - g(a)}{h} + g(a) \cdot \frac{f(a + h) - f(a)}{h} \right]$$

$$= f(a)g'(a) + g(a)f'(a).$$

Polynomials and rational functions are differentiated in the same way as in elementary calculus. As an example, let $f(z) = z^n$ with n a positive integer. Using the binomial theorem, we have

$$f'(z) = \lim_{h \to 0} \frac{f(z + h) - f(z)}{h} = \lim_{h \to 0} \frac{(z + h)^n - z^n}{h}$$

$$= \lim_{h \to 0} \frac{(z^n + nz^{n-1}h + \ldots + h^n) - z^n}{h} = nz^{n-1}.$$

In particular, it follows that every polynomial

$$P(z) = a_0 + a_1 z + a_2 z^2 + \ldots + a_n z^n$$

is entire, since at each point z in $\mathbb{C}$ it has the derivative

$$P'(z) = a_1 + 2a_2 z + \ldots + na_n z^{n-1}.$$

In spite of these similarities, there is a fundamental difference between differentiation for functions of real variables and differentiation for functions of a complex variable. Let $z = (x, y)$ and suppose that h is real. Then

$$f'(z) = \lim_{h \to 0} \frac{f(x + h, y) - f(x, y)}{h} = \frac{\partial f}{\partial x}(z) = f_x(z).$$

But if $h = ik$ is purely imaginary, then

$$f'(z) = \lim_{k \to 0} \frac{f(x, y + k) - f(x, y)}{ik} = \frac{1}{i}\frac{\partial f}{\partial y}(z) = -if_y(z).$$

Thus, the existence of a complex derivative forces the function to satisfy the partial differential equation

$$f_x = -if_y.$$

Writing $f(z) = u(z) + iv(z)$, where u and v are real-valued functions of a complex variable, and equating the real parts and imaginary parts of

$$u_x + iv_x = f_x = -if_y = v_y - iu_y,$$

we obtain the **Cauchy-Riemann** differential equations

$$\boxed{u_x = v_y, \qquad v_x = -u_y.}$$

We have proved:

THEOREM: If the function $f(z) = u(z) + iv(z)$ has a derivative at the point z, then the first partial derivatives of u and v, with respect to x and y, exist and satisfy the Cauchy–Riemann equations.

EXAMPLE.

Let $f(z) = z^2 = (x^2 - y^2) + 2xyi$. Since f is entire, $u = x^2 - y^2$ and $v = 2xy$ must satisfy the Cauchy-Riemann equations. Observe that

$$u_x = 2x = v_y \qquad \text{and} \qquad -u_y = 2y = v_x.$$

On the other hand, if $f(z) = |z|^2 = x^2 + y^2$, then $u = x^2 + y^2$, $v = 0$ and $u_x = 2x$, $u_y = 2y$, $v_x = 0 = v_y$, so f satisfies the Cauchy-Riemann equations only at 0. Moreover, f has a derivative when $z = 0$, since

$$f'(0) = \lim_{h \to 0} \frac{|h|^2}{h} = \lim_{h \to 0} \bar{h} = 0.$$

EXERCISES

In Exercises 1-4, prove that each function satisfies the Cauchy-Riemann equations.

1. $f(z) = e^x (\cos y + i \sin y)$
2. $f(z) = \cos x \cosh y - i \sin x \sinh y$
3. $f(z) = \sin x \cosh y + i \cos x \sinh y$
4. $f(z) = e^{x^2 - y^2} (\cos 2xy + i \sin 2xy)$

Using the rules for differentiation find the (complex) derivatives of the functions in Exercises 5–8.

5. $f(z) = 18z^3 - \dfrac{z^2}{4} + 4z + 8$

6. $f(z) = (2z^3 + 1)^5$

7. $f(z) = \dfrac{z + 1}{z - 1}, \qquad z \neq 1$

8. $f(z) = z^3 (z^2 + 1)^{-2}, \qquad z \neq \pm i$

Let f and g be analytic functions defined on the region G. Prove the rules of differentiation stated in Exercises 9 and 10.

9. $(f \pm g)' = f' \pm g'$

10. $\left(\dfrac{f}{g} \right)' = \dfrac{gf' - fg'}{g^2}, \quad$ with $g(z) \neq 0$ for any z in G

11. Show that the quotient $P(z)/Q(z)$ of two polynomials has a derivative at every point z where $Q(z) \neq 0$.

Using the Cauchy-Riemann equations, prove that the functions in Exercises 12–15 do not have a derivative at any point in $\mathbb{C}$.

12. $f(z) = \bar{z}$ 13. $f(z) = \mathrm{Re}\, z$
14. $f(z) = \mathrm{Im}\, z$ 15. $f(z) = |z|$

Using the Cauchy-Riemann equations and the definition of a derivative, determine where the functions in Exercises 16–19 have a derivative.

16. $f(z) = \bar{z}^2$ 17. $f(z) = (\mathrm{Re}\, z)^2$
18. $f(z) = \bar{z}\, \mathrm{Re}\, z$ 19. $f(z) = z\, \mathrm{Im}\, z$

20. Prove the chain rule of differentiation $[f(g(z))]' = f'(g(z))g'(z)$, assuming f and g are entire.

21. Using the chain rule, prove that an entire function of an entire function is entire.

*22. If all the zeros of a polynomial $P(z)$ have negative real parts, prove that the same is true for all the zeros of $P'(z)$. (*Hint:* Factor $P(z)$ and consider $P'(z)/P(z)$.)

23. If u and v are expressed in terms of polar coordinates (r, θ), show that the Cauchy-Riemann equations can be written in the form

$$\frac{\partial u}{\partial r} = \frac{1}{r} \frac{\partial v}{\partial \theta}, \qquad \frac{1}{r} \frac{\partial u}{\partial \theta} = -\frac{\partial v}{\partial r}, \qquad r \neq 0.$$

24. Prove that the function

$$f(z) = r^5 (\cos 5\theta + i \sin 5\theta)$$

satisfies the Cauchy–Riemann equations in polar form for all $z \neq 0$.

1.6 SUFFICIENT CONDITIONS FOR ANALYTICITY

At this point one might ask whether the Cauchy–Riemann equations are enough to guarantee the existence of a derivative at a given point. The following example by D. Menchoff shows that this is not the case. Let

$$f(z) = \begin{cases} \dfrac{z^5}{|z|^4}, & z \neq 0, \\ 0, & z = 0. \end{cases}$$

Then

$$\frac{f(z)}{z} = \left(\frac{z}{|z|} \right)^4, \qquad z \neq 0,$$

which has value 1 on the real axis and value -1 on the line $y = x$. Thus, f does not have a derivative at $z = 0$; but expanding the expression for f yields

$$u(x, 0) = x, \qquad u(0, y) = 0 = v(x, 0), \qquad v(0, y) = y.$$

Hence,

$$u_x(0, 0) = 1 = v_y(0, 0), \qquad -u_y(0, 0) = 0 = v_x(0, 0),$$

and the Cauchy–Riemann equations hold.

However, we do have the following theorem.

THEOREM

Let $f(z) = u(x, y) + iv(x, y)$, defined in some region G containing the point z_0, have *continuous* first partial derivatives, with respect to x and y, satisfying the Cauchy–Riemann equations at z_0. Then $f'(z_0)$ exists.

PROOF† Assuming $x \neq x_0$ and $y \neq y_0$, the difference quotient may be written in the form

$$\frac{f(z) - f(z_0)}{z - z_0} = \frac{u(x, y) - u(x_0, y_0)}{z - z_0} + i \; \frac{v(x, y) - v(x_0, y_0)}{z - z_0}$$

$$= \frac{x - x_0}{z - z_0} \left[\frac{u(x, y) - u(x_0, y)}{x - x_0} + i \; \frac{v(x, y) - v(x_0, y)}{x - x_0} \right]$$

$$+ \frac{y - y_0}{z - z_0} \left[\frac{u(x_0, y) - u(x_0, y_0)}{y - y_0} + i \; \frac{v(x_0, y) - v(x_0, y_0)}{y - y_0} \right]$$

$$= \frac{x - x_0}{z - z_0} \left\{ u_x \left(x_0 + t_1 (x - x_0), y \right) + i v_x \left(x_0 + t_2 (x - x_0), y \right) \right\}$$

$$+ \frac{y - y_0}{z - z_0} \left\{ u_y \left(x_0, y_0 + t_3 (y - y_0) \right) + i v_y \left(x_0, y_0 + t_4 (y - y_0) \right) \right\},$$

where $0 < t_k < 1$, $k = 1, 2, 3, 4$, by the mean value theorem of differential calculus. This result also holds if $x = x_0$ or $y = y_0$. Since the partials are continuous at z_0, we may write

$$\frac{f(z) - f(z_0)}{z - z_0} = \frac{x - x_0}{z - z_0} \left[u_x(z_0) + i v_x(z_0) + \epsilon_1 \right]$$

$$+ \frac{y - y_0}{z - z_0} \left[u_y(z_0) + i v_y(z_0) + \epsilon_2 \right],$$

where $\epsilon_1, \epsilon_2 \to 0$ as $z \to z_0$. Applying the Cauchy-Riemann equations to the last term, we may combine the terms, obtaining

$$\frac{f(z) - f(z_0)}{z - z_0} = u_x(z_0) + i v_x(z_0) + \frac{(x - x_0)\epsilon_1 + (y - y_0)\epsilon_2}{z - z_0}.$$

Since $|x - x_0|, |y - y_0| \leqslant |z - z_0|$, the triangle inequality yields

$$\left| \frac{(x - x_0)\epsilon_1 + (y - y_0)\epsilon_2}{z - z_0} \right| \leqslant |\epsilon_1| + |\epsilon_2| \to 0, \qquad \text{as } z \to z_0.$$

Hence, the last term tends to 0 as z tends to z_0; so taking the limit, we have

$$f'(z_0) = \lim_{z \to z_0} \frac{f(z) - f(z_0)}{z - z_0} = u_x(z_0) + i v_x(z_0)$$

In particular, if the hypotheses in the theorem hold at all points of the region G, then f is analytic in G. ∎

EXAMPLE 1.

Show that the function
$$f(z) = e^{x^2 - y^2} \left(\cos 2xy + i \sin 2xy \right)$$
is entire.

SOLUTION: We must check that the first partials of
$$u = e^{x^2 - y^2} \cos 2xy \text{ and } v = e^{x^2 - y^2} \sin 2xy$$
are continuous and satisfy the Cauchy–Riemann equations at all points of $\mathbb{C}$. Clearly,
$$u_x = 2e^{x^2 - y^2} \left(x \cos 2xy - y \sin 2xy \right) = v_y$$
and
$$-u_y = 2e^{x^2 - y^2} \left(y \cos 2xy + x \sin 2xy \right) = v_x$$
are continuous functions in $\mathbb{C}$, so $f(z)$ is entire.

EXAMPLE 2.

Describe the region of analyticity of the function
$$f(z) = \frac{(x - 1) - iy}{(x - 1)^2 + y^2}.$$

SOLUTION: The first partials of $u = \text{Re } f$ and $v = \text{Im } f$ satisfy
$$u_x = \frac{y^2 - (x - 1)^2}{[(x - 1)^2 + y^2]^2} = v_y$$
and
$$u_y = \frac{-2y(x - 1)}{[(x - 1)^2 + y^2]^2} = -v_x.$$
These functions are continuous for all $z \neq 1$. Note that $f(z)$ is not defined at $z = 1$. Hence, $f(z)$ is analytic for all $z \neq 1$.

In the real-variable case of elementary calculus, we know that when the derivative of a function is zero in some interval, then the function is constant in that interval. A similar result holds for complex variables.

ZERO DERIVATIVE THEOREM

Let f be analytic on a region G and $f'(z) = 0$ at each z in G. Then f is constant on G. The same conclusion holds if either Re f, Im f, $|f|$, or arg f is constant in G.

PROOF: Since $f'(z) = u_x(z) + iv_x(z)$, the vanishing of the derivative implies $u_x = v_y$, $v_x = -u_y$ are all zero. Thus, u and v are constant on lines parallel to the coordinate axes, and since G is polygonally connected (see the theorem and the remarks following its proof in Section 1.3), $f = u + iv$ is constant on G.

If u (or v) is constant, $v_x = -u_y = 0 = u_x = v_y$, implying $f'(z) = u_x(z) + iv_x(z) = 0$ and f is constant.

If $|f|$ is constant, so is $|f|^2 = u^2 + v^2$, implying that

$$uu_x + vv_x = 0, \qquad uu_y + vv_y = vu_x - uv_x = 0.$$

Solving these two equations for u_x, v_x, we have $u_x = v_x = 0$ unless the determinant $u^2 + v^2$ vanishes. Since $|f|^2 = u^2 + v^2$ is constant, if $u^2 + v^2 = 0$ at a single point, then it is constantly zero and f vanishes identically. Otherwise the derivative vanishes and f is constant.

If arg $f = c$, then $f(G)$ lies on the line

$$v = (\tan c) \cdot u,$$

unless $u \equiv 0$, in which case we are done. But $(1 - i \tan c) f$ is analytic and

$$\text{Im}(1 - i \tan c) f = v - (\tan c)u = 0,$$

implying $(1 - i \tan c) f$ is constant. Thus, so is f. ∎

EXERCISES

Prove that each of the functions in Exercise 1–5 is entire.

1. $f(z) = e^x (\cos y + i \sin y)$
2. $f(z) = \cos x \cosh y - i \sin x \sinh y$
3. $f(z) = \sin x \cosh y + i \cos x \sinh y$
4. $f(z) = (x^3 - 3xy^2) + i(3x^2 y - y^3)$
5. $f(z) = \sin(x^2 - y^2)\cosh(2xy) + i \cos(x^2 - y^2)\sinh(2xy)$

For Exercises 6–8, state where the functions are analytic.

6. $f(z) = \dfrac{x}{x^2 + y^2} - i \dfrac{y}{x^2 + y^2}$

7. $f(\bar{z}) = \sin\left(\dfrac{x}{x^2 + y^2}\right) \cosh\left(\dfrac{y}{x^2 + y^2}\right)$

$\qquad - i \cos\left(\dfrac{x}{x^2 + y^2}\right) \sinh\left(\dfrac{y}{x^2 + y^2}\right)$

8. $f(\bar{z}) = \frac{1}{2} \log(x^2 + y^2) + i \arctan \dfrac{y}{x}$

9. Show that at $z = 0$ the function

$$f(z) = \begin{cases} \dfrac{\bar{z}^3}{|z|^2}, & z \neq 0, \\ 0 & z = 0. \end{cases}$$

satisfies the Cauchy–Riemann equations but does not have a derivative.

10. Show that the function

$$f(z) = \begin{cases} e^{-1/z^4}, & z \neq 0, \\ 0, & z = 0, \end{cases}$$

satisfies the Cauchy–Riemann equations at $z = 0$ but does not have a derivative at that point.

11. If $f(z) = u + iv$ and $\bar{f} = u - iv$ are both analytic, prove f is constant.

12. Let $f(z) = u + iv$ be entire and suppose $u \cdot v$ is constant. Prove f is constant.

13. If $f(z) = u + iv$ is entire and $v = u^2$, then show that f is constant.

14. If $f = u + iv$ is entire and $u^2 = v^2$, then prove that f is constant.

15. Suppose the analytic function f is real-valued on the region G. Prove that f is constant on G.

16. Let $f(z) = z^3$, $z_1 = 1$, and $z_2 = i$. Prove there is no point z_0 on the line segment from z_1 to z_2 such that

$$f(z_2) - f(z_1) = f'(z_0)(z_2 - z_1).$$

This shows that the mean value theorem for real functions does not extend to complex functions.

17. If $z = x + iy$, show that no entire function has the function $f(z) = x$ as its derivative.

1.7 THE COMPLEX EXPONENTIAL

We have seen in Section 1.4 that polynomials and rational functions in a real variable yield analytic functions when the real variable is replaced by z. This is by no means an isolated example. In fact, all elementary functions in calculus, such as exponentials, logarithms, and trigonometric functions, give rise to analytic functions when suitably extended to the complex plane. In the next three sections, we shall define extensions of these elementary functions and indicate some of their properties.

We begin with the exponential e^x. We wish to define a function $f(z) = e^z$ that is analytic and coincides with the real exponential function when z is real. Recalling that the real exponential is determined by the differential equation

$$f'(x) = f(x), \qquad f(0) = 1,$$

we ask if there is an analytic solution of the equation

$$f'(z) = f(z), \qquad f(0) = 1.$$

If such a solution exists, it will necessarily coincide with e^x when $z = x$, as it will satisfy the determining equation on the real axis. By the definition of f', we have

$$u_x + iv_x = u + iv, \qquad u(0) = 1, \qquad v(0) = 0.$$

Since $u_x = u$, $v_x = v$, separating variables, we have

$$u(x, y) = p(y)e^x, \qquad v(x, y) = q(y)e^x,$$

with $p(0) = 1$, $q(0) = 0$ by the initial conditions. Differentiating these two equations with respect to y and applying the Cauchy-Riemann equations, we obtain

$$p'(y)e^x = u_y = -v_x = -q(y)e^x, \qquad q'(y)e^x = v_y = u_x = p(y)e^x.$$

Hence, $p' = -q$, $q' = p$, so that

$$q'' = p' = -q, \qquad p'' = -q' = -p,$$

and p, q are solutions of the real differential equation $\phi''(y) + \phi(y) = 0$. All solutions of this equation are of the form $A \cos y + B \sin y$, with A and B constants. Since $q'(0) = p(0) = 1$, $p'(0) = -q(0) = 0$, we must have $p(y) = \cos y$, $q(y) = \sin y$. Hence, we obtain the function

$$f(z) = e^x \cos y + ie^x \sin y = e^x (\cos y + i \sin y),$$

which coincides with e^x when $z = x$ and is analytic since the construc-

tion automatically guarantees that the partials are continuous and
satisfy the Cauchy–Riemann equations.

DEFINITION

The *complex exponential* given by

$$e^z = e^x(\cos y + i \sin y)$$

is a nonzero entire function satisfying the differential equation

$$f'(z) = f(z), \qquad f(0) = 1.$$

That $e^z \neq 0$ follows, since neither e^x nor $\cos y + i \sin y$ vanishes.
Observe further that since $z = x + iy$, the notation yields

$$e^{iy} = \cos y + i \sin y, \qquad |e^{iy}| = 1.$$

Thus, the polar representation of a complex number becomes (see
Section 1.2)

$$z = |z|e^{i \arg z}.$$

If $z_1 = x_1 + iy_1$ and $z_2 = x_2 + iy_2$, then the addition formulas of
trigonometry imply that

$$\begin{aligned}
e^{z_1} e^{z_2} &= e^{x_1} e^{x_2} (\cos y_1 + i \sin y_1)(\cos y_2 + i \sin y_2) \\
&= e^{x_1 + x_2} [(\cos y_1 \cos y_2 - \sin y_1 \sin y_2) + \\
&\qquad i(\sin y_1 \cos y_2 + \cos y_1 \sin y_2)] \\
&= e^{x_1 + x_2} [\cos(y_1 + y_2) + i \sin(y_1 + y_2)] \\
&= e^{x_1 + x_2} e^{i(y_1 + y_2)} = e^{z_1 + z_2}.
\end{aligned}$$

Since

$$e^{z_1 - z_2} e^{z_2} = e^{z_1 - z_2 + z_2} = e^{z_1},$$

it follows that

$$e^{z_1 - z_2} = e^{z_1}/e^{z_2}.$$

Using the sum of exponents formula repeatedly, we obtain $e^{nz} = (e^z)^n$.
This identity provides a quick proof of De Moivre's theorem by
letting $z = e^{i\theta}$:

$$(\cos \theta + i \sin \theta)^n = (e^{i\theta})^n = e^{in\theta} = \cos n\theta + i \sin n\theta,$$

for $n = 0, \pm 1, \pm 2, \ldots$.

Using this version of De Moivre's theorem we have

$$(1 - i)^{23} = (\sqrt{2}\, e^{-\pi i/4})^{23} = 2^{23/2}\, e^{-23\pi i/4}$$
$$= 2^{23/2}\, e^{\pi i/4} = 2^{11}(\sqrt{2}\, e^{\pi i/4})$$
$$= 2^{11}(1 + i).$$

The complex exponential will play a key role in applications. In order to understand the complex exponential thoroughly, we need to discuss its properties as a mapping. To visualize the mapping

$$w = e^{z} = e^{x}(\cos y + i \sin y),$$

observe that the infinite strip $-\pi \leqslant y < \pi$ is mapped onto $\mathcal{C} - \{0\}$: the points on the line segment $x = 0,\ -\pi \leqslant y < \pi$ are mapped one-to-one onto the circle $|w| = 1$, vertical lines left of the imaginary axis are mapped onto circles of radius $r < 1$, vertical lines right of the imaginary axis onto circles of radius $r > 1$, the left half of the strip in

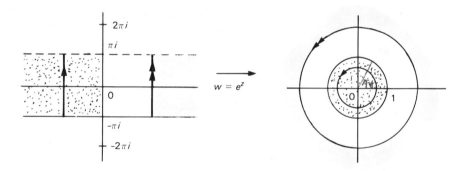

FIGURE 1.21. The exponential

Figure 1.21 is mapped onto $0 < |w| < 1$, and the right half goes onto $|w| > 1$. Observe that e^{z} has period $2\pi i$, because

$$e^{z+2\pi i} = e^{x+(2\pi+y)i} = e^{x}[\cos(2\pi + y) + i \sin(2\pi + y)] = e^{z},$$

so the complex values e^{z} and $e^{z+2\pi i k}$, k an integer, are identical. Hence, each infinite strip $-\pi \leqslant y - 2\pi k < \pi$, $k = 0, \pm 1, \pm 2, \ldots$, is also mapped onto $\mathcal{C} - \{0\}$, and the mapping $e^{z}: \mathcal{C} \to \mathcal{C} - \{0\}$ sends infinitely many points in $\mathcal{C}$ to the same point in $\mathcal{C} - \{0\}$. This is an undesirable development, since it prevents the discussion of an inverse function except on each of the infinite strips described above. The inverse function is certain to be important because the inverse of the real exponential is the logarithm. To eliminate this difficulty, imagine the range of the mapping to consist of infinitely many copies of $\mathbb{C} - \{0\}$ stacked as layers one upon another, each cut along the negative real axis with the upper edge of one layer "glued" to the lower edge of the layer above, yielding a set $\mathfrak{R}$ resembling an infinite spiraling ramp (see Figure 1.22). The set $\mathfrak{R}$ differs from $\mathbb{C} - \{0\}$ in that each point on $\mathfrak{R}$ is uniquely determined in polar coordinates, whereas points in $\mathbb{C} - \{0\}$ cannot be uniquely determined in polar

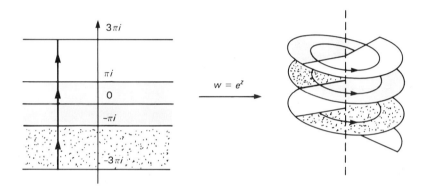

FIGURE 1.22. The Riemann surface of $w = e^z$

coordinates because the argument is multivalued. Using $\mathfrak{R}$ as the range of the function e^z and measuring short distances in $\mathfrak{R}$ in the obvious fashion, we observe that e^z maps $\mathbb{C}$ continuously onto $\mathfrak{R}$ and that the mapping is one-to-one. Thus, $e^z : \mathbb{C} \to \mathfrak{R}$ has an inverse, which we will study in Section 1.9. The analyticity of e^z is not affected by making this change in the range set because

$$\frac{e^{z+h} - e^z}{h} = e^z \left(\frac{e^h - e^0}{h} \right)$$

and the quantity in parentheses tends to e^0 as $h \to 0$, when e^h lies in the same layer of $\mathfrak{R}$ as e^0. Alternatively, if Im $z \neq (2k + 1)\pi$ and h is small, z and $z + h$ will lie in the same strip, so e^z and e^{z+h} lie on the same copy of $\mathbb{C} - \{0\}$.

The set $\mathfrak{R}$ is called a **Riemann surface**; the cut lines on each copy of $\mathbb{C} - \{0\}$ are called **branch cuts**; the ends of the branch cuts $0, \infty$ are called **branch points**; and each copy of $\mathbb{C} - \{0\}$ is called a **branch** of $\mathfrak{R}$.

EXERCISES

In Exercises 1-7, express each number in the form $x + iy$.

1. $e^{i\pi}$
2. $e^{(1+\pi i)/2}$
3. $e^{-1+(\pi i/4)}$
4. $e^{(-1+\pi i)/4}$
5. $e^{3i\pi/2}$
6. $e^{-i\pi/2}$
7. $e^{7\pi i/2}$.

In Exercises 8-10, find all the complex numbers z that satisfy the given conditions.

8. $e^{2z} = -1$ 9. $e^{iz} = 2$ 10. $e^{iz} = -1$
11. Obtain all values of $e^{\pi i k/2}$, k an integer
12. Show that $\overline{(e^z)} = e^{\bar{z}}$.

In Exercises 13-20, calculate each number using De Moivre's theorem.

13. $(1 + i)^{29}$
14. $(-1 + i)^{17}$
15. $(-1 - i)^{36}$
16. $(2 + 2i)^{12}$
17. $(\sqrt{3} + i)^{15}$
18. $(-\sqrt{3} + i)^{13}$
19. $(1 - \sqrt{3}i)^{14}$
20. $\left(\dfrac{1}{\sqrt{2}} - \dfrac{1}{\sqrt{2}}i\right)^{19}$

In Exercises 21-24, find the sums using De Moivre's theorem.

21. $1 + \cos x + \cos 2x + \ldots + \cos nx$
22. $\cos x + \cos 3x + \cos 5x + \ldots + \cos(2n - 1)x$
23. $\sin x + \sin 2x + \sin 3x + \ldots + \sin nx$
24. $\sin x + \sin 3x + \sin 5x + \ldots + \sin(2n - 1)x$
25. If $f(z)$ is entire, show that $e^{f(z)}$ is entire and find its derivative.
26. Prove that e^z is the only analytic solution to the complex differential equation $f'(z) = f(z), f(0) = 1$.

What is the image of the set $\{z: |x| < 1, |y| < 1\}$ under the mappings given in Exercises 27 and 28?

27. $w = e^{\pi z}$
28. $w = e^{\pi z/2}$

29. Find an analytic function mapping $\{z: 0 < x < 1, 0 \leqslant y < 1\}$ one-to-one and onto $1 < |w| < e^{2\pi}$.

1.8 THE COMPLEX TRIGONOMETRIC AND HYPERBOLIC FUNCTIONS

The complex exponential can be used to define the complex trigono-metric functions. Since $e^{iy} = \cos y + i \sin y$ and $e^{-iy} = \cos y - i \sin y$, it follows that

$$\cos y = \frac{e^{iy} + e^{-iy}}{2}, \qquad \sin y = \frac{e^{iy} - e^{-iy}}{2i}.$$

We extend these definitions to the complex planes as follows.

DEFINITION

$$\cos z = \frac{e^{iz} + e^{-iz}}{2}, \qquad \sin z = \frac{e^{iz} - e^{-iz}}{2i}.$$

These functions are entire as they are sums of entire functions and satisfy

$$(\cos z)' = \frac{ie^{iz} - ie^{-iz}}{2} = -\frac{e^{iz} - e^{-iz}}{2i} = -\sin z,$$

$$(\sin z)' = \frac{ie^{iz} + ie^{-iz}}{2i} = \frac{e^{iz} + e^{-iz}}{2} = \cos z.$$

The other four trigonometric functions, defined in terms of the sine and cosine function by the usual relations

$$\tan z = \frac{\sin z}{\cos z}, \qquad \cot z = \frac{\cos z}{\sin z},$$

$$\sec z = \frac{1}{\cos z}, \qquad \csc z = \frac{1}{\sin z},$$

are analytic except where their denominators vanish, and satisfy the standard rules of differentiation (see Exercise 22)

$$(\tan z)' = \sec^2 z, \qquad (\sec z)' = \sec z \tan z,$$

$$(\cot z)' = -\csc^2 z, \qquad (\csc z)' = -\csc z \cot z.$$

All the usual trigonometric identities are still valid in complex variables, the proofs depending on properties of the exponential. For example,

$$\cos^2 z + \sin^2 z = \tfrac{1}{4} \left[(e^{iz} + e^{-iz})^2 - (e^{iz} - e^{-iz})^2 \right] = 1,$$

and

$$\cos z_1 \cos z_2 - \sin z_1 \sin z_2$$

$$= \frac{e^{iz_1} + e^{-iz_1}}{2} \frac{e^{iz_2} + e^{-iz_2}}{2} - \frac{e^{iz_1} - e^{-iz_1}}{2i} \frac{e^{iz_2} - e^{-iz_2}}{2i}$$

$$= \frac{2e^{iz_1} e^{iz_2} + 2e^{-iz_1} e^{-iz_2}}{4} = \cos(z_1 + z_2).$$

From the definition of $\cos z$ we have

$$\cos z = \cos(x + iy) = \frac{e^{-y + ix} + e^{y - ix}}{2}$$

$$= \tfrac{1}{2}e^{-y}(\cos x + i \sin x) + \tfrac{1}{2}e^{y}(\cos x - i \sin x)$$

$$= \left(\frac{e^{y} + e^{-y}}{2}\right) \cos x - i \left(\frac{e^{y} - e^{-y}}{2}\right) \sin x.$$

Thus,

$$\cos z = \cos x \cosh y - i \sin x \sinh y.$$

Similarly, we find

$$\sin z = \sin x \cosh y + i \cos x \sinh y.$$

THEOREM

The real zeros of sin z and cos z are their only zeros.

PROOF: If $\sin z = 0$, the last equation shows we must have

$$\sin x \cosh y = 0, \qquad \cos x \sinh y = 0.$$

But $\cosh y \geqslant 1$, implying the first term vanishes only when $\sin x = 0$, that is, $x = 0, \pm\pi, \pm2\pi, \ldots$ However, for these values $\cos x$ does not vanish. Hence, we must have $\sinh y = 0$, or $y = 0$. Thus

$$\sin z = 0 \quad \text{implies} \quad z = n\pi, \qquad n \text{ an integer.}$$

This statement also applies to $\tan z$, and in like manner we find

$$\cos z = 0 \quad \text{implies} \quad z = (n + \tfrac{1}{2})\pi, \qquad n \text{ an integer.} \blacksquare$$

The complex hyperbolic functions are defined by extending the real definitions to the complex plane.

DEFINITION

$$\sinh z = \frac{e^z - e^{-z}}{2}, \qquad \cosh z = \frac{e^z + e^{-z}}{2}.$$

Again, all the usual identities and rules of differentiation apply to the complex hyperbolic functions (see Exercises 23–30). Note, moreover, that

$$\sinh iz = \frac{e^{iz} - e^{-iz}}{2} = i \sin z$$

and

$$\cosh iz = \frac{e^{iz} + e^{-iz}}{2} = \cos z.$$

Thus, the complex hyperbolic functions are intimately related to the complex trigonometric functions, as multiplying by i simply rotates every vector in $\mathcal{C}$ counterclockwise by $90°$. Hence, the zeros of $\sinh z$ and $\cosh z$ are pure imaginary.

EXERCISES

In Exercises 1–8, express each of the numbers in the form $x + iy$.

1. $\sin i$ 2. $\cos (-i)$
3. $\cosh (1 + i)$ 4. $\sinh \pi i$
5. $\cos (1 + i)$ 6. $\tan 2i$
7. $\sinh (1 + \pi i)$ 8. $\cosh (\pi i/4)$

In Exercises 9–12, find all complex numbers z such that the given conditions are met.

9. $\cos z = \sin z$ 10. $\cos z = -i \sin z$
11. $\cosh z = 2$ 12. $\cosh z = i$
13. Are there any points z where $\sinh z = \cosh z$?
14. Show that $\overline{\sin z} = \sin \bar{z}$.
15. Show that $\overline{\cos z} = \cos \bar{z}$.

In Exercises 16–21, prove the identities.

16. $\sin(z_1 \pm z_2) = \sin z_1 \cos z_2 \pm \cos z_1 \sin z_2$
17. $\cos(z_1 - z_2) = \cos z_1 \cos z_2 + \sin z_1 \sin z_2$
18. $\sin(-z) = -\sin z, \qquad \cos(-z) = \cos z$

19. $\sin 2z = 2 \sin z \cos z, \qquad \cos 2z = \cos^2 z - \sin^2 z,$

$$\tan 2z = \frac{2 \tan z}{1 - \tan^2 z}$$

20. $|\sin z|^2 = \sin^2 x + \sinh^2 y$
21. $|\cos z|^2 = \cos^2 x + \sinh^2 y$
22. Prove that the rules of differentiation for the functions $\tan z$, $\cot z$, $\sec z$, and $\csc z$ are valid as stated.

In Exercises 23–27, prove the identities.

23. $\cosh^2 z - \sinh^2 z = 1$, $\cosh(-z) = \cosh z$, $\sinh(-z) = -\sinh z$
24. $\sinh(z_1 + z_2) = \sinh z_1 \cosh z_2 + \cosh z_1 \sinh z_2$
25. $\cosh(z_1 + z_2) = \cosh z_1 \cosh z_2 + \sinh z_1 \sinh z_2$
26. $i \sinh z = \sin iz, \qquad \cosh z = \cos iz, \qquad i \tanh z = \tan iz$
27. $|\sinh z|^2 = \sinh^2 x + \sin^2 y$, $|\cosh z|^2 = \sinh^2 x + \cos^2 y$

Prove the rules of differentiation given in Exercises 28–30.

28. $(\sinh z)' = \cosh z, \qquad (\cosh z)' = \sinh z$
29. $(\tanh z)' = \mathrm{sech}^2 z, \qquad (\coth z)' = -\mathrm{csch}^2 z$
30. $(\mathrm{sech}\, z)' = -\mathrm{sech}\, z \tanh z, \qquad (\mathrm{csch}\, z)' = -\mathrm{csch}\, z \coth z$
31. Find all the zeros of $\sinh z$ and $\cosh z$
32. Verify that $e^z = \cosh z + \sinh z$.
33. Verify that $e^{iz} = \cos z + i \sin z$.

Show that the function $w = \sin z$ maps each of the strips in Exercises 34–36 onto the given set by indicating what happens to horizontal and vertical line segments under the transformation

$$w = \sin z = \sin x \cosh y + i \cos x \sinh y.$$

34. The strip $|x| < \pi/2$ onto $\mathbb{C} - \{z: y = 0, |x| \geq 1\}$
35. The semiinfinite strip $|x| < \pi/2, y > 0$ onto the upper half plane
36. The semiinfinite strip $0 < x < \pi/2, y > 0$ onto the first quadrant
37. Describe the function $w = \cos z$ by considering the mapping of vertical and horizontal line segments under the transformation

$$w = \cos z = \cos x \cosh y - i \sin x \sinh y.$$

1.9 THE COMPLEX LOGARITHM AND COMPLEX POWER FUNCTIONS

Since $e^z: \mathbb{C} \to \mathcal{R}$ is one-to-one, where $\mathcal{R}$ is the Riemann surface defined in Section 1.7, we can define its inverse function mapping $\mathcal{R}$

onto $\mathbb{C}$. Imitating the real-valued case, we call this inverse mapping the **logarithm** and we denote it by

$$\log z : \mathfrak{R} \to \mathbb{C}.$$

Since the complex exponential and the logarithm are inverse functions, it follows that

$$\log e^z = z, \qquad \text{for any } z \text{ in } \mathbb{C},$$

and

$$e^{\log z} = z, \qquad \text{for any } z \text{ in } \mathfrak{R}.$$

The only remaining task is to obtain an expression for $\log z$ in terms of known functions. One complication is that the logarithm is defined on the Riemann Surface $\mathfrak{R}$ illustrated in Figure 1.22. Since $\mathfrak{R}$ consists of infinitely many copies of $\mathbb{C} - \{0\}$ stacked to resemble an infinite spiraling ramp, we must find a way of identifying the points on each branch of the Riemann surface. It is at this point that a previous complication becomes an asset: Although the argument arg z is multivalued on $\mathbb{C} - \{0\}$, it is *single-valued* on $\mathfrak{R}$. Thus, we can distinguish between different branches of $\mathfrak{R}$ by using the polar representation $z = |z|e^{i \arg z}$ for each z in $\mathfrak{R}$. The polar representation and the inverse nature of the logarithm and exponential functions provide a natural definition for the complex logarithm:

$$\log z = \log(|z|e^{i \arg z}) = \log(e^{\log |z| + i \arg z})$$
$$= \log |z| + i \arg z,$$

where $\log |z|$ is the natural logarithm of elementary calculus.

To complete the description of the Riemann surface $\mathfrak{R}$ we define the ϵ-neighborhoods for the points on $\mathfrak{R}$. If z lies on a branch of $\mathfrak{R}$ and $|z| > \epsilon$, then the set of all points on that branch whose distance from z is less than ϵ constitute an ϵ-neighborhood of z. This concept is important because limits are defined in terms of ϵ-neighborhoods. By defining an ϵ-neighborhood on a Riemann surface, we can extend the notions of continuity, differentiability, and analyticity of a function defined on that Riemann surface, since these notions depend only on the local behavior of the function. That is, continuity at z depends only on the difference $f(z) - f(w)$ for any w in any ϵ-neighborhood of z, while the derivative at z depends only on the difference quotient $[f(z) - f(w)]/(z - w)$. Using these concepts it is not hard to verify that $\log z$ is continuous, since

$$\log z - \log w = \log |z| + i \arg z - \log |w| - i \arg w$$
$$= [\log |z| - \log |w|] + i[\arg z - \arg w],$$

and the natural logarithm and argument function are continuous.

THEOREM

The function $\log z = \log |z| + i \arg z$ is analytic for all z in $\mathcal{R}$.

PROOF: Since $u = \log |z| = \frac{1}{2} \log(x^2 + y^2)$, $v = \arg z = \tan^{-1} y/x + \pi n$

$$u_x = \frac{x}{x^2 + y^2}, \quad u_y = \frac{y}{x^2 + y^2}, \quad v_x = \frac{-y}{x^2 + y^2}, \quad v_y = \frac{x}{x^2 + y^2},$$

the Cauchy–Riemann equations hold and the partial derivatives are all continuous in $\mathcal{R}$. Because analyticity is a local property and the proof of the theorem on sufficient conditions for analyticity in Section 1.6 relies on a local argument, $\log z$ is analytic in $\mathcal{R}$. ∎

The complex logarithm has the usual properties of a logarithm:

$$\log z_1 z_2 = \log z_1 + \log z_2,$$

$$\log \frac{z_1}{z_2} = \log z_1 - \log z_2.$$

Note that in these two identities, we are assuming that z_1 and z_2 are points of the Riemann surface $\mathcal{R}$. Since

$$z = e^{\log z}$$

for any z in $\mathcal{R}$, we apply the chain rule of differentiation, obtaining

$$1 = e^{\log z} (\log z)'$$

or

$$(\log z)' = 1/z, \qquad \text{for } z \text{ in } \mathcal{R}.$$

Thus, the usual differentiation formula holds on $\mathcal{R}$.

Just as it is convenient to define the principal value Arg z of the argument arg z, we can extend this concept to the logarithm. Visualizing the logarithm as the inverse mapping of the exponential, we call the branch of $\mathcal{R}$ cut along the negative real axis, which is mapped onto the infinite strip $-\pi \leqslant y < \pi$, the **principal branch** of the logarithm (see Figure 1.23). We denote $\log z$ when restricted to the principal branch by

$$\text{Log } z = \log |z| + i \text{ Arg } z,$$

and we call this the **principal value** of $\log z$.

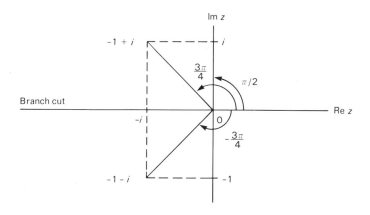

FIGURE 1.23. Principal branch of $\mathcal{R}$

Note that the principal value $\text{Log } z$ is defined only on the branch of $\mathcal{R}$ for which $\text{Arg } z$ exists. Care must be taken when working with the principal branch of the logarithm $\text{Log } z$, as the usual properties of logarithms may not apply. For example,

$$\text{Log } i = \log |i| + i \text{ Arg } i = i\pi/2,$$
$$\text{Log } (-1 + i) = \log |-1 + i| + i \text{ Arg } (-1 + i)$$
$$= \log\sqrt{2} + i \ \frac{3\pi}{4},$$

but

$$\text{Log } [i(-1 + i)] = \text{Log}(-1 - i)$$
$$= \log |-1 - i| + i \text{ Arg } (-1 - i)$$
$$= \log\sqrt{2} - i \ \frac{3\pi}{4},$$

so that

$$\text{Log } [i(-1 + i)] \neq \text{Log } i + \text{Log}(-1 + i).$$

Instead, the two expressions differ by a multiple of $2\pi i$. (Why?)

The complex logarithm and exponential functions may be used to define the power functions.

DEFINITION

$$z^a = e^{a \log z}, \qquad a \text{ complex} \neq 0, \qquad z \neq 0.$$

The function $z^a: \mathcal{R} \to \mathcal{R}$ is analytic and one-to-one, as it is the composition of such functions. By the chain rule,

$$\left(z^a\right)' = e^{a \log z} \cdot \frac{a}{z} = az^{a-1}.$$

The *principal value* of the power function is given by

$$z^a = e^{a \operatorname{Log} z}.$$

We are often interested in the case where $a = m/n > 0$, m, n positive integers with no common factors. Consider the set of numbers $e^{\operatorname{Log}(z) + 2\pi k i}$, $k = 0, \pm 1, \pm 2, \ldots$, that is, those points in $\mathcal{R}$ lying directly "above" and "below" the point $e^{\operatorname{Log} z}$. Then $\left(e^{\operatorname{Log}(z) + 2\pi k i}\right)^{m/n} = e^{(m/n)\operatorname{Log} z} e^{(m/n)2\pi k i}$ and writing $k = pn + q$ with p and q integers, $0 \leqslant q < n$, we have

$$e^{(m/n)2\pi k i} = e^{2\pi p m i} e^{2\pi i q m/n} = e^{2\pi i q m/n},$$

so there are only n different complex-valued answers. Thus, the mapping $z^{m/n}: \mathcal{R} \to \mathcal{R}$ takes every n copies of $\mathcal{C} - \{0\}$ onto one copy of $\mathcal{C} - \{0\}$ and repeats itself thereafter. This fact makes it possible to simplify the model used in describing the mapping $w = z^{m/n}$. For simplicity, suppose $m = 1$. Then

$$w = z^{1/n} = e^{(1/n)\operatorname{Log} z} e^{2\pi i q/n}, \qquad q = 0, 1, \ldots, n-1,$$

may be visualized as a mapping of $[\mathcal{C} - \{0\}]^n$ onto $[\mathcal{C} - \{0\}]$, where $[\mathcal{C} - \{0\}]^n$ consists of n copies of $\mathcal{C} - \{0\}$ "glued" one after another along the negative real axis, as in $\mathcal{R}$, except that the upper edge of the top branch is "glued" to the lower edge of the bottom branch.

EXAMPLE 1.

Describe the modified Riemann surface of the function

$$w = \sqrt{z}.$$

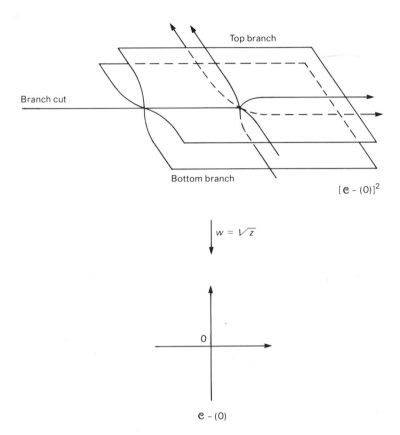

FIGURE 1.24. The Riemann surface for $w = z^{1/2}$

SOLUTION: From the discussion above, the function maps $[\mathbb{C} - \{0\}]^2$ onto $[\mathbb{C} - \{0\}]$, as illustrated in Figure 1.24. We may view the top branch as mapping onto the right half plane and the bottom branch as mapping onto the left half plane.

The mapping $z^m = [\mathbb{C} - \{0\}] \to [\mathbb{C} - \{0\}]^m$ is the inverse of the mapping $z^{1/m}$. Hence, the composite function

$$(z^{1/n})^m = z^{m/n} : [\mathbb{C} - \{0\}]^n \to [\mathbb{C} - \{0\}]^m$$

is analytic and one-to-one on the modified Riemann surfaces described above.

The logarithm can also be used to define the inverse trigonometric functions.

EXAMPLE 2.

Show that

$$\sin^{-1} z = -i \log \left[iz + (1 - z^2)^{\frac{1}{2}} \right].$$

SOLUTION. The function $w = \sin^{-1} z$ reverses the action of the mapping

$$z = \sin w = \frac{e^{iw} - e^{-iw}}{2i}.$$

Multiplying both ends of this equation by $2ie^{iw}$, we have

$$e^{2iw} - 2ize^{iw} - 1 = 0.$$

Using the quadratic formula to solve for e^{iw}, we obtain

$$e^{iw} = iz + (1 - z^2)^{\frac{1}{2}},$$

where the square root maps $[\mathbb{C} - \{0\}]^2$ onto $[\mathbb{C} - \{0\}]$ (or is two-valued on $\mathbb{C}$). The result now follows by taking the logarithm of both sides.

The usual identities and rules for differentiation for inverse trigonometric and hyperbolic functions apply here too. In fact, most of the mathematical functions arising in physical and engineering problems are analytic. Thus, the concept of analyticity applies to a large, useful class of functions.

EXERCISES

In Exercises 1–6, find all the values of the given expressions.

1. $\log i$ 2. $\log (1 + i)$
3. $\log (-1)$ 4. 1^i
5. i^i 6. $(1 + i)^{1+i}$

In Exercises 7–10, find the principal values of the given expressions

7. $\log i$ 8. $\log(1 + i)$
9. i^i 10. $(1 + i)^{1+i}$

11. For what complex numbers a can z^a be extended continuously at $z = 0$? When is the resulting function entire?

12. Prove that $\log z$ is the only analytic solution of the differential equation

$$f'(z) = \frac{1}{z}, \qquad f(1) = 0,$$

in the disk $|z - 1| < 1$.

13. Show that $\log z_1 + \log z_2 = \log z_1 z_2$.

14. Show that $\log z_1 - \log z_2 = \log \dfrac{z_1}{z_2}$.

15. Show that $z^a z^b = z^{a+b}$.

16. Show that $\dfrac{z^a}{z^b} = z^{a-b}$.

17. Show that $\text{Log}(-1 - i) - \text{Log}\, i \neq \text{Log} \left(\dfrac{-1 - i}{i} \right)$.

18. Show that $\text{Log}\,(i^3) \neq 3\,\text{Log}\, i$.

19. Prove that $\log z^a = a \log z$, a complex $\neq 0$, $z \neq 0$.

20. Is 1 raised to any power always equal to 1?

21. Prove that $\cos^{-1} z = -i \log [z + (z^2 - 1)^{\frac{1}{2}}]$.

22. Prove that $\tan^{-1} z = \dfrac{i}{2} \log \left(\dfrac{i + z}{i - z} \right)$, $z \neq \pm i$.

23. Prove that $\cot^{-1} z = \dfrac{i}{2} \log \left(\dfrac{z - i}{z + i} \right)$, $z \neq \pm i$.

24. Prove that $\sinh^{-1} z = \log [z + (z^2 + 1)^{\frac{1}{2}}]$.

25. Prove that $\cosh^{-1} z = \log \left[z + (z^2 - 1)^{\frac{1}{2}} \right]$.

26. Prove that $\tanh^{-1} z = \frac{1}{2} \log \left(\dfrac{1+z}{1-z} \right)$, $z \neq \pm 1$.

27. Prove that $\left(\sin^{-1} z \right)' = (1 - z^2)^{-\frac{1}{2}}$, $z \neq \pm 1$.

28. Prove that $\left(\cos^{-1} z \right)' = -(1 - z^2)^{-\frac{1}{2}}$, $z \neq \pm 1$.

29. Prove that $\left(\tan^{-1} z \right)' = \dfrac{1}{1+z^2}$, $z \neq \pm i$.

30. Prove that $\left(\sinh^{-1} z \right)' = (1 + z^2)^{-\frac{1}{2}}$, $z \neq \pm i$.

31. Prove that $\left(\cosh^{-1} z \right)' = (z^2 - 1)^{-\frac{1}{2}}$, $z \neq \pm 1$.

32. Prove that $\left(\tanh^{-1} z \right)' = \dfrac{1}{1-z^2}$, $z \neq \pm 1$.

33. Find the flaw in the following argument:

$$i = (-1)^{1/2} = [(-1)^3]^{1/2} = (-1)^{3/2} = i^3 = -i.$$

1.10 APPLICATIONS IN OPTICS (Optional)

One of the models that has been suggested for interpreting the empirical properties of light assumes that a light source creates a disturbance that produces spherical waves in a homogeneous medium. This model is analogous to the ever-widening circle of ripples that occur when the surface of a body of water is disturbed. A mathematical analysis of this model (using James Maxwell's equations of electromagnetism) leads to the one-dimensional **wave equation**

$$\frac{\partial^2 E}{\partial x^2} = \frac{1}{c^2} \frac{\partial^2 E}{\partial t^2},$$

where E is the optical disturbance, x is the direction of propagation of the wave, c is the velocity of the propagation of light, and t is time (see Exercise 4). It is easy to prove that any function of the form $E = f(ct - x)$ is a solution of the wave equation, as

$$\frac{\partial^2 E}{\partial x^2} = \frac{\partial}{\partial x} \left[-f'(ct - x) \right] = f''(ct - x)$$

and

$$\frac{\partial^2 E}{\partial t^2} = \frac{\partial}{\partial t} \left[cf'(ct - x) \right] = c^2 f''(ct - x).$$

Observation of the **interference** effects that occur when two beams of light from a common source arrive at the same point along different paths suggests that the optical disturbance consists of a sum of nearly sinusoidal functions. That is, E can be closely approximated by a sum of **sinusoidal waves** of the form

$$A \cos \left[\omega \left(t - \frac{x}{c} \right) + \phi \right],$$

where A is the **amplitude**, $\omega/2\pi$ is the **frequency**, and $\alpha = \phi - \omega x/c$ is the **phase shift** of the wave.

We can easily add sinusoidal waves of the same frequency by using the complex exponential:

$$A_1 \cos \left(\omega t + \alpha_1 \right) + \ldots + A_n \cos \left(\omega t + \alpha_n \right)$$
$$= \mathrm{Re} \left[A_1 e^{i(\omega t + \alpha_1)} + \ldots + A_n e^{i(\omega t + \alpha_n)} \right]$$
$$= \mathrm{Re} \, A e^{i(\omega t + \alpha)} = A \cos \left(\omega t + \alpha \right),$$

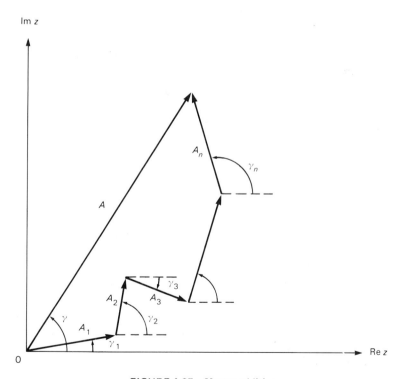

FIGURE 1.25. Vector addition

where we obtain

$$A_1 e^{i\alpha_1} + \ldots + A_n e^{i\alpha_n} = A e^{i\alpha}$$

graphically by using the parallelogram law of vector addition (see Figure 1.25).

Telescopic images are affected by **interference fringes** which occur when entering rays of light are transmitted and reflected at the surfaces of glass plates and air spaces in the telescope. Consider Figure 1.26, where a ray of light from a distant source S reaches the

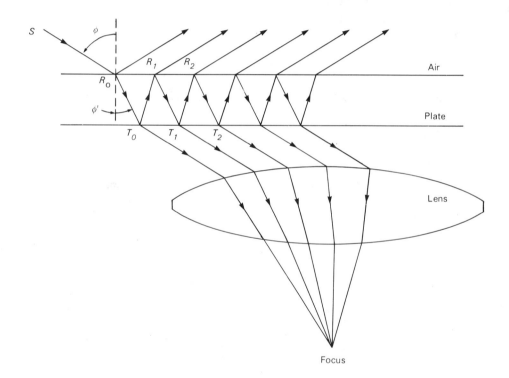

FIGURE 1.26. Multiple wave interference

glass plate at R_0. Part of the incident ray is reflected, while the rest is transmitted through the plate. At T_0, part is reflected to R_1 and part is transmitted and focused by the lens. On reaching R_1, part of the ray is reflected to T_1 and another part is transmitted. And so on.

Let $r(s)$ be the ratio of the reflected (transmitted) to incident amplitude, and suppose the initial incident ray has amplitude A. Then the optical disturbance at T_0 is given by

$$E_{T_0} = s^2 A \text{ Re } e^{i\omega t},$$

because the ray has been transmitted through two surfaces whereas the optical disturbance at T_1 is given by

$$E_{T_1} = s^2 r^2 A \text{ Re } e^{i(\omega t - \alpha)}$$

where α is the phase shift that arises from the rays traveling the extra distance during the two reflections at T_0 and R_1 (see Exercise 2). Similarly

$$E_{T_2} = s^2 r^4 A \text{ Re } e^{i(\omega t - 2\alpha)}$$

and

$$E_{T_n} = s^2 r^{2n} A \text{ Re } e^{i(\omega t - n\alpha)}.$$

To determine the resulting optical disturbance, we add all these disturbances together, obtaining

$$E = \sum_{n=0}^{\infty} E_{T_n} = s^2 A \text{ Re } \left[e^{i\omega t} \sum_{n=0}^{\infty} (r^2 e^{-i\alpha})^n \right]$$

$$= s^2 A \text{ Re } \left(\frac{e^{i\omega t}}{1 - r^2 e^{-i\alpha}} \right).$$

The last equality follows by observing that the *geometric series*

$$G = \sum_{n=0}^{\infty} z^n = 1 + z + z^2 + \ldots + z^n + \ldots$$

satisfies the identity $zG = G - 1$ or $(1 - z)G = 1$. More will be said about this series in Chapter 3. But

$$\frac{e^{i\omega t}}{1 - r^2 e^{-i\alpha}} = \frac{e^{i\omega t}}{1 - r^2 e^{-i\alpha}} \cdot \frac{1 - r^2 e^{i\alpha}}{1 - r^2 e^{i\alpha}}$$

$$= \frac{e^{i\omega t}(1 - r^2 e^{i\alpha})}{1 + r^4 - 2r^2 \cos \alpha},$$

so that

$$E = \frac{s^2 A}{1 + r^4 - 2r^2 \cos \alpha} \cdot \text{Re } [e^{it\omega} - r^2 e^{i(\omega t + \alpha)}]$$

$$= \frac{s^2 A [\cos \omega t - r^2 \cos (\omega t + \alpha)]}{1 + r^4 - 2r^2 \cos \alpha}$$

$$= \frac{s^2 A [(1 - r^2 \cos \alpha) \cos \omega t + (r^2 \sin \alpha) \sin \omega t]}{1 + r^4 - 2r^2 \cos \alpha}.$$

Since

$$(1 - r^2 \cos \alpha)^2 + (r^2 \sin \alpha)^2 = 1 + r^4 - 2r^2 \cos \alpha,$$

letting

$$\cos \beta = \frac{1 - r^2 \cos \alpha}{\sqrt{1 + r^4 - 2r^2 \cos \alpha}}$$

yields the transmitted optical disturbance

$$E = \frac{s^2 A \cos (\omega t - \beta)}{\sqrt{1 + r^4 - 2r^2 \cos \alpha}}.$$

The law of conservation of energy implies that $s + r = 1$ and

$$E = \frac{(1 - r)^2 A}{\sqrt{1 + r^4 - 2r^2 \cos \alpha}} \cdot \cos (\omega t - \beta).$$

The phase angle α depends on the length of the path the light travels during the reflections at T_0 and R_1 (see Exercise 2). Since $\omega = 2\pi c/\lambda$, where the **wavelength** λ is the distance between successive maxima of the wave, we obtain

$$\alpha = \frac{2\pi (\ell_2 - \ell_1)}{\lambda}.$$

If α is an integer multiple of 2π, then the amplitude of the optical disturbance is

$$\frac{(1 - r)^2 A}{\sqrt{(1 - r^2)^2}} = \frac{1 - r}{1 + r} A,$$

while odd multiples of π yield an amplitude equal to

$$\frac{(1 - r)^2 A}{\sqrt{(1 + r^2)^2}} = \frac{(1 - r)^2}{1 + r^2} A.$$

Small changes in the angle of incidence ϕ in Figure 1.26 may produce substantial changes in the phase angle α. Thus, neighboring rays of light with identical incidence amplitudes will produce images with *different* amplitudes. If r is close to 1 (large reflectance), the change in amplitude

$$\frac{\dfrac{1-r}{1+r}A}{\dfrac{(1-r)^2}{1+r^2}A} = \frac{1+r^2}{1-r^2}$$

will be large, yielding an image that consists of narrow bright lines on a dark background. This halolike effect is called the *interference fringe*. If r is close to zero (small reflectance), the change of amplitude will be small and the interference fringes will be broad and faint.

EXERCISES

1. Prove that the reflected optical disturbance in Figure 1.26 also has the form $E = A^* \cos(\omega t - \gamma)$. Find A^* and γ.
2. Consider Figure 1.27.

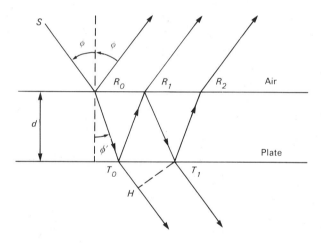

FIGURE 1.27. Snell's law

Let ℓ_1 be the distance from the source S to H and ℓ_2 the distance from S to T_1. **Snell's law** of refraction states that

$$v \sin \phi = v' \sin \phi',$$

where ϕ and ϕ' are the angles of incidence and refraction, respectively, and v and v' are the indices of refraction for air and the plate, respectively. Prove that if d is the thickness of the plate, then

$$\ell_2 - \ell_1 = 2dv' \cos \phi'.$$

3. Show that $E = f(ct + x)$ is also a solution of the wave equation.

*4. Let E and H be, respectively, the electric and magnetic intensities at any point in an electromagnetic field. Show that Maxwell's equations

$$\text{div } H = 0, \qquad\qquad \text{div } E = 0,$$

$$\text{curl } E = -\mu_0 \frac{\partial H}{\partial t}, \qquad \text{curl } H = \epsilon \frac{\partial E}{\partial t}$$

yield the wave equation if we assume that $c = 1/\sqrt{\epsilon\mu_0}$ and that E and H depend only on the time t and the coordinate x. (This is a good approximation when the source is located extremely far away in the x-direction from a neighborhood of the point where we are deriving the wave equation.)

NOTES

SECTION 1.1

Formulas relating z to Z in the stereographic projection are easy to compute: [A, pp. 18-20] or [H, pp. 38-44].

SECTION 1.5

Other synonyms for *analytic* are *holomorphic, monogenic,* and *regular.*

SECTION 1.6

Far weaker sufficient conditions are known for analyticity. The best such result appears to be in [S, pp. 197-199], where the *Looman–Menchoff* theorem states that if u and v are continuous in G, have first partials at all except an enumerable number of points in G, and satisfy the Cauchy–Riemann equations almost everywhere in G, then

$f = u + iv$ is analytic in G. Other examples showing that the Cauchy–Riemann equations alone are insufficient for analyticity may be found in [T, pp. 67, 70].

SECTIONS 1.7 and 1.9

For a more detailed elementary development of Riemann surfaces, see [Kn, Part II, pp. 100-146]. Tables of elementary mappings of domains can be found in the appendix and in [Ko]. Since the derivative of a function at a point is obtained by considering difference quotients of nearby points, the definition of analyticity extends to any Riemann surface.

2 COMPLEX INTEGRATION

Integration is an important and useful concept in elementary calculus. The two-dimensional nature of the complex plane suggests the consideration of integrals along arbitrary curves in $\mathbb{C}$ instead of only on segments of the real axis. These "line integrals" have interesting and unusual properties when the function being integrated is analytic. Complex integration is one of the most beautiful and elegant theories in mathematics.

2.1 LINE INTEGRALS

The properties of analytic functions discussed in the preceding chapter were all consequences of the differentiability of the function. In real calculus, the fundamental theorem of calculus reveals a surprising and useful connection between derivatives and definite integrals.

FUNDAMENTAL THEOREM OF CALCULUS

If a real-valued function $f(x)$ is continuous in an interval $a \leqslant x \leqslant b$, then $f(x)$ possesses antiderivatives in that interval. If $F(x)$ is any antiderivative of $f(x)$ in $a \leqslant x \leqslant b$, then

$$\int_a^b f(x)\, dx = F(b) - F(a).$$

70

One of the main goals of this chapter is to prove a similar theorem for line integrals of an analytic function in the complex plane. At first glance this appears to be a very difficult job, as there is an infinity of curves joining two given points, but the proof is easy and the applications are very useful.

An *arc* γ in the plane is any set of points that can be described in parametric form by

$$\gamma: x = x(t), \qquad y = y(t), \qquad \alpha \leqslant t \leqslant \beta,$$

with $x(t)$, $y(t)$ continuous functions of the real variable t in the closed real interval $[\alpha, \beta]$. In the complex plane we describe the arc γ by the continuous complex-valued function of a real variable

$$\gamma: z = z(t) = x(t) + iy(t), \qquad \alpha \leqslant t \leqslant \beta.$$

The arc γ is said to be **smooth** if the function $z'(t) = x'(t) + iy'(t)$ is nonzero and continuous on $\alpha \leqslant t \leqslant \beta$. A **piecewise smooth (pws)** arc is an arc consisting of a finite number of smooth arcs joined end to end. If γ is a pws arc, then $x(t)$ and $y(t)$ are continuous, but their derivatives $x'(t)$ and $y'(t)$ are piecewise continuous. An arc is a **simple**, or **Jordan**, arc if $z(t_1) = z(t_2)$ only if $t_1 = t_2$, that is, if it is non-self-intersecting. An arc is a **closed curve** if $z(\alpha) = z(\beta)$ and a **Jordan curve** if it is closed and simple except at the endpoints α and β. Figure 2.1 illustrates some of these concepts.

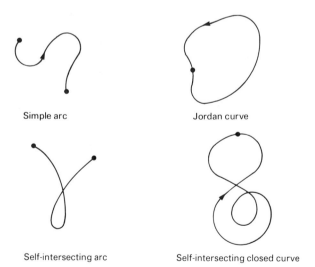

Simple arc   Jordan curve

Self-intersecting arc   Self-intersecting closed curve

FIGURE 2.1. Arcs and curves

EXAMPLE 1.

Sketch the arcs given by the parametrizations

(a) $\gamma: z(t) = e^{it}, \qquad 0 \leqslant t \leqslant 2\pi,$

(b) $\gamma^*: z(t) = \begin{cases} 1 - i(1 - t), & 0 \leqslant t \leqslant 1, \\ 1 + t - i, & -1 \leqslant t \leqslant 0. \end{cases}$

SOLUTION: (a) Since $(e^{it})' = ie^{it} \neq 0$, the arc γ is smooth. Note that $|e^{it}| = 1$ and $e^0 = e^{2\pi i} = 1$. Hence, γ is a parametrization of the unit circle traversed in a counterclockwise direction. Clearly, γ is a Jordan curve (see Figure 2.2a). (b) γ^* is not a smooth arc because $z'(t)$ is not defined at $t = 0$. However, $z(t)$ is a smooth

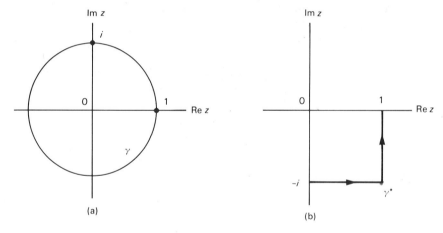

(a)   (b)

FIGURE 2.2. Unit circle and pws arc γ^*

arc on each of the intervals $[-1, 0]$ and $[0, 1]$. Hence γ^* is pws. We see from the graph in Figure 2.2b that γ^* is a simple arc.

Jordan curves satisfy the following property.

JORDAN CURVE THEOREM

A Jordan curve separates the extended plane into two simply connected regions, both having the curve as their boundary.

The region containing the point at infinity is called the **outside** of the curve; the other region is called the **inside**. Although this theorem seems obvious, its proof is difficult, so we shall accept its validity on intuitive grounds. A Jordan curve is parametrized in its **positive sense** if its interior is kept to the left as the curve is traversed. For example, the parametrization $z(t) = e^{it} = \cos t + i \sin t$, $0 \leqslant t \leqslant 2\pi$, parametrizes $|z| = 1$ in its positive sense, whereas $z(t) = e^{-it}$, $0 \leqslant t \leqslant 2\pi$, does not.

Let γ be a smooth arc in $\mathbb{C}$ and let the complex function $f(z)$ be continuous on γ. We use the parametrization of γ to define the line integral of f on γ in terms of two real integrals. If the two real integrals can be evaluated, a value can be assigned to the line integral.

DEFINITION

Let $\gamma: z = z(t)$, $\alpha \leqslant t \leqslant \beta$, be a smooth arc and $f(z) = u + iv$ be continuous on γ. Then, the **line integral of f on γ is given by**

$$\int_{\gamma} f(z)\, dz = \int_{\alpha}^{\beta} f(z(t)) z'(t)\, dt$$

$$= \int_{\alpha}^{\beta} [u(z(t)) + iv(z(t))] [x'(t) + iy'(t)]\, dt$$

$$= \int_{\alpha}^{\beta} [u(z(t)) x'(t) - v(z(t)) y'(t)]\, dt$$

$$+ i \int_{\alpha}^{\beta} [u(z(t)) y'(t) + v(z(t)) x'(t)]\, dt.$$

The line integral over a pws arc γ is obtained by applying the definition above to each of the finitely many closed intervals on which $z(t)$ is smooth and summing the results. If you are unfamiliar with line integrals, you may wish to read Appendix A.3.

EXAMPLE 2.

To evaluate $\int_{\gamma} x\, dz$ along the pws arc γ shown in Figure 2.3, parametrize γ by

$$\gamma \colon z(t) = \begin{cases} 1 + it, & 0 \leqslant t \leqslant 1, \\ (2 - t) + i, & 1 \leqslant t \leqslant 2. \end{cases}$$

Then

$$z'(t) = \begin{cases} i, & 0 \leqslant t \leqslant 1, \\ -1, & 1 \leqslant t \leqslant 2, \end{cases}$$

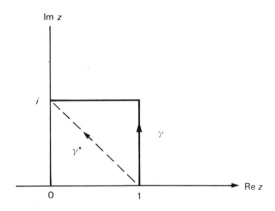

FIGURE 2.3. pws arc γ

with unequal left and right derivatives at $t = 1$. By definition, integrating over each of the intervals $0 \leqslant t \leqslant 1$ and $1 \leqslant t \leqslant 2$, we get

$$\int_{\gamma} x \, dz = \int_{0}^{1} i \, dt + \int_{1}^{2} (2 - t)(-1) dt = -\tfrac{1}{2} + i,$$

since $x(t) = 1$ on $0 \leqslant t \leqslant 1$ and $x(t) = 2 - t$ on $1 \leqslant t \leqslant 2$.

If we choose a different parametrization for γ, say

$$\gamma \colon z(t) = \begin{cases} 1 + i \log t, & 1 \leqslant t \leqslant e, \\ 2 - \dfrac{t}{e} + i, & e \leqslant t \leqslant 2e, \end{cases}$$

we have

$$z'(t) = \begin{cases} i/t, & 1 \leqslant t \leqslant e, \\ -1/e, & e \leqslant t \leqslant 2e, \end{cases}$$

and

$$\int_{\gamma} x \, dz = \int_{1}^{e} \frac{i}{t} \, dt + \int_{e}^{2e} \left(2 - \frac{t}{e}\right) \left(\frac{-1}{e}\right) dt = -\tfrac{1}{2} + i.$$

Hence, *the line integral is independent of the two parametrizations of* γ. This is always the case when the change of parameters is piecewise differentiable as can easily be seen using the change of variable formula of integral calculus.

A different value is obtained if we integrate over the line segment γ^* joining 1 to i. Here

$$\gamma^*: z(t) = (1 - t) + it, \; 0 \leqslant t \leqslant 1,$$

so that

$$\int_{\gamma^*} x \, dz = \int_0^1 (1 - t)(-1 + i) \, dt = \frac{-1 + i}{2}.$$

This example shows that we cannot obtain a theorem similar to the fundamental theorem of calculus for *all* continuous complex-valued functions $f(z)$. Suppose, instead, that we consider only those continuous functions $f(z)$ that are derivatives of an analytic function $F = U + iV$ in some region G containing the smooth arc γ. Then, by definition,

$$\int_\gamma f(z) \, dz = \int_\gamma F'(z) \, dz = \int_\alpha^\beta F'(z(t)) z'(t) \, dt.$$

By the chain rule of calculus, we have

$$\int_\alpha^\beta F'(z(t)) z'(t) \, dt = \int_\alpha^\beta \frac{d}{dt} [F(z(t))] \, dt$$

$$= \int_\alpha^\beta \frac{d}{dt} [U(z(t))] \, dt + i \int_\alpha^\beta \frac{d}{dt} [V(z(t))] \, dt.$$

Applying the fundamental theorem of calculus to each of these real integrals, we obtain

$$\int_\gamma f(z) \, dz = [U(z(\beta)) - U(z(\alpha))] + i [V(z(\beta)) - V(z(\alpha))]$$

$$= F(z(\beta)) - F(z(\alpha)).$$

We can easily extend this result to pws arcs by adding the results obtained from the smooth subarcs. Since the result depends only on the endpoints of each smooth subarc, we have proved the following theorem.

FUNDAMENTAL THEOREM (OF CALCULUS)

If $F(z)$ is an analytic function with a continuous derivative $f(z) = F'(z)$ in a region G containing the pws arc $\gamma: z = z(t)$, $\alpha \leqslant t \leqslant \beta$, then

$$\int_\gamma f(z)\, dz = F(z(\beta)) - F(z(\alpha)).$$

Since the integral depends only on the endpoints of the arc γ, *the integral is independent of path.* Thus, the same result is obtained for *any* pws arc in G with these endpoints. *For pws closed curves γ, the fundamental theorem yields*

$$\int_\gamma f(z)\, dz = 0,$$

since $F(z(\beta)) = F(z(\alpha))$.

EXAMPLE 3.

Calculate

$$\int_\gamma z\, dz \quad \text{and} \quad \int_{\gamma^*} z\, dz,$$

where γ and γ^* are the two arcs shown in Figure 2.3.

SOLUTION: The continuous function $f(z) = z$ is the derivative of the entire function $F(z) = z^2/2$. Parametrizing γ as in Example 2, we have

$$\int_\gamma z\, dz = \int_0^1 (1 + it)i\, dt + \int_1^2 [(2 - t) + i]\,(-1)\, dt$$

$$= i \int_0^1 dt - \int_0^1 t\, dt + \int_1^2 (t - 2)\, dt - i \int_1^2 dt$$

$$= -1.$$

Using the parametrization of γ^* from Example 2, we get

$$\int_{\gamma^*} z\, dz = \int_0^1 [(1 - t) + it]\,(-1 + i)\, dt$$

$$= - \int_0^1 dt + i \int_0^1 (1 - 2t) \, dt = -1.$$

By the fundamental theorem, any pws arc γ beginning at 1 and ending at i satisfies

$$\int_\gamma z \, dz = \frac{z^2}{2} \bigg|_1^i = \frac{i^2 - 1}{2} = -1.$$

EXAMPLE 4.

Show that

$$\int_{|z|=1} \frac{dz}{z} = 2\pi i.$$

SOLUTION: It might appear as if this result contradicts the fundamental theorem, since $|z| = 1$ is a Jordan curve. However, the antiderivatives of the continuous function $f(z) = 1/z$ are logarithms and are analytic on the Riemann surface $\mathfrak{R}$ described in Sections 1.7 and 1.9. The curve $|z| = 1$ is *not* a closed curve in $\mathfrak{R}$. Two methods for obtaining this integral will be illustrated. First note that unless the contrary is stated, *integrations over Jordan curves are assumed to be carried out in the positive sense.* Thus, parametrizing $|z| = 1$ by $z(t) = e^{it}$, $0 \leqslant t \leqslant 2\pi$, we have

$$z'(t) = ie^{it}.$$

Then the integral becomes

$$\int_{|z|=1} \frac{dz}{z} = \int_0^{2\pi} \frac{z'(t) \, dt}{z(t)} = \int_0^{2\pi} \frac{ie^{it}}{e^{it}} \, dt = i \int_0^{2\pi} dt = 2\pi i.$$

To use the fundamental theorem in evaluating this integral, we select *any* branch of the Riemann surface $\mathfrak{R}$ of the analytic function

$$F(z) = \log z = \log |z| + i \arg z.$$

For example, beginning at $-i$ on the principal branch yields

$$\int_{|z|=1} \frac{dz}{z} = \log |z| + i \arg z \,\bigg|_{e^{-\pi i/2}}^{e^{3\pi i/2}} = i(3\pi/2) - i(-\pi/2)$$

$$= 2\pi i.$$

EXAMPLE 5.

Let $P(z)$ be any polynomial and γ be a pwś arc. Show that:
(a) $\int_\gamma P(z) \, dz = 0$ if γ is a closed curve,
(b) $\int_\gamma P(z) \, dz$ depends only on the endpoints of γ.

SOLUTION: Every polynomial $P(z)$ is continuous in $\mathbb{C}$. Futhermore, if

$$P(z) = a_n z^n + a_{n-1} z^{n-1} + \ldots + a_1 z + a_0,$$

then $P(z)$ is the derivative of the analytic polynomial

$$Q(z) = \frac{a_n z^{n+1}}{n+1} + \frac{a_{n-1} z^n}{n} + \ldots + \frac{a_1 z^2}{2} + a_0 z.$$

Thus, the fundamental theorem is satisfied and parts (a) and (b) hold.

EXAMPLE 6.

Since $\cos z$ is entire and has antiderivative $\sin z$, we have

$$\int_{-i}^{i} \cos z \, dz = \sin z \Big|_{-i}^{i} = 2 \sin i = 2i \sinh(1),$$

and along any pws closed curve γ,

$$\int_\gamma \cos z \, dz = 0.$$

EXERCISES

1. Show that the parametrization γ: $z(t) = a \cos t + ib \sin t$, $0 \leqslant t \leqslant 2\pi$, describes the ellipse

$$\frac{x^2}{a^2} + \frac{y^2}{b^2} = 1.$$

In Exercises 2-5, determine pws parametrizations for the indicated arcs or curves.

2. Semicircle from 1 to −1

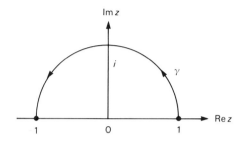

3. Triangle

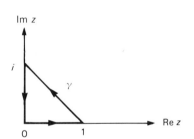

4. Square

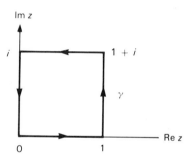

5. Barbell beginning at 1

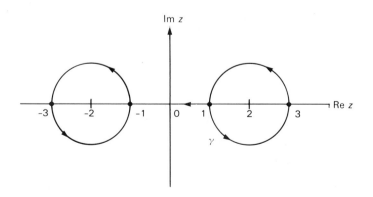

6. Show that $z'(t)$ can be interpreted as a vector tangent to the arc $\gamma: z = z(t)$ at all points where $z'(t)$ is nonzero.

Evaluate the integrals

$$\int x \, dz, \qquad \int y \, dz, \qquad \int \bar{z} \, dz$$

along the paths given in Exercises 7-9.

7. The directed line segment from 0 to $1 - i$
8. Around the circle $|z| = 1$
9. Around the circle $|z - a| = R$
10. Evaluate $\int_\gamma y \, dz$, where γ is the straight line joining 1 to i.
11. Evaluate $\int_\gamma y \, dz$, where γ is the arc in the first quadrant along $|z| = 1$ joining 1 to i.
12. Evaluate $\int_\gamma y \, dz$, where γ is the arc along the coordinate axes joining 1 to i.
13. Since all three values in Exercises 10-12 are different, why doesn't this result violate the fundamental theorem?

Use parametrizations of the arcs to evaluate the integrals in Exercises 14-24. Confirm your answer by using the fundamental theorem.

14. Evaluate the integral $\int (z - a)^n \, dz$, n an integer, around the circle $|z - a| = R$. (The answer for $n = -1$ differs from the rest.)
15. Evaluate $\int_\gamma e^z \, dz$, where γ is the straight-line path joining 1 to i.
16. Evaluate $\int_\gamma e^z \, dz$, where γ is the path in the first quadrant along the circle $|z| = 1$ joining 1 to i.
17. Evaluate $\int_\gamma e^z \, dz$, where γ is the path along the coordinate axes joining 1 to i.
18. Evaluate $\int_{-i}^{i} e^{\pi z} \, dz$.
19. Evaluate $\int_{-1}^{i} \sinh(az) \, dz$.
20. Evaluate $\int_{1}^{i} (z - 1)^3 \, dz$.
*21. If γ is the ellipse $z(t) = a \cos t + ib \sin t$, $0 \leqslant t \leqslant 2\pi$, $a^2 - b^2 = 1$, show that

$$\int_\gamma \frac{dz}{\sqrt{1 - z^2}} = \pm 2\pi,$$

depending on which value of the radical is taken.
(*Hint:* $1 - z^2(t) = [z'(t)]^2$.)
22. Let $\gamma_1: z(t) = e^{it}$ and $\gamma_2: z(t) = e^{-it}$, $0 \leqslant t \leqslant \pi$. Evaluate
$$\int \frac{dz}{z^2}$$
along each curve.

23. Evaluate $\int \text{Log } z \, dz$ along each curve given in Exercise 22.
24. Evaluate $\int \sqrt{z} \, dz$ along each curve given in Exercise 22. (Use the principal branch of $\sqrt{z}$.)

2.2 GREEN'S THEOREM AND ITS CONSEQUENCES

In Examples 4 and 5 of the previous section, we found that the line integral of a polynomial along a pws closed curve vanishes but that

$$\int_{|z|=1} \frac{dz}{z} = 2\pi i.$$

Note that the function $1/z$ is not analytic at the origin. Could it be that the line integral of a function along a pws Jordan curve vanishes when the function is analytic on and *inside* the curve? Surprisingly, that is correct.

It is easy to prove that the line integral along a pws Jordan curve is zero if we assume that the derivative of the analytic function in the integrand is continuous inside the pws Jordan curve. This is not an unreasonable requirement, since the derivative of every analytic function we have encountered is analytic. The proof is based on the following result found in most elementary calculus textbooks, as well as in Appendix A.3.

GREEN'S THEOREM

Let G be the region inside of a pws Jordan curve γ and suppose the real-valued functions p and q are continuous on $G \cup \gamma$ and have continuous first partials in G. Then

$$\iint_G (p_x + q_y) \, dx \, dy = \int_\gamma p \, dy - q \, dx.$$

Let $f = u + iv$ be analytic on and inside the pws Jordan curve γ, and rewrite the integral of f along γ in the form

$$\int_\gamma f(z)\,dz = \int_\gamma (u + iv)\,(dx + i\,dy) = \int_\gamma u\,dx - v\,dy + i\int_\gamma v\,dx + u\,dy.$$

If f' is continuous on G, then the first partials u_x, u_y, v_x, v_y are continuous. Applying Green's theorem to the two line integrals on the right, we get

$$\int_\gamma f(z)\,dz = -\iint_G (v_x + u_y)\,dx\,dy + i\iint_G (u_x - v_y)\,dx\,dy.$$

The first partials satisfy the Cauchy–Riemann equations, $u_x = v_y$ and $u_y = -v_x$, since f is analytic. Hence, both integrands on the right side are zero. Under the assumption that $f'(z)$ is continuous on G, we have proved the following theorem.

CAUCHY'S THEOREM

Let the function $f(z)$ be analytic on and inside the pws Jordan curve γ. Then

$$\int_\gamma f(z)\,dz = 0.$$

The drawback to this proof is the assumption that $f'(z)$ is continuous on G. In this section we will verify this condition before we use Cauchy's theorem. However, in Section 2.5 we will prove that this condition is unnecessary. Indeed, we will show that analytic functions have *analytic* derivatives.

EXAMPLE 1.

Evaluate $\displaystyle\int_{|z|=1} \frac{e^z}{z^2 + 4}\,dz.$

SOLUTION: The notation employed signifies that the integration is taken over the unit circle in its positive sense. The function $f(z) = e^z/(z^2 + 4)$ and its derivative

$$f'(z) = \frac{(z^2 - 2z + 4)}{(z^2 + 4)^2}\,e^z$$

are both analytic on and inside $|z| = 1$. As the derivative is analytic, it is continuous. Hence, Cauchy's theorem applies and

$$\int_{|z|=1} \frac{e^z}{z^2 + 4}\,dz = 0.$$

EXAMPLE 2.

Show that

$$\frac{1}{2\pi} \int_0^{2\pi} \frac{R^2 - r^2}{R^2 - 2Rr \cos \theta + r^2} \, d\theta = 1, \qquad 0 < r < R.$$

The integrand appearing in this integral is called the **Poisson kernel**. It has many useful properties, which we will study in Chapter 6.

SOLUTION: The Poisson kernel is equal to the real part of the quotient

$$\frac{R + re^{i\theta}}{R - re^{i\theta}} = \frac{(R + re^{i\theta})(R - re^{-i\theta})}{(R - re^{i\theta})(R - re^{-i\theta})} = \frac{R^2 - r^2 + 2irR \sin \theta}{R^2 - 2rR \cos \theta + r^2}.$$

Letting $z = re^{i\theta}$ with r fixed, we have

$$\frac{dz}{d\theta} = rie^{i\theta} = iz$$

so that

$$\frac{1}{2\pi} \int_0^{2\pi} \frac{R^2 - r^2}{R^2 - 2rR \cos \theta + r^2} \, d\theta = \text{Re} \left(\frac{1}{2\pi i} \int_{|z|=r} \frac{R + z}{R - z} \frac{dz}{z} \right).$$

But by partial fractions,

$$\frac{1}{2\pi i} \int_{|z|=r} \frac{R + z}{z(R - z)} \, dz = \frac{1}{2\pi i} \int_{|z|=r} \left(\frac{1}{z} + \frac{2}{R - z} \right) dz$$

$$= \frac{1}{2\pi i} \int_{|z|=r} \frac{dz}{z} + \frac{1}{2\pi i} \int_{|z|=r} \frac{2dz}{R - z}.$$

We can show that the first of the integrals on the right side equals 1 by the methods used in Example 4 of Section 2.1 with $z = re^{it}$ and $z' = ire^{it}$, $0 \leqslant t \leqslant 2\pi$. The last integral of the right side is zero by Cauchy's theorem, since $f(z) = 2/(R - z)$ and $f'(z) = 2/(R - z)^2$ are both analytic on $|z| \leqslant r$.

EXAMPLE 3.

Show that

$$\int_{-\infty}^{\infty} e^{-x^2} \cos 2bx \, dx = \sqrt{\pi} e^{-b^2}.$$

SOLUTION: Applying Cauchy's theorem to the function $f(z) = e^{-z^2}$ analytic on a region containing the rectangle $|x| \leqslant a$, $0 \leqslant y \leqslant b$ (see Figure 2.4), we have

$$0 = \int_{-a}^{a} e^{-x^2} \, dx + \int_{0}^{b} e^{-(a+iy)^2} i \, dy + \int_{a}^{-a} e^{-(x+ib)^2} \, dx + \int_{b}^{0} e^{-(-a+iy)^2} i \, dy$$

$$= \int_{-a}^{a} e^{-x^2} \, dx - e^{b^2} \int_{-a}^{a} e^{-x^2} (\cos 2bx - i \sin 2bx) \, dx$$

$$- i e^{-a^2} \int_{0}^{b} e^{y^2} (e^{2iay} - e^{-2iay}) \, dy$$

$$= \int_{-a}^{a} e^{-x^2} \, dx - e^{b^2} \int_{-a}^{a} e^{-x^2} \cos 2bx \, dx + 2e^{-a^2} \int_{0}^{b} e^{y^2} \sin 2ay \, dy, \quad (1)$$

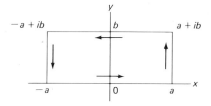

FIGURE 2.4.   Rectangle of integration

since the imaginary part of the middle integral vanishes. But using polar coordinates.

$$\left(\int_{-\infty}^{\infty} e^{-x^2} \, dx \right)^2 = \int_{-\infty}^{\infty} e^{-x^2} \, dx \cdot \int_{-\infty}^{\infty} e^{-y^2} \, dy$$

$$= \int_{-\infty}^{\infty} \int_{-\infty}^{\infty} e^{-(x^2 + y^2)} \, dx \, dy$$

$$= \int_{0}^{2\pi} \int_{0}^{\infty} e^{-r^2} r \, dr \, d\theta = \pi, \quad (2)$$

so the first two integrals in (1) are convergent as $a \to \infty$. Letting $a \to \infty$, the last term in (1) vanishes and

$$\int_{-\infty}^{\infty} e^{-x^2} \cos 2bx \, dx = e^{-b^2} \int_{-\infty}^{\infty} e^{-x^2} \, dx = \sqrt{\pi} e^{-b^2}.$$

EXAMPLE 4.

Prove that

$$\int_{0}^{\infty} \frac{\sin(x^2)}{x} \, dx = \frac{\pi}{4}.$$

SOLUTION: Integrating e^{iz^2}/z along the boundary of $r \leqslant |z| \leqslant R$, $0 \leqslant \arg z \leqslant \pi/2$ (see Figure 2.5), Cauchy's theorem yields

$$\int_{r}^{R} \frac{e^{ix^2}}{x} \, dx + i \int_{0}^{\pi/2} e^{i(Re^{i\theta})^2} \, d\theta - \int_{r}^{R} \frac{e^{-iy^2}}{y} \, dy - i \int_{0}^{\pi/2} e^{i(re^{i\theta})^2} \, d\theta = 0. \quad (3)$$

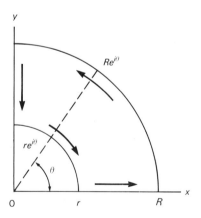

FIGURE 2.5. Region of integration

Using the inequality $|\int_{a}^{b} f(\theta) \, d\theta| \leqslant \int_{a}^{b} |f(\theta)| \, d\theta$ that we will prove for complex-valued integrands in Section 2.3,

$$\left| i \int_0^{\pi/2} e^{i(Re^{i\theta})^2} \, d\theta \right| \leq \int_0^{\pi/2} e^{-R^2 \sin 2\theta} \, d\theta$$

$$= 2 \int_0^{\pi/4} e^{-R^2 \sin 2\theta} \, d\theta$$

$$< 2 \int_0^{\pi/4} e^{-4R^2 \theta/\pi} \, d\theta$$

$$= \frac{\pi}{2R^2} [1 - e^{-R^2}],$$

since $h(\theta) = \sin 2\theta - (4\theta/\pi)$ vanishes at $\theta = 0, \pi/4$ and satisfies $h''(\theta) < 0$ for $0 < \theta < \pi/4$, implying that h is concave down and $\sin 2\theta \geqslant 4\theta/\pi$. Therefore, the second integral in (3) vanishes as $R \to \infty$. Given $\epsilon > 0$, there exists an $r > 0$ such that $|e^{iz^2} - 1| < \epsilon$ whenever $|z| < r$. Then

$$\left| i \int_0^{\pi/2} e^{i(re^{i\theta})^2} \, d\theta - \frac{i\pi}{2} \right| = \left| i \int_0^{\pi/2} (e^{i(re^{i\theta})^2} - 1) \, d\theta \right| < \epsilon \frac{\pi}{2},$$

so the last integral in (3) approaches $i\pi/2$ as $r \to 0$. Adding the first and third integrals in (3) and letting $R \to \infty$ and $r \to 0$, we have

$$0 = \int_0^\infty \frac{e^{ix^2} - e^{-ix^2}}{x} \, dx - \frac{i\pi}{2} = 2i \int_0^\infty \frac{\sin(x^2)}{x} \, dx - \frac{i\pi}{2}.$$

EXERCISES

1. Show that

$$\int_{|z|=1} \frac{\text{Log } z}{z} \, dz = 0,$$

even though $(\text{Log } z)/z$ is not analytic on $|z| \leqslant 1$. What result is obtained if we integrate

$$\int_\gamma \frac{\log z}{z} \, dz$$

on $\gamma: z(t) = e^{it}, 0 \leqslant t \leqslant 2\pi$? Explain.

Use Green's theorem for Exercises 2–4, where A equals the area of G and ∂G is the boundary of G.

2. Show that $\int_{\partial G} x \, dz = iA$.

3. Show that $\int_{\partial G} y \, dz = -A$.

4. Show that $\int_{\partial G} \bar{z} \, dz = 2iA$.

5. Prove that

$$\int_0^{\pi/2} e^{a \cos t} \cos (t + a \sin t) \, dt = \frac{\sin a}{a}, \quad a > 0,$$

by integrating e^z along the Jordan curve in the first quadrant composed of the quarter circle of $|z| = a$ and the line segments from ia to 0 and 0 to a.

6. Show that

$$\int_0^T e^{at} \cos bt \, dt = \frac{e^{aT}(a \cos bT + b \sin bT) - a}{a^2 + b^2},$$

$$\int_0^T e^{at} \sin bt \, dt = \frac{e^{aT}(a \sin bT - b \cos bT) + b}{a^2 + b^2},$$

by integrating $f(z) = e^z$ along the line segment joining 0 to $(a + ib)T$.

7. Show that

$$\int_0^T \sin at \cosh bt \, dt = \frac{b \sin aT \sinh bT - a \cos aT \cosh bT + a}{a^2 + b^2},$$

$$\int_0^T \cos at \sinh bt \, dt = \frac{b \cos aT \cosh bT + a \sin aT \sinh bT - b}{a^2 + b^2},$$

by integrating $f(z) = \sin z$ along the line segment from 0 to $(a + ib) T$.

8. Obtain the integrals

$$\int_0^T \cos at \cosh bt \, dt = \frac{a \sin aT \cosh bT + b \cos aT \sinh bT}{a^2 + b^2},$$

$$\int_0^T \sin at \sinh bt \, dt = \frac{b \sin aT \cosh bT - a \cos aT \sinh bT}{a^2 + b^2},$$

by integrating $f(z) = \cos z$ along the line segment from 0 to $(a + ib)T$.

9. Assuming $0 < b < 1$ and applying Cauchy's theorem to the function $f(z) = (1 + z^2)^{-1}$ along the boundary of the rectangle in

Figure 2.4, show that

$$\int_{-\infty}^{\infty} \frac{(1 - b^2 + x^2)\, dx}{(1 - b^2 + x^2)^2 + 4x^2 b^2} = \pi.$$

10. Prove that

$$\int_{-\infty}^{\infty} e^{-kx^2} \cos ax\, dx = \sqrt{\frac{\pi}{k}}\, e^{-a^2/4k}, \qquad k > 0, \quad a \text{ real},$$

by using the same procedure as in Example 3 with the function $f(z) = e^{-kz^2}$. Check your answer by changing variables.

11. Prove that

$$\int_{-\infty}^{\infty} \frac{(1 - b^2 + x^2)\cos kx + 2xb \sin kx}{(1 - b^2 + x^2)^2 + 4x^2 b^2}\, dx = e^{kb} \int_{-\infty}^{\infty} \frac{\cos kx}{1 + x^2}\, dx,$$

$$\int_{-\infty}^{\infty} \frac{(1 - b^2 + x^2)\sin kx - 2xb \cos kx}{(1 - b^2 + x^2)^2 + 4x^2 b^2}\, dx = 0.$$

with $0 < b < 1$ and k real.

12. Let $0 < b < 1$ and show that

$$\int_{-\infty}^{\infty} \frac{\text{Re}(1 + (x - ib)^4)}{|1 + (x + ib)^4|^2}\, dx = \int_{-\infty}^{\infty} \frac{dx}{1 + x^4}.$$

13. Prove that

$$\int_{0}^{\infty} e^{-x^2} \sin 2xb\, dx = e^{-b^2} \int_{0}^{b} e^{x^2}\, dx, \qquad b > 0,$$

by integrating around a suitable rectangle.

14. Prove the equalities

$$\int_{0}^{\infty} \cos x^2\, dx = \int_{0}^{\infty} \sin x^2\, dx = \frac{\sqrt{\pi}}{2\sqrt{2}}, \text{ Fresnel's integrals},$$

by applying Cauchy's theorem to the function $f(z) = e^{-z^2}$ along the boundary of the sector $0 \leqslant |z| \leqslant R$, $0 \leqslant \arg z \leqslant \pi/4$.

15. Show that

$$\int_{0}^{\infty} e^{-x^2} \cos(x^2)\, dx = \frac{\sqrt{\pi}}{4} \sqrt{\sqrt{2} + 1},$$

$$\int_{0}^{\infty} e^{-x^2} \sin(x^2)\, dx = \frac{\sqrt{\pi}}{4} \sqrt{\sqrt{2} - 1},$$

by integrating e^{-z^2} along the boundary of the sector $0 \leqslant |z| \leqslant R$, $0 \leqslant \arg z \leqslant \pi/8$.

16. Prove Dirichlet's integral

$$\int_0^\infty \frac{\sin x}{x} \, dx = \frac{\pi}{2},$$

by integrating $f(z) = e^{iz}/z$ along the boundary of the set $r \leqslant |z| \leqslant R$, $0 \leqslant \arg z \leqslant \pi$. Check your answer by changing variables in Example 4.

2.3 **THE CAUCHY INTEGRAL FORMULA**

We shall need the following facts about integrals

THEOREM

(i) $\int_\gamma [\alpha f_1(z) + \beta f_2(z)] \, dz = \alpha \int_\gamma f_1(z) \, dz + \beta \int_\gamma f_2(z) \, dz$.

(ii) $\int_{\gamma_1 + \gamma_2} f(z) \, dz = \int_{\gamma_1} f(z) \, dz + \int_{\gamma_2} f(z) \, dz$, where $\gamma_1 + \gamma_2$ is the path consisting of traversing first γ_1 followed by γ_2.

(iii) $\int_{-\gamma} f(z) \, dz = -\int_\gamma f(z) \, dz$, where $-\gamma$ is the path traversing the arc γ in the reverse direction,

(iv) $|\int_\gamma f(z) \, dz| \leqslant \int_\gamma |f(z)| \, |dz|$, where we define $|dz|$ to be the differential with respect to *arc length*, with

$$|dz| = |dx + idy| = \sqrt{(dx)^2 + (dy)^2} = ds.$$

PROOF: To prove (iv), notice that for any real constant θ,

$$\text{Re}\left(e^{-i\theta} \int_\gamma f(z) \, dz \right) = \int_\alpha^\beta \text{Re}(e^{-i\theta} f(z(t)) z'(t)) \, dt \leqslant \int_\alpha^\beta |f(z(t))| \, |z'(t)| \, dt,$$

since the real part of a complex number cannot exceed its absolute value. Writing $\int_\gamma f(z) \, dz$ in polar form and setting $\theta = \arg [\int_\gamma f(z) \, dz]$, the expression on the left reduces to the absolute value of the integral and the inequality holds. The remaining proofs are immediate consequences of the definition of a line integral in Section 2.1. Their straightforward and somewhat tedious verification will be left for the exercises. ∎

If $|f(z)| \leqslant M$ at every point z on an arc γ of length L, then part (iv) of the theorem yields the inequality

$$\left| \int_\gamma f(z) \, dz \right| \leqslant M \int_\gamma |dz| = ML.$$

EXAMPLE 1.

$$\left| \int_{|z|=1} e^z \, dz \right| \leqslant 2\pi e.$$

SOLUTION: From part (iv) we have

$$\left| \int_{|z|=1} e^z \, dz \right| \leqslant \int_{|z|=1} |e^z| \, |dz|.$$

Since $|e^z| = e^x \leqslant e$ for all points $z = x + iy$ on the unit circle.

$$\int_{|z|=1} |e^z| \, |dz| \leqslant e \int_{|z|=1} |dz| = 2\pi e,$$

and the inequality is verified. In fact, it is clear that

$$\left| \int_{|z|=1} e^z \, dz \right| < 2\pi e,$$

since $|e^z|$ attains the value e only at $z = 1$.

In many applications it is necessary to consider regions that are not simply connected. We shall generalize Cauchy's theorem to the case of a *multiply connected* region.

THEOREM

Let the inside of the pws Jordan curve γ_0 contain the disjoint pws Jordan curves $\gamma_1, \ldots, \gamma_n$, none of which is contained inside another. Suppose $f(z)$ is analytic in a region G containing the set S consisting of all points on and inside γ_0 but not inside γ_k, $k = 1, \ldots, n$. Then

$$\int_{\gamma_0} f(z) \, dz = \sum_{k=1}^n \int_{\gamma_k} f(z) \, dz.$$

PROOF: We can always find disjoint pws arcs L_k, $k = 0, \ldots, n$, joining γ_k to γ_{k+1} (with L_n joining γ_n to γ_0) such that two pws Jordan curves are formed, each lying in some simply connected subregion of G. (We omit a proof on intuitive grounds. See Figure 2.6.) By

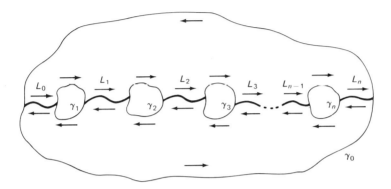

FIGURE 2.6. A multiply connected domain

Cauchy's theorem, the integral of $f(z)$ on these curves, each traversed in the positive sense, vanishes. But the total contribution of these two curves is equivalent to traversing γ_0 in the positive sense, $\gamma_1, \ldots,$ γ_n in the negative (opposite) sense, and $L_0, \ldots, L_n$ in opposite directions. Thus, the integrals on the arcs L_k cancel out and

$$0 = \int_{\gamma_0 - \sum_{k=1}^{n} \gamma_k} f(z)\, dz = \int_{\gamma_0} f(z)\, dz - \sum_{k=1}^{n} \int_{\gamma_k} f(z)\, dz. \quad \blacksquare$$

We next prove the surprising result that the values of an analytic function inside a pws Jordan curve are completely determined by its values on the curve.

THE CAUCHY INTEGRAL FORMULA

Let $f(z)$ be analytic on a simply connected region containing the pws Jordan curve γ. Then

$$f(\zeta) = \frac{1}{2\pi i} \int_{\gamma} \frac{f(z)}{z - \zeta}\, dz,$$

for all points ζ inside γ.

PROOF: Fix ζ, then given $\epsilon > 0$, there exists a closed disk $|z - \zeta| \leqslant r$ lying inside γ for which $|f(z) - f(\zeta)| < \epsilon$. (See Figure 2.7.) Since

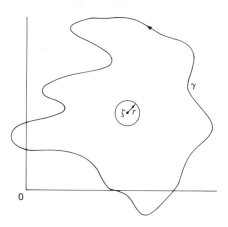

FIGURE 2.7. The Cauchy integral formula

$f(z)/(z - \zeta)$ is analytic on a region containing the points on and inside γ satisfying $|z - \zeta| \geqslant r$, the Cauchy theorem on multiply connected regions implies.

$$\frac{1}{2\pi i} \int_\gamma \frac{f(z)}{z - \zeta} \, dz = \frac{1}{2\pi i} \int_{|z-\zeta|=r} \frac{f(z)}{z - \zeta} \, dz.$$

But

$$\int_{|z-\zeta|=r} \frac{f(z)}{z - \zeta} \, dz = f(\zeta) \int_{|z-\zeta|=r} \frac{dz}{z - \zeta} + \int_{|z-\zeta|=r} \frac{f(z) - f(\zeta)}{z - \zeta} \, dz.$$

By Example 4 or Exercise 14 of Section 2.1, the first integral on the right-hand side equals $2\pi i$, so

$$\left| \int_{|z-\zeta|=r} \frac{f(z)}{z - \zeta} \, dz - 2\pi i f(\zeta) \right| \leqslant \int_{|z-\zeta|=r} \frac{|f(z) - f(\zeta)|}{|z - \zeta|} \, |dz| < 2\pi\epsilon.$$

Since ϵ can be chosen arbitrarily close to 0, the proof is complete. ∎

EXAMPLE 2.

Integrate

$$\int_\gamma \frac{\cos z}{z^3 + z} \, dz$$

over the given curves: (a) γ: $|z| = 2$, (b) γ: $|z| = \frac{1}{2}$, and (c) γ: $|z - i/2| = 1$.

SOLUTION: (a) γ: $|z| = 2$. Decomposing the integral by partial fractions, we obtain

$$\int_\gamma \frac{\cos z}{z^3 + z} \, dz = \int_\gamma \frac{\cos z}{z} \, dz - \frac{1}{2} \int_\gamma \frac{\cos z}{z + i} \, dz - \frac{1}{2} \int_\gamma \frac{\cos z}{z - i} \, dz$$

$$= 2\pi i \left[\cos(0) - \frac{1}{2} \cos(-i) - \frac{1}{2} \cos i \right] = 2\pi i \left[1 - \cosh(1) \right].$$

(b) γ: $|z| = \frac{1}{2}$. As $\cos z/(z^2 + 1)$ is analytic on and inside γ, the integral equals $2\pi i$ times its value at $z = 0$, that is,

$$\int_\gamma \frac{\cos z}{z^3 + z} \, dz = 2\pi i.$$

(c) γ: $|z - i/2| = 1$. Since $\cos z/(z + i)$ is analytic on and inside γ, by partial fractions we have

$$\frac{1}{z(z - i)} = i \left(\frac{1}{z} - \frac{1}{z - i} \right),$$

so that

$$\int_\gamma \frac{\cos z}{z^3 + z} \, dz = 2\pi i \left[i \left(\frac{\cos(0)}{i} \right) - i \left(\frac{\cos i}{2i} \right) \right] = 2\pi i \left[1 - \frac{1}{2} \cosh(1) \right].$$

Of course, we can do all three examples utilizing the partial fraction decomposition in part (a), since the corresponding integrals vanish when the points 0 or $\pm i$ lie outside γ.

If the Cauchy integral formula is differentiated formally with respect to ζ inside the integral sign, an expression for the derivative at all points inside γ is obtained:

$$f'(\zeta) = \frac{1}{2\pi i} \int_\gamma \frac{f(z)}{(z-\zeta)^2}\, dz.$$

To verify this equation, we use the Cauchy integral formula, rewriting

$$\frac{f(\zeta + h) - f(\zeta)}{h} - \frac{1}{2\pi i} \int_\gamma \frac{f(z)}{(z-\zeta)^2}\, dz$$

$$= \frac{1}{2\pi i} \int_\gamma f(z) \left[\frac{1}{h}\left(\frac{1}{z-\zeta-h} - \frac{1}{z-\zeta}\right) - \frac{1}{(z-\zeta)^2}\right] dz$$

$$= \frac{h}{2\pi i} \int_\gamma \frac{f(z)\, dz}{(z-\zeta)^2 (z-\zeta-h)}.$$

Let d be the shortest distance from ζ to γ, M the maximum value of $|f(z)|$ on γ, and L the length of γ, and assume that $|h| \leqslant d/2$. Then

$$|z - \zeta - h| \geqslant |z - \zeta| - |h| \geqslant d - \frac{d}{2} = \frac{d}{2}$$

so that

$$\left| \frac{h}{2\pi i} \int_\gamma \frac{f(z)\, dz}{(z-\zeta)^2 (z-\zeta-h)} \right| \leqslant \frac{ML|h|}{\pi d^3}.$$

Letting $h \to 0$, it follows that

$$f'(\zeta) = \lim_{h \to 0} \frac{f(\zeta + h) - f(\zeta)}{h} = \frac{1}{2\pi i} \int_\gamma \frac{f(z)}{(z-\zeta)^2}\, dz.$$

In Section 2.5 we shall generalize this procedure and prove **Cauchy's theorem for derivatives:**

$$f^{(n)}(\zeta) = \frac{n!}{2\pi i} \int_\gamma \frac{f(z)}{(z-\zeta)^{n+1}}\, dz, \quad n = 1, 2, \ldots,$$

valid for all points ζ inside the pws Jordan curve γ contained in a simply connected region G on which $f(z)$ is analytic. Observe that this formula implies that $f(z)$ possesses derivatives of all orders on G. Hence, the derivative of an analytic function is also analytic. Assuming this fact, we obtain a converse to Cauchy's theorem that is often useful in establishing the analyticity of a function.

MORERA'S THEOREM

If $f(z)$ is continuous in a simply connected region G and satisfies

$$\int_\gamma f(z) \, dz = 0,$$

for all pws closed curves γ in G, then $f(z)$ is analytic in G.

PROOF: Select a point z_0 in G and define

$$F(z) = \int_{z_0}^z f(\zeta) \, d\zeta,$$

for all z in G. Then $F(z)$ is well defined because it is independent of path: If γ_1 and γ_2 are both pws curves in G from z_0 to z, then $\gamma = \gamma_1 - \gamma_2$ is a pws closed curve in G and

$$0 = \int_\gamma f(\zeta) \, d\zeta = \int_{\gamma_1} f(\zeta) \, d\zeta - \int_{\gamma_2} f(\zeta) \, d\zeta.$$

Since f is continuous, for any point z in G and $\epsilon > 0$ there is a disk $|\zeta - z| < \delta$ in G such that $|f(\zeta) - f(z)| < \epsilon$. If $|h| < \delta$, we have

$$\frac{F(z+h) - F(z)}{h} = \frac{1}{h} \left[\int_{z_0}^{z+h} f(\zeta) \, d\zeta - \int_{z_0}^z f(\zeta) \, d\zeta \right] = \frac{1}{h} \int_z^{z+h} f(\zeta) \, d\zeta,$$

where the integration can be taken over the line from z to $z + h$. Since

$$f(z) = \frac{f(z)}{h} \int_z^{z+h} d\zeta,$$

we get by subtraction

$$\left| \frac{F(z+h) - F(z)}{h} - f(z) \right| = \left| \frac{1}{h} \int_z^{z+h} [f(\zeta) - f(z)] \, d\zeta \right|$$

$$\leqslant \frac{1}{|h|} \int_z^{z+h} |f(\zeta) - f(z)| \, |d\zeta| < \epsilon.$$

Hence, $F'(z) = f(z)$, so that F is analytic on G. But then F has an analytic derivative, implying that f is also analytic on G.

EXAMPLE 3.

Integrate

$$\int_\gamma \frac{\cos z}{z^2 (z - 1)} \, dz$$

over (a) γ: $|z| = \frac{1}{3}$, (b) γ: $|z - 1| = \frac{1}{3}$, and (c) γ: $|z| = 2$.

SOLUTION: (a) γ: $|z| = \frac{1}{3}$. In this case $\cos z/(z - 1)$ is analytic on and inside γ, so by Cauchy's theorem for derivatives we obtain

$$\int_\gamma \frac{\left(\dfrac{\cos z}{z - 1}\right)}{z^2}\, dz = 2\pi i \left(\frac{\cos z}{z - 1}\right)' \Bigg|_{z=0} = -2\pi i.$$

(b) γ: $|z - 1| = \frac{1}{3}$. Now $z^{-2} \cos z$ is analytic on and inside γ, so the integral equals $2\pi i$ times the value of $z^{-2} \cos z$ at $z = 1$, that is,

$$\int_\gamma \frac{z^{-2} \cos z}{z - 1}\, dz = 2\pi i \cos(1).$$

(c) γ: $|z| = 2$. By the Cauchy theorem on multiply connected regions, we may replace γ by the circles in parts (a) and (b). Hence, the integral equals $2\pi i [\cos(1) - 1]$. Alternatively, decomposing the integrand by partial fractions, we obtain

$$\int_\gamma \frac{\cos z}{z^2 (z - 1)}\, dz = \int_\gamma \cos z \left(\frac{1}{z - 1} - \frac{1}{z} - \frac{1}{z^2}\right) dz$$

$$= 2\pi i [\cos(1) - \cos(0) + \sin(0)] = 2\pi i [\cos(1) - 1],$$

by Cauchy's theorem for derivatives.

EXERCISES

In Exercises 1–3, evaluate the integral

$$\int_\gamma \frac{dz}{(z - a)(z - b)}$$

by decomposing the integrand into partial fractions.

1. If a and b lie inside γ
2. If a lies inside and b outside γ
3. If b lies inside and a outside γ

Let γ: $z(t) = 2e^{it} + 1, 0 \leqslant t \leqslant 2\pi$. Evaluate the integrals in Exercises 4–7.

4. $\displaystyle\int_\gamma \frac{e^z}{z}\, dz$

5. $\displaystyle\int_\gamma \frac{\cos z}{z - 1}\, dz$

6. $\displaystyle\int_\gamma \frac{\sin z}{z^2 + 1}\, dz$ **7.** $\displaystyle\int_\gamma \frac{\sin z}{z^2 - z}\, dz$

Let $\gamma: z(t) = 2e^{it} + 1, 0 \leqslant t \leqslant 2\pi$. Evaluate the integrals in Exercises 8–11.

8. $\displaystyle\int_\gamma \frac{e^z}{z^2}\, dz$ **9.** $\displaystyle\int_\gamma \frac{\cos z}{(z - 1)^2}\, dz$

10. $\displaystyle\int_\gamma \frac{\sin z}{(z^2 + 1)^2}\, dz$ **11.** $\displaystyle\int_\gamma \frac{\sin z}{(z - 1)^3}\, dz$

Prove the integral identities in Exercises 12–14.

12. $\int_\gamma [\alpha f_1(z) + \beta f_2(z)]\, dz = \alpha \int_\gamma f_1(z)\, dz + \beta \int_\gamma f_2(z)\, dz$

13. $\int_{\gamma_1 + \gamma_2} f(z)\, dz = \int_{\gamma_1} f(z)\, dz + \int_{\gamma_2} f(z)\, dz$

14. $\int_{-\gamma} f(z)\, dz = -\int_\gamma f(z)\, dz$

15. Without computing the integral, show that

$$\left| \int_{|z|=2} \frac{dz}{z^2 + 1} \right| \leqslant \frac{4\pi}{3}.$$

16. If γ is the semicircle $|z| = R$, $|\arg z| \leqslant \pi/2, R > 1$, show that

$$\left| \int_\gamma \frac{\text{Log}\, z}{z^2}\, dz \right| \leqslant \frac{\pi}{R}\left(\text{Log}\, R + \frac{\pi}{2}\right)$$

and hence that the value of the integral tends to zero as $R \to \infty$.

17. Evaluate $\int_{|z|=1} |z + 1|\, |dz|$.

18. Let $f(z)$ be analytic and bounded by M in $|z| \leqslant R$. Prove that

$$|f^{(n)}(z)| \leqslant \frac{MRn!}{(R - |z|)^{n+1}}, \qquad |z| < R.$$

19. If $f(z)$ is analytic in $|z| < 1$ and $|f(z)| \leqslant (1 - |z|)^{-1}$, show that

$$|f^{(n)}(0)| \leqslant (n + 1)!\left(1 + \frac{1}{n}\right)^n.$$

20. Can an analytic function $f(z)$ satisfy $|f^{(n)}(z)| > n!\, n^n$, for all positive integers n, at some point z?

21. Compute

$$\int_{|z|=1} \frac{e^{kz^n}}{z}\, dz,$$

for n a positive integer. Then show that

$$\int_0^{2\pi} e^{k\cos n\theta} \cos(k\sin n\theta)\, d\theta = 2\pi.$$

22. The Legendre polynomial $P_n(z)$ is defined by

$$P_n(z) = \frac{1}{2^n n!} \frac{d^n}{dz^n}\, [(z^2 - 1)^n].$$

Using Cauchy's formula for derivatives, show that

$$P_n(z) = \frac{1}{2\pi i} \int_\gamma \frac{(\zeta^2 - 1)^n\, d\zeta}{2^n (\zeta - z)^{n+1}},$$

where z is inside the pws Jordan curve γ.

23. Prove the following extension of Morera's theorem: Let $f(z)$ be continuous in the region G (possibly multiply connected). Suppose that for each ζ in G there is a disk D, containing ζ, in G such that

$$\int_\gamma f(z)\, dz = 0,$$

for all pws closed curves γ in D. Then $f(z)$ is analytic in G.

24. Let $P(z)$ be a polynomial none of whose roots lies on the pws Jordan curve γ. Show that

$$\frac{1}{2\pi i} \int_\gamma \frac{P'(z)}{P(z)}\, dz$$

equals the number of roots of $P(z)$ inside γ including multiplicities.

2.4 LIOUVILLE'S THEOREM AND THE MAXIMUM PRINCIPLE

In this section we present three useful consequences of the Cauchy integral formula and its extension to higher derivatives.

GAUSS'S MEAN VALUE THEOREM

Let $f(z)$ be analytic in $|z - \zeta| < R$. Then

$$f(\zeta) = \frac{1}{2\pi} \int_0^{2\pi} f(\zeta + re^{i\theta})\, d\theta, \qquad 0 < r < R.$$

PROOF: The Cauchy integral formula states that

$$f(\zeta) = \frac{1}{2\pi i} \int_{|z-\zeta|=r} \frac{f(z)}{z-\zeta}\, dz,$$

for all $0 < r < R$. If $z = \zeta + re^{i\theta}$, then $z'(\theta) = ire^{i\theta}$, from which the desired identity follows. ∎

CAUCHY'S ESTIMATE

Let $f(z)$ be analytic and satisfy $|f(z)| \leqslant M$ in $|z - \zeta| \leqslant r$. Then

$$|f^{(n)}(\zeta)| \leqslant \frac{Mn!}{r^n}.$$

PROOF: By Cauchy's theorem for derivatives, we have

$$|f^{(n)}(\zeta)| = \left| \frac{n!}{2\pi i} \int_{|z-\zeta|=r} \frac{f(z)}{(z-\zeta)^{n+1}}\, dz \right|$$

$$\leqslant \frac{Mn!}{r^n} \left| \frac{1}{2\pi i} \int_{|z-\zeta|=r} \frac{dz}{z-\zeta} \right| = \frac{Mn!}{r^n}. \quad ∎$$

LIOUVILLE'S THEOREM

An entire function cannot be bounded on all of $\mathcal{C}$ unless it is a constant.

PROOF: Suppose $f(z)$ is entire and bounded by M. Then at any point ζ in $\mathcal{C}$, Cauchy's estimate implies that $|f'(\zeta)| \leqslant M/r$. But r can be made arbitrarily large, so $f'(\zeta) = 0$ at all ζ in $\mathcal{C}$. Hence, $f(z)$ is constant in $\mathcal{C}$. ∎

We next prove one of the most useful theorems in the theory of analytic functions.

MAXIMUM PRINCIPLE

If $f(z)$ is analytic and nonconstant in a region G, then $|f(z)|$ has no maximum in G.

PROOF: Suppose there is a point z_0 in G satisfying $|f(z)| \leqslant |f(z_0)|$ for all z in G. Since z_0 is an interior point, there exists a number

$r > 0$ such that $|z - z_0| \leqslant r$ lies in G. Then by Gauss's mean value theorem,

$$f(z_0) = \frac{1}{2\pi} \int_0^{2\pi} f(z_0 + re^{it}) \, dt,$$

that is, the value at the center of the circle equals the integral average of its values on the circle. By assumption, $|f(z_0 + re^{it})| \leqslant |f(z_0)|$, and if strict inequality holds for some value t, it must hold, by the continuity of $|f(z)|$, on an arc of the circle. But then

$$|f(z_0)| \leqslant \frac{1}{2\pi} \int_0^{2\pi} |f(z_0 + re^{it})| \, dt < \frac{1}{2\pi} \int_0^{2\pi} |f(z_0)| \, dt = |f(z_0)|,$$

a contradiction. So $|f(z_0 + re^{it})| = |f(z_0)|$ for $0 \leqslant t \leqslant 2\pi$, and since the procedure holds on all circles $|z - z_0| = s$, $0 < s \leqslant r$, $|f(z)|$ is constant on the disk $|z - z_0| \leqslant r$. Let S be the set of all points z in G satisfying $|f(z)| = |f(z_0)|$. The argument above shows that each such point is an interior point of S, so S is open. But any point in $T = G - S$ is also an interior point by the continuity of $|f(z)|$. Neither T nor S contains a boundary point of the other, as both are open. Since G is connected, T must be empty. Hence, $S = G$ and, by the zero derivative theorem in Section 1.6, $f(z)$ is constant in G, contradicting the hypothesis. Thus, $|f(z)|$ has no maximum in G, and the proof is complete. ■

We denote by $\bar{G}$ the set consisting of G together with its boundary. Since the exterior of G is open, $\bar{G}$ is closed. We can now reformulate the maximum principle in the following way.

COROLLARY

Let $f(z)$ be analytic in a bounded region G and continuous on $\bar{G}$. Then $|f(z)|$ attains its maximum on the boundary of G.

PROOF: Since $\bar{G}$ is closed and bounded and $|f(z)|$ is continuous on $\bar{G}$, a theorem of ordinary calculus states that $|f(z)|$ attains a maximum somewhere on $\bar{G}$. By the maximum principle, it cannot be in G, so it must be on the boundary of G. ■

MINIMUM PRINCIPLE

Let $f(z)$ be analytic in a bounded region G and continuous and nonzero on $\bar{G}$. Then $|f(z)|$ attains its minimum on the boundary of G.

PROOF: Let $g(z) = 1/f(z)$. Then g is analytic in G and continuous on $\bar{G}$. By the corollary above, $|g(z)|$ attains its maximum (and hence $|f(z)|$ attains its minimum) on the boundary of G. ∎

Liouville's theorem yields an easy verification of a very important theorem of elementary algebra that is usually stated without proof.

FUNDAMENTAL THEOREM OF ALGEBRA

Every polynomial of degree greater than zero has a root.

PROOF: Suppose $P(z) = a_n z^n + a_{n-1} z^{n-1} + \ldots + a_1 z + a_0$ is not zero for any value z. Then the function $f(z) = 1/P(z)$ is entire. Furthermore, $|f(z)|$ approaches zero as $|z|$ tends to infinity because

$$|f(z)| = \frac{1}{|z|^n \left| a_n + \dfrac{a_{n-1}}{z} + \ldots + \dfrac{a_0}{z^n} \right|}.$$

Therefore, $|f(z)|$ is bounded for all z. By Liouville's theorem, $f(z)$ and consequently $P(z)$ are constant, contradicting the hypothesis that $n > 0$. Thus $P(z)$ has at least one root.

To show $P(z)$ actually has n roots (including multiple roots), we observe that by the fundamental theorem of algebra, it has at least one root, say, ζ_0. Thus,

$$P(z) = P(z) - P(\zeta_0)$$
$$= a_n (z^n - \zeta_0^n) + a_{n-1} (z^{n-1} - \zeta_0^{n-1}) + \ldots + a_1 (z - \zeta_0)$$
$$= (z - \zeta_0) Q(z),$$

where $Q(z)$ is a polynomial in z of degree $n - 1$. If $n - 1 > 0$, then $Q(z)$ has a root. Continuing in this fashion, we can extract n factors from $P(z)$; thus $P(z)$ has exactly n roots. ∎

EXERCISES

1. Prove that an entire function satisfying $|f(z)| < |z|^n$ for some n and all sufficiently large $|z|$ must be a polynomial. [*Hint:* Apply the inequalities in Exercise 18, Section 2.3 to either $f^{(n+1)}(z)$ or $f^{(n)}(z)$.]

2. Let $f(z)$ be analytic in $|z| < 1$ and satisfy $f(0) = 0$. Define $F(z) = f(z)/z$ for all z in $0 < |z| < 1$. What value can be given to

$F(0)$ to make $F(z)$ analytic in $|z| < 1$? (*Hint:* Apply Cauchy's theorem for derivatives to $F(z)$ on $|z| = r < 1$. Then show that the resulting function, analytic in $|z| < r$, coincides with F on $0 < |z| < r$. Use partial fractions.)

3. Using the results in the exercise above and the maximum principle, prove **Schwarz's lemma:** Let $f(z)$ be analytic for $|z| < 1$ and satisfy the conditions $f(0) = 0$ and $|f(z)| \leqslant 1$. Then $|f(z)| \leqslant |z|$ and $|f'(0)| \leqslant 1$, with equality only if $f(z) = e^{i\theta} z$ for some fixed real θ.

4. Show that in Schwarz's lemma, $|f(z)| \leqslant 1$ for $|z| < 1$ implies that $|f'(0)| \leqslant 1$ regardless of the value of $f(0)$.

5. Give an example to show why the nonzero condition is necessary for the validity of the minimum principle.

6. Let $f(z)$ be analytic and nonconstant in $|z| < R$ and denote by $M(r)$ the maximum of $|f(z)|$ on $|z| = r$. Prove that $M(r)$ is strictly increasing for $0 \leqslant r < R$.

7. Prove that if $f(z)$ is analytic and nonconstant in the bounded region G, is continuous in $\bar{G}$, and has constant absolute value on the boundary of G, then it must have at least one zero in G.

*8. Prove the **three-circles theorem:** If $f(z)$ is analytic in a region containing the annulus $0 < r_1 \leqslant |z| \leqslant r_2$ and satisfies the inequalities $|f(z)| \leqslant M_1$ on $|z| = r_1$ and $|f(z)| \leqslant M_2$ on $|z| = r_2$, then the maximum of $|f(z)|$ on $|z| = r$, $r_1 \leqslant r \leqslant r_2$, is at most equal to

$$M_1^{(\log r_2/r)/(\log r_2/r_1)} \cdot M_2^{(\log r/r_1)/(\log r_2/r_1)}.$$

9. *Fundamental theorem of algebra* (alternate proof): Show that any nonconstant polynomial

$$P(z) = a_n z^n + \ldots + a_1 z + a_0, \qquad a_n \neq 0,$$

has at least one root, by assuming that $P(z)$ does not vanish and by integrating $a_0/zP(z)$ over $|z| = R$ with $R \to \infty$.

2.5 THE CAUCHY-GOURSAT THEOREM (Optional)

The fundamental theorem and Cauchy's theorem, proved in Sections 2.1 and 2.2, provide conditions guaranteeing that

$$\int_\gamma f(z) \, dz = 0,$$

where γ is a pws closed curve. The fundamental theorem requires that $f(z)$ be the continuous derivative of a function $F(z)$ analytic in a region G containing γ, whereas Cauchy's theorem requires that $f(z)$ be analytic and have a continuous derivative on and inside the pws Jordan curve γ. In this section we will prove that both hypotheses are satisfied when $f(z)$ is analytic. Furthermore, we will be able to extend Cauchy's theorem to any pws closed curve γ.

The next theorem provides the first step in proving that analytic functions have analytic antiderivatives. Note that this result is very similar to Cauchy's theorem in Section 2.2, except that $f'(z)$ is *not* assumed to be continuous inside the rectangle R shown in Figure 2.8.

CAUCHY-GOURSAT THEOREM

Let the function $f(z)$ be analytic in a region containing the rectangle R, given by the inequalities $a \leqslant x \leqslant b$, $c \leqslant y \leqslant d$. Then

$$\int_{\partial R} f(z)\, dz = 0,$$

where ∂R is the boundary of R.

PROOF: To simplify the notation, let

$$I(R) = \int_{\partial R} f(z)\, dz$$

for any rectangle R. Bisect R into four congruent rectangles R^1, R^2, R^3, R^4, and observe that

$$I(R) = I(R^1) + I(R^2) + I(R^3) + I(R^4),$$

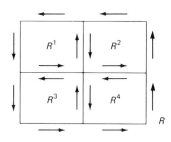

FIGURE 2.8. Bisecting the rectangle

because the integrals over the common sides cancel each other, by part (iii) of the first theorem in Section 2.3, as they have opposite orientation (see Figure 2.8.) By the triangle inequality, we have

$$|I(R)| \leqslant |I(R^1)| + |I(R^2)| + |I(R^3)| + |I(R^4)|,$$

so at least one R^j satisfies $|I(R^j)| \geqslant |I(R)|/4$. More than one R^j may have this property; select the one with smallest superscript and call it R_1. Repeating the above process indefinitely, we obtain a nested sequence of rectangles $R \supset R_1 \supset \cdots \supset R_n \supset R_{n+1} \supset \cdots$ satisfying

$$|I(R_n)| \geqslant \frac{|I(R_{n-1})|}{4},$$

implying that

$$|I(R_n)| \geqslant \frac{|I(R)|}{4^n}$$

(see Figure 2.9). Denote by $z_n^* = x_n^* + iy_n^*$ the lower left corner of the rectangle R_n. From the construction of the rectangles R_n, it is clear that the sequences $\{x_n^*\}$ and $\{y_n^*\}$ of real numbers are nondecreasing and bounded above by b and d, respectively. Thus, their limits x^* and y^* exist. We will show that the point $z^* = x^* + iy^*$ belongs to all

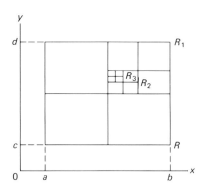

FIGURE 2.9. $R \supset R_1 \supset R_2 \supset R_3 \supset \ldots$

the rectangles R_n. If $z_n = x_n + iy_n$ is the upper right corner of R_n, then x_n and y_n are upper bounds for the sequences $\{x_n^*\}$ and $\{y_n^*\}$, implying that $x_n^* \leqslant x^* \leqslant x_n$, $y_n^* \leqslant y^* \leqslant y_n$. Thus z^* lies in R_n, for all n. Moreover, no other point lies in all the rectangles R_n, since $|z_n - z_n^*| \to 0$ as $n \to \infty$.

Given $\epsilon > 0$, we can find a $\delta > 0$ such that $f(z)$ is analytic and

$$\left| \frac{f(z) - f(z^*)}{z - z^*} - f'(z^*) \right| < \epsilon,$$

whenever $|z - z^*| < \delta$. For sufficiently large n, we have R_n contained in $|z - z^*| < \delta$. Since $z^*, f(z^*), f'(z^*)$ are constants, Example 5 of Section 2.1 implies that

$$\int_{\partial R_n} f(z^*)\, dz = 0 = \int_{\partial R_n} f'(z^*)(z - z^*)\, dz.$$

Thus, adding zero to the integral $I(R_n)$, we have

$$|I(R_n)| = \left| \int_{\partial R_n} [f(z) - f(z^*) - f'(z^*)(z - z^*)]\, dz \right|.$$

By part (iv) of the first theorem in Section 2.3 and the conditions above, we have

$$|I(R_n)| \leqslant \int_{\partial R_n} |f(z) - f(z^*) - f'(z^*)(z - z^*)|\, |dz|$$

$$< \epsilon \int_{\partial R_n} |z - z^*|\, |dz| \leqslant \epsilon D_n L_n,$$

where

$$D_n = |z_n - z_n^*| \qquad \text{and} \qquad L_n = \int_{\partial R_n} |dz|$$

are the diagonal and length of the perimeter of R_n, respectively. But

$$D_n = \tfrac{1}{2} D_{n-1} = \ldots = 2^{-n} D, \qquad L_n = \tfrac{1}{2} L_{n-1} = \ldots = 2^{-n} L,$$

where D and L are the diagonal and length of the perimeter of R, so

$$4^{-n}|I(R)| \leqslant |I(R_n)| \leqslant \epsilon D_n L_n = 4^{-n} \epsilon DL.$$

Thus, $|I(R)| \leqslant \epsilon DL$, and since ϵ is arbitrary, we can only have $I(R) = 0$, and the proof is complete. ∎

The next step is to show that any function analytic in a disk has an antiderivative in that disk.

THEOREM

If $f(z)$ is analytic in the disk $|z - z_0| < r$, then there is a function $F(z)$ analytic in $|z - z_0| < r$ satisfying $F'(z) = f(z)$.

PROOF: For any z in the disk $|z - z_0| < r$, let γ_z be the arc consisting of the line segments joining z_0 to $x + iy_0$ and $x + iy_0$ to z, where $z = x + iy$ and $z_0 = x_0 + iy_0$ (see Figure 2.10). Define

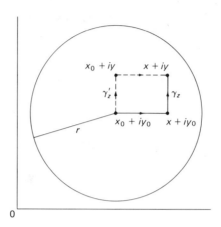

FIGURE 2.10.   Arcs γ_z and γ_z'

$$F(z) = \int_{\gamma_z} f(z)\, dz = \int_{x_0}^{x} f(t + iy_0)\, dt + i \int_{y_0}^{y} f(x + it)\, dt. \quad (1)$$

If γ_z' is the arc consisting of the line segments joining z_0 to $x_0 + iy$ and $x_0 + iy$ to z, then $\gamma_z - \gamma_z'$ is the boundary of a rectangle, and by the Cauchy–Goursat theorem,

$$0 = \int_{\gamma_z - \gamma_z'} f(z)\, dz = \int_{\gamma_z} f(z)\, dz - \int_{\gamma_z'} f(z)\, dz.$$

Thus we also can compute $F(z)$ along the path γ_z',

$$F(z) = \int_{\gamma_z'} f(z)\, dz = i \int_{y_0}^{y} f(x_0 + it)\, dt + \int_{x_0}^{x} f(t + iy)\, dt. \quad (2)$$

The partial derivative of (1) with respect to y is given by

$$F_y(z) = i\, \frac{\partial}{\partial y} \int_{y_0}^{y} f(x + it)\, dt = if(x + iy) = if(z),$$

since the first integral in (1) is independent of y. Similarly, taking the partial derivative of (2) with respect to x yields $F_x(z) = f(z)$. Thus, $F(z)$ satisfies the Cauchy–Riemann equations

$$F_x(z) = f(z) = -iF_y(z),$$

and since $f(z)$ is continuous, we have sufficient conditions for the analyticity of $F(z)$ in $|z - z_0| < r$. Finally, $F'(z) = F_x(z) = f(z)$. ∎

As in real calculus, *two antiderivatives of the same function differ at most by a constant:* If $F(z)$ and $H(z)$ are both antiderivatives of the function $f(z)$, then

$$[F(z) - H(z)]' = f(z) - f(z) = 0,$$

implying $F(z) - H(z)$ is a constant, by the zero derivative theorem of Section 1.6.

The proof in the theorem above may be extended to any simply connected region.

ANTIDERIVATIVE THEOREM

Let $f(z)$ be analytic in a simply connected region G. Then there is a function $F(z)$ analytic in G such that $F'(z) = f(z)$.

PROOF: Fix a point z_0 in G. Then by the theorem on polygonal paths in Section 1.3, we can find a polygon, with sides parallel to the axes, joining z_0 to any point z in G. Suppose γ and γ' are two such polygons; then $\gamma - \gamma'$ consists of a finite number of boundaries of rectangles in G (possibly some of them degenerate) traversed alternatively in the positive and negative direction (see Figure 2.11). This fact requires a delicate proof, utilizing the simple connectedness of the region G, and is omitted as it is intuitively clear (see the Notes at the end of the chapter). By the Cauchy-Goursat theorem,

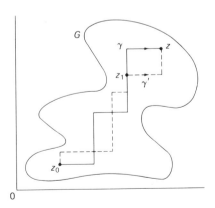

FIGURE 2.11. The curve $\gamma - \gamma'$ consisting of the boundaries of rectangles in G

$$0 = \int_{\gamma - \gamma'} f(z) \, dz = \int_{\gamma} f(z) \, dz - \int_{\gamma'} f(z) \, dz.$$

Thus

$$F(z) = \int_{\gamma} f(z) \, dz = \int_{\gamma'} f(z) \, dz$$

is independent of the choice of path. Let the last line segment of $\gamma(\gamma')$ be horizontal (vertical) and $z_1 = x_1 + iy_1$ be the last point of inter-section. Then

$$F(z) = i \int_{y_1}^{y} f(x_1 + it) \, dt + \int_{x_1}^{x} f(t + iy) \, dt + C$$

$$= \int_{x_1}^{x} f(t + iy_1) \, dt + i \int_{y_1}^{y} f(x + it) \, dt + C,$$

where $z = x + iy$ and the constant $C = F(z_1)$. Taking the partial derivatives, the first equation yields $F_x(z) = f(z)$, the second $F_y(z) = if(z)$. Since $f(z)$ is continuous and $F_x = -iF_y$, $F(z)$ is analytic in G and $F'(z) = f(z)$.

It is essential that G be simply connected, for otherwise the poly-gons γ and γ' might form a rectangle containing a "hole" in its interior. The function $f(z)$ would not be analytic in a region contain-ing that rectangle, so the Cauchy–Goursat theorem would not apply. ∎

EXAMPLE.

The function $f(z) = 1/z$, analytic in $\mathbb{C} - \{0\}$, has $F(z) = \log z$ as an antiderivative. If we traverse the top half of the unit circle starting at 1 on the principal branch, we obtain

$$F(e^{i\pi/2}) = \frac{i\pi}{2},$$

whereas traversing the bottom half of the unit circle yields

$$F(e^{-i\pi/2}) = \frac{-i\pi}{2}.$$

Hence, in this case the value of the antiderivative at $z = -1$ depends on the path chosen.

The antiderivative theorem provides an immediate simplification of the hypotheses of the following two theorems.

FUNDAMENTAL THEOREM

Let $f(z)$ be analytic in a simply connected region G. Then, for any pws arc $\gamma\colon z = z(t),\ \alpha \leqslant t \leqslant \beta$,

$$\int_{\gamma} f(z)\, dz = F(z(\beta)) - F(z(\alpha)),$$

where $F(z)$ is any antiderivative of $f(z)$ in G.

CAUCHY'S THEOREM

If $f(z)$ is analytic and γ is a pws closed curve in the simply connected region G, then

$$\int_{\gamma} f(z)\, dz = 0.$$

PROOF: If $f(z)$ is analytic in the simply connected region G, there is an analytic function $F(z)$ in G such that $F' = f$. Thus, the fundamental theorem of Section 2.1 holds, implying that for any pws arc $\gamma\colon z = z(t),\ \alpha \leqslant t \leqslant \beta$,

$$\int_{\gamma} f(z)\, dz = F(z(\beta)) - F(z(\alpha)).$$

If $z(\beta) = z(\alpha)$, we obtain Cauchy's theorem. ∎

We now consider the properties possessed by integrals of the type found in the Cauchy integral formula.

RIEMANN'S THEOREM

Let $g(\zeta)$ be continuous on the pws arc γ. Then the function

$$F_n(z) = \int_{\gamma} \frac{g(\zeta)\, d\zeta}{(\zeta - z)^n}, \qquad n = 1, 2, 3, \ldots,$$

is analytic at all z in the complement of γ, and its derivative satisfies $F_n'(z) = n F_{n+1}(z)$.

PROOF: Select a point z_0 not on γ and a disk $|z - z_0| < \delta$ disjoint from γ. For z in the disk $|z - z_0| < \delta/2$, we have

$$|F_1(z) - F_1(z_0)| = \left| \int_\gamma g(\zeta) \left(\frac{1}{\zeta - z} - \frac{1}{\zeta - z_0} \right) d\zeta \right|$$

$$\leq |z - z_0| \int_\gamma \frac{|g(\zeta)| \, |d\zeta|}{|\zeta - z| |\zeta - z_0|}.$$

The arc γ has finite length L, thus is a closed and bounded set of points. A theorem of ordinary calculus states that continuous real-valued functions attain a maximum on any closed and bounded set. Thus, $|g(\zeta)|$ is bounded by M on γ. Since $|\zeta - z| > \delta/2$ for all ζ on γ.

$$|F_1(z) - F_1(z_0)| \leq \frac{2ML}{\delta^2} \cdot |z - z_0|,$$

which proves the continuity of $F_1(z)$ at z_0. Applying this fact to the functions

$$G_n(z) = \int_\gamma \frac{\dfrac{g(\zeta)}{(\zeta - z_0)}}{(\zeta - z)^n} \, d\zeta,$$

we find that $G_1(z)$ is continuous at z_0 since $g(\zeta)/(\zeta - z_0)$ is continuous on γ. Then, since the difference quotient of $F_1(z)$ equals $G_1(z)$,

$$F_2(z_0) = G_1(z_0) = \lim_{z \to z_0} G_1(z) = \lim_{z \to z_0} \frac{F_1(z) - F_1(z_0)}{z - z_0} = F_1'(z_0).$$

Suppose it were true that $F_{n-1}'(z) = (n - 1)F_n(z)$ (and since $g(\zeta)$ is arbitrary, also that $G_{n-1}'(z) = (n - 1)G_n(z)$). Then

$$\frac{1}{\zeta - z} = \frac{1}{\zeta - z_0} + \frac{z - z_0}{(\zeta - z)(\zeta - z_0)}$$

implies that

$$F_n(z) - F_n(z_0) = [G_{n-1}(z) - G_{n-1}(z_0)] + (z - z_0)G_n(z).$$

Since $G_{n-1}(z)$ is differentiable, it is continuous, and

$$|G_n(z)| = \left| \int_\gamma \frac{g(\zeta) \, d\zeta}{(\zeta - z)^n (\zeta - z_0)} \right| \leq \frac{2^n ML}{\delta^{n+1}},$$

for $|z - z_0| < \delta/2$. By the triangle inequality,

$$0 \leqslant \lim_{z \to z_0} |F_n(z) - F_n(z_0)| \leqslant \frac{2^n ML}{\delta^{n+1}} \lim_{z \to z_0} |z - z_0| = 0,$$

implying that $F_n(z)$ (and hence $G_n(z)$) is continuous at z_0. Thus,

$$F_n'(z_0) = \lim_{z \to z_0} \left[\frac{G_{n-1}(z) - G_{n-1}(z_0)}{z - z_0} \right] + G_n(z)$$

$$= G_{n-1}'(z_0) + G_n(z_0)$$

$$= nG_n(z_0) = nF_{n+1}(z_0).$$

The proof now follows by induction. ∎

Riemann's theorem provides a proof for Cauchy's theorem for derivatives and the remarkable fact that analytic functions have analytic derivatives.

CAUCHY'S THEOREM FOR DERIVATIVES

Let $f(z)$ be analytic in a simply connected region containing the pws Jordan curve γ. Then, for all points ζ inside γ,

$$f^{(n)}(\zeta) = \frac{n!}{2\pi i} \int_\gamma \frac{f(z)}{(z - \zeta)^{n+1}} \, dz, \qquad n = 0, 1, 2, \ldots .$$

PROOF: Set $g(z) = f(z)$ in Riemann's theorem. Then

$$F_1(\zeta) = \int_\gamma \frac{f(z)}{z - \zeta} \, dz = 2\pi i f(\zeta)$$

by the Cauchy integral formula for all points ζ inside γ. Applying Riemann's theorem repeatedly, we have

$$F_{n+1}(\zeta) = \frac{F_n'(\zeta)}{n} = \frac{F_{n-1}''(\zeta)}{n(n-1)} = \ldots = \frac{F_1^{(n)}(\zeta)}{n!} = \frac{2\pi i f^{(n)}(\zeta)}{n!},$$

so

$$f^{(n)}(\zeta) = \frac{n!}{2\pi i} F_{n+1}(\zeta) = \frac{n!}{2\pi i} \int_\gamma \frac{f(z)}{(z - \zeta)^{n+1}} \, dz,$$

and the result follows. Adopting the convention that $f^{(0)} = f$ and $0! = 1$, note that the equation above reduces to the Cauchy integral formula when $n = 0$. ∎

COROLLARY

If $f(z)$ is analytic in a region G, then so is its derivative $f'(z)$. Furthermore, $f(z)$ possesses derivatives of all orders on G.

PROOF:　Since analyticity need only be proved in a neighborhood of a point, for each ζ we can find a disk $|z - \zeta| \leqslant r$ contained in G. Let γ be the circle $|z - \zeta| = r$. Then $f^{(n)}(\zeta)$ exists for all positive integers n, so $f'(z)$ has a derivative at ζ and is thus analytic. ∎

This corollary completes the task of showing that analytic functions have analytic derivatives and allows the removal of all unnecessary hypotheses in the versions of the fundamental theorem, Cauchy's theorem, and Morera's theorem that were proved in Sections 2.1–2.3.

EXERCISES

1. The **Laguerre polynomials** $L_n(z)$ are given by

$$L_n(z) = e^z \, \frac{d^n}{dz^n} \, (z^n e^{-z}).$$

Show that for all z inside the pws Jordan curve γ,

$$L_n(z) = \frac{n!}{2\pi i} \int_\gamma \frac{\zeta^n e^{-(\zeta - z)}}{(\zeta - z)^{n+1}} \, d\zeta.$$

2. Derive **Wallis's formula**

$$\int_0^{\pi/2} \cos^{2n} \theta \, d\theta = \frac{(2n)!}{(2^n \, n!)^2} \cdot \frac{\pi}{2}$$

by integrating $f(z) = (z + 1/z)^{2n}/z$ over $|z| = 1$.

3. Let $f(z)$ be continuous in the region $\operatorname{Re} z \geqslant \sigma$ and suppose $\lim_{z \to \infty} f(z) = 0$. Then, for every negative number t, prove that

$$\lim_{R \to \infty} \int_{\Gamma_R} e^{zt} f(z) \, dz = 0,$$

where $\Gamma_R = \{|z| = R\} \cap \{\operatorname{Re} z \geqslant \sigma\}$.

*4. Let $f(z)$ be analytic on the set R^* obtained by omitting finitely many interior points $z_1, z_2, \ldots, z_n$ of the rectangle R. Prove that

$$\int_{\partial R} f(z) \, dz = 0,$$

provided

$$\lim_{z \to z_k} (z - z_k)f(z) = 0,$$

for all $k = 1, 2, \ldots, n$.

*5. Let $f(z)$ be analytic on the set D obtained by omitting the points $z_1, z_2, \ldots, z_n$ in $|z - z_0| < r$. Prove that

$$\int_\gamma f(z) \, dz = 0$$

for any closed curve γ in $|z - z_0| < r$, provided that

$$\lim_{z \to z_k} (z - z_k)f(z) = 0, \qquad k = 1, 2, \ldots, n.$$

*6. Show that the statement in Exercise 5 remains true when D is obtained by omitting infinitely many points $z_1, z_2, \ldots$, having no accumulation points in $|z - z_0| < r$.

NOTES

SECTION 2.1

A proof of the Jordan curve theorem is found in [W, p. 30].

SECTION 2.4

Generalizations of Schwarz's lemma may be found in [A, p. 136].

SECTION 2.5

The Cauchy-Goursat theorem may be shown to hold on R with weaker hypotheses. [V, p. 76] proves it is valid for $f(z)$ analytic inside ∂R and continuous on R. Verification in the antiderivative theorem that $\gamma - \gamma'$ consists of the boundaries of finitely many rectangles may be found in [A, pp. 141–143] or [L, pp. 128–131]. A proof of Cauchy's theorem bypassing the topology has recently been given by J. D. Dixon. It may be found in [L, pp. 148–150] or in the original article in *Proc. Amer. Math. Soc.* 29 (1971): 625–626. Further generalizations of Cauchy's theorem may be found in [A,

p. 144] and [Ho, pp. 3, 26]. Riemann's theorem holds whenever $\int_\gamma |g(\zeta)| \, |d\zeta| < \infty$. The proof, using this weaker hypothesis, is essentially the same. A proof of analyticity of the derivative independent of integration is given by [W, p. 77].

3 INFINITE SERIES

3.1 TAYLOR SERIES

DEFINITION

An infinite series of complex numbers

$$a_1 + a_2 + \ldots + a_n + \ldots$$

converges to the sum A if the partial sums

$$S_n = a_1 + a_2 + \ldots + a_n$$

satisfy $S_n \to A$ as $n \to \infty$, and in this case we write $\Sigma_1^\infty \, a_n = A$. Otherwise, we say the series **diverges**. A series with the property that the absolute values of its terms form a convergent series is said to be **absolutely convergent**.

Since $a_{n+1} = S_{n+1} - S_n$, if the series converges, we have

$$\lim_{n \to \infty} a_{n+1} = \lim_{n \to \infty} (S_{n+1} - S_n) = A - A = 0.$$

Thus, the general term in a convergent series tends to zero. This condition is necessary but not sufficient, as the following example shows.

EXAMPLE 1.

The series

$$1 + \tfrac{1}{2} + \tfrac{1}{2} + \tfrac{1}{3} + \tfrac{1}{3} + \tfrac{1}{3} + \tfrac{1}{4} + \tfrac{1}{4} + \tfrac{1}{4} + \tfrac{1}{4} + \ldots$$

diverges, because if like terms are added,

$$1 + (\tfrac{1}{2} + \tfrac{1}{2}) + (\tfrac{1}{3} + \tfrac{1}{3} + \tfrac{1}{3}) + (\tfrac{1}{4} + \tfrac{1}{4} + \tfrac{1}{4} + \tfrac{1}{4}) + \ldots,$$

the partial sums grow without bound.

An absolutely convergent series must converge. The proof is left as an exercise, as it is identical to that given in any calculus text.

Frequently, we are interested in an infinite series of functions defined on a region G,

$$\sum_{n=1}^{\infty} f_n(z) = f_1(z) + f_2(z) + \ldots + f_n(z) + \ldots.$$

The series is said to converge on G if it converges for each z_0 in G. We write

$$f(z) = \sum_{1}^{\infty} f_n(z)$$

and call $f(z)$ the **sum** of the series.

EXAMPLE 2.

Show that the **geometric series**

$$\sum_{n=0}^{\infty} z^n$$

converges to

$$\frac{1}{1-z}$$

for $|z| < 1$.

SOLUTION: Using polynomial long division, we have

$$\frac{1}{1-z} = 1 + z + z^2 + \ldots + z^{n-1} + \frac{z^n}{1-z} = S_{n-1} + \frac{z^n}{1-z}.$$

Since $|z| < 1$, it follows that $z^n \to 0$ as $n \to \infty$. Thus,

$$\frac{1}{1-z} = \sum_{n=0}^{\infty} z^n, \qquad |z| < 1.$$

We will now show that every analytic function can be expressed as a convergent **Taylor series**.

TAYLOR'S THEOREM

Let $f(z)$ be analytic in the region G containing the point z_0. Then the representation

$$f(z) = f(z_0) + \frac{f'(z_0)}{1!}(z - z_0) + \ldots + \frac{f^{(n)}(z_0)}{n!}(z - z_0)^n + \ldots$$

holds in all disks $|z - z_0| < r$ contained in G.

PROOF: Let z be any point of the closed disk $|\zeta - z_0| \leqslant r$ contained in G, and use the Cauchy integral formula to express $f(z)$ as an integral.

$$f(z) = \frac{1}{2\pi i} \int_{|\zeta - z_0| = r} \frac{f(\zeta)}{\zeta - z} \, d\zeta.$$

Since

$$\zeta - z = (\zeta - z_0)\left[1 - \frac{z - z_0}{\zeta - z_0}\right] \quad \text{and} \quad \left|\frac{z - z_0}{\zeta - z_0}\right| < 1,$$

we use the partial sums of the geometric series to rewrite the integral in the form

$$f(z) = \frac{1}{2\pi i} \int_{|\zeta - z_0| = r} \frac{f(\zeta)}{\zeta - z_0} \, \frac{1}{\left[1 - \frac{z - z_0}{\zeta - z_0}\right]} \, d\zeta$$

$$= \frac{1}{2\pi i} \int_{|\zeta - z_0| = r} \frac{f(\zeta)}{\zeta - z_0}\left[1 + \frac{z - z_0}{\zeta - z_0} + \ldots + \left(\frac{z - z_0}{\zeta - z_0}\right)^{n-1} + Q_n\right] d\zeta,$$

where

$$Q_n = \frac{(z - z_0)^n / (\zeta - z_0)^n}{1 - (z - z_0)/(\zeta - z_0)} = \frac{(z - z_0)^n}{(\zeta - z)(\zeta - z_0)^{n-1}}.$$

Using Cauchy's theorem for derivatives, we obtain

$$f(z) = f(z_0) + (z - z_0)\frac{f'(z_0)}{1!} + \ldots + (z - z_0)^{n-1}\frac{f^{(n-1)}(z_0)}{(n-1)!} + R_n,$$

where

$$R_n = \frac{(z - z_0)^n}{2\pi i} \int_{|\zeta - z_0| = r} \frac{f(\zeta)}{(\zeta - z)(\zeta - z_0)^n} \, d\zeta.$$

If we select z inside $|\zeta - z_0| = r$, let $|z - z_0| = \rho$, and observe that $|\zeta - z| \geq r - \rho$ for all ζ on $|\zeta - z_0| = r$, we have

$$|R_n| \leq \frac{\rho^n}{2\pi} \frac{2\pi r M}{(r - \rho) r^n} = \frac{rM}{r - \rho} \left(\frac{\rho}{r} \right)^n,$$

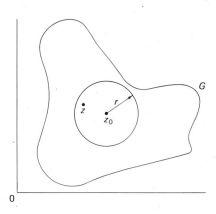

FIGURE 3.1.   The disk $|\zeta - z_0| < r$ contained in G

with M the maximum of $|f(\zeta)|$ on $|\zeta - z_0| = r$ (see Figure 3.1). But $\rho/r < 1$, hence $R_n \to 0$ as $n \to \infty$. Therefore, $f(z)$ is represented by the Taylor series for all such z. ∎

This theorem allows us to obtain Taylor series for analytic functions in the same manner as was done in ordinary calculus. For example, if $f(z) = e^z$, then $f^{(n)}(z) = e^z$ and $f^{(n)}(0) = 1$, and we have the **Maclaurin series**

$$e^z = \sum_{n=0}^{\infty} \frac{z^n}{n!}, \qquad |z| < \infty,$$

valid for all z in $\mathbb{C}$, since $f(z)$ is entire.

The next two theorems are useful consequences of Taylor's theorem.

THEOREM

If $f(z)$ is analytic in a region G containing the point z_0, and $f^{(n)}(z_0) = 0$ for $n = 1, 2, \ldots$, then $f(z)$ is constant in G.

PROOF: By Taylor's theorem, $f(z) = f(z_0)$ for all z in any disk $|\zeta - z_0| < r$ contained in G. Let $g(z) = f(z) - f(z_0)$. Then g is analytic in G and $g^{(n)}(z) = 0$, $n = 0, 1, 2, \ldots$, for all z in this disk. Let S be the set of all points z in G at which $g^{(n)}(z) = 0$, $n = 0, 1, 2, \ldots$, and let $T = G - S$. If z_1 is in S, then by Taylor's theorem, g has the series representation $g(z) = 0$ in all disks $|z - z_1| < r$ contained in G. Hence, by the argument above, S is open, as all its points are interior points. If z_1 is in T, there is an integer $n \geqslant 0$ for which $g^{(n)}(z_1) \neq 0$. Thus, in a disk centered on z_1 lying in G, the Taylor series of $g(z)$ does not vanish, implying that z_1 is an interior point of T. Thus, T is open. Neither T nor S contains a boundary point of the other, as both are open. Since G is connected, T must be empty; hence $g(z) = f(z) - f(z_0) = 0$ for all z in G. ∎

This theorem implies that if a nonconstant function $f(z)$ analytic in a region G vanishes at a point z_0 in G, then there is a positive integer n for which $f^{(n)}(z_0) \neq 0$. The least such integer determines the *order of the zero of f at z_0* and allows us to write

$$f(z) = (z - z_0)^n f_n(z), \qquad f_n(z) = \frac{1}{2\pi i} \int_{|\zeta - z_0| = r} \frac{f(\zeta)}{(\zeta - z)(\zeta - z_0)^n} \, d\zeta,$$

with $f_n(z)$ analytic inside the disk $|\zeta - z_0| \leqslant r$ contained in G by Cauchy's theorem for derivatives (or by Riemann's theorem in Section 2.5.) Furthermore,

$$f_n(z_0) = \frac{1}{2\pi i} \int_{|\zeta - z_0| = r} \frac{f(\zeta)}{(\zeta - z_0)^{n+1}} \, d\zeta = \frac{f^{(n)}(z_0)}{n!} \neq 0.$$

Thus, there is an ϵ-neighborhood of z_0 lying in G on which $f_n(z)$ does not vanish, since f_n is continuous. This shows that z_0 is the only zero of f in $|z - z_0| < \epsilon$. We have proved the following theorem.

THEOREM

The zeros of a nonconstant analytic function are isolated.

EXAMPLE 3.

Obtain the Maclaurin series of $f(z) = (1 - z)^{-2}$.

SOLUTION: As $f^{(n)}(z) = (n + 1)!(1 - z)^{-(n+2)}$, for $n = 0, 1, 2, \ldots$, we have $f^{(n)}(0) = (n + 1)!$ and

$$\frac{1}{(1 - z)^2} = \sum_{n=0}^{\infty} (n + 1)z^n, \qquad |z| < 1,$$

since $f(z)$ is not analytic at $z = 1$. The Taylor series of $f(z)$ centered at the point $z_0 = -1$ is

$$\frac{1}{(1 - z)^2} = \sum_{n=0}^{\infty} \frac{(n + 1)}{2^{n+2}} (z + 1)^n, \qquad |z| < 2,$$

since $f^{(n)}(-1) = (n + 1)!/2^{n+2}$.

EXAMPLE 4.

Find the order of the zero of $f(z) = 2z(e^z - 1)$ at $z = 0$.

SOLUTION: First calculate $f^{(n)}(0)$ for $n = 0, 1, 2, \ldots$. As

$$f'(z) = 2ze^z + 2(e^z - 1), \qquad f''(z) = 2ze^z + 4e^z,$$

we find $f''(0) = 4$, so the order is 2. This is also clear from the Maclaurin series of $f(z)$

$$2z(e^z - 1) = 2z \sum_{n=1}^{\infty} \frac{z^n}{n!} = 2z^2 \left(1 + \frac{z}{2!} + \frac{z^2}{3!} + \ldots \right).$$

We must be especially careful when seeking the Taylor series of analytic functions defined on a Riemann surface. The next example illustrates this situation.

EXAMPLE 5.

Recall that the function $\log z$ is defined on the Riemann surface $\Re$ described in Section 1.9. In constructing the Taylor series of

$f(z) = \log z$, it is essential to specify the branch under consideration. If we seek the Taylor series about $z = 1$, a point on the principal branch, we have $f^{(n)}(z) = (-1)^{n-1}(n-1)!z^{-n}$, $n = 1, 2, 3, \ldots$, and $\log 1 = \mathrm{Log}\, 1 = 0$, so that

$$\mathrm{Log}\, z = \sum_{n=1}^{\infty} \frac{f^{(n)}(1)}{n!}(z-1)^n = -\sum_{n=1}^{\infty} \frac{(1-z)^n}{n}.$$

On the other hand, the Taylor series about $z = e^{2\pi i}$ on the next branch of $\mathfrak{R}$ is

$$\log z = 2\pi i - \sum_{n=1}^{\infty} \frac{(1-z)^n}{n},$$

since the functions $f^{(n)}(z)$ are analytic on $\mathbb{C} - \{0\}$ and

$$f(e^{2\pi i}) = \log e^{2\pi i} = 2\pi i,$$
$$f^{(n)}(e^{2\pi i}) = (-1)^{n-1}(n-1)!, \qquad n = 1, 2, \ldots .$$

Similar remarks would apply to any other branch of $\mathfrak{R}$. The formulas are valid in $|z - 1| < 1$, since $f^{(n)}(z)$ is not analytic at $z = 0$, for $n = 0, 1, 2, \ldots$.

EXAMPLE 6.

Show that there is no function analytic in $|z| < 2$ satisfying the condition

$$f\left(\frac{1}{n}\right) = \frac{(-1)^n}{n}, \qquad n = 1, 2, 3, \ldots .$$

SOLUTION: Otherwise, $F(z) = z - f(z)$ is a nonconstant analytic function satisfying $F(1/2m) = 0$, for $m = 1, 2, 3, \ldots$. Hence, $z = 0$ is a zero of $F(z)$ that is not isolated, contradicting the last theorem.

EXERCISES

1. Show that the series $\sum_{n=1}^{\infty} 1/n$ diverges.

2. Prove that an absolutely convergent series must converge.

Obtain the Maclaurin series given in Exercises 3–7.

3. $\sin z = \sum\limits_{n=1}^{\infty} (-1)^{n-1} \dfrac{z^{2n-1}}{(2n-1)!}$, $|z| < \infty$

4. $\cos z = \sum\limits_{n=0}^{\infty} (-1)^{n} \dfrac{z^{2n}}{(2n)!}$, $|z| < \infty$

5. $\sinh z = \sum\limits_{n=1}^{\infty} \dfrac{z^{2n-1}}{(2n-1)!}$, $|z| < \infty$

6. $\cosh z = \sum\limits_{n=0}^{\infty} \dfrac{z^{2n}}{(2n)!}$, $|z| < \infty$

7. $\dfrac{1}{1 - z^2} = \sum\limits_{n=0}^{\infty} z^{2n}$, $|z| < 1$

In Exercises 8–15, expand the given functions in a Taylor series about z_0. Indicate the largest disk where the representation is valid.

8. $f(z) = \dfrac{1}{1 - z}$, $z_0 = -1$ 9. $f(z) = \dfrac{1}{1 - z}$, $z_0 = i$

10. $f(z) = \cos z$, $z_0 = \dfrac{\pi}{2}$ 11. $f(z) = \sin z$, $z_0 = \dfrac{\pi}{2}$

12. $f(z) = \dfrac{1}{z}$, $z_0 = 1$ 13. $f(z) = \text{Log } z$, $z_0 = i$

14. $f(z) = \log z$, $z_0 = -1$ 15. $f(z) = \log z$, $z_0 = 2e^{3\pi i}$

Find the order of the zero at $z = 0$ for the functions given in Exercises 16–19.

16. $z^2 (\cos z - 1)$ 17. $6 \sin z^2 + z^2 (z^4 - 6)$
18. $z - \tan z$ 19. $z^2 - \sinh z^2$

20. Prove that if two functions analytic on a region G coincide on a subset of G that has an accumulation point in G, then they coincide everywhere in G.

Determine if there exists a function analytic in $|z| < 2$ assuming at the points $z = 1/n$, $n = 1, 2, 3, \ldots$, the values given in Exercises 21–24.

21. $0, 1, 0, -1, 0, 1, 0, -1, \ldots$
22. $1, 0, \frac{1}{3}, 0, \frac{1}{5}, 0, \frac{1}{7}, 0, \frac{1}{9}, 0, \ldots$
23. $1, \frac{2}{3}, \frac{3}{5}, \frac{4}{7}, \frac{5}{9}, \frac{6}{11}, \frac{7}{13}, \frac{8}{15}, \ldots$

24. $\frac{1}{2}, -\frac{1}{2}, \frac{1}{3}, -\frac{1}{3}, \frac{1}{4}, -\frac{1}{4}, \frac{1}{5}, -\frac{1}{5}, \ldots$

25. Give an example of two functions that agree at infinitely many points in a region G, yet are different.

26. Show that if f is a nonconstant analytic function in G, the set of points z satisfying $f(z) = \alpha$, α in $\mathbb{C}$, does not have an accumulation point in G.

27. Prove the **binomial theorem** for complex α:

$$(1 + z)^\alpha = 1 + \frac{\alpha}{1} z + \frac{\alpha(\alpha - 1)}{1 \cdot 2} z^2 + \frac{\alpha(\alpha - 1)(\alpha - 2)}{1 \cdot 2 \cdot 3} z^3 + \ldots, \quad |z| < 1.$$

3.2 UNIFORM CONVERGENCE OF SERIES

In this section we will prove the converse of Taylor's theorem, namely, that convergent power series are in fact analytic functions in their region of convergence.

DEFINITION

The series $f(z) = \Sigma_1^\infty f_n(z)$ is **uniformly convergent** on G if for every $\epsilon > 0$ there exists a positive *number K* such that

$$\left| f(z) - \sum_{n=1}^k f_n(z) \right| < \epsilon,$$

for any $k > K$ and z in G.

Uniform convergence differs from convergence in that for convergence we need only show the existence of a positive *function K(z)*, such that for each z_0 in G,

$$\left| f(z_0) - \sum_{n=1}^k f_n(z_0) \right| < \epsilon,$$

whenever $k > K(z_0)$. The importance of the concept of uniform convergence comes from the following result.

WEIERSTRASS'S THEOREM

The sum of a uniformly convergent series of *analytic* functions is analytic and may be differentiated or integrated term by term.

PROOF: Let $f(z) = \sum_{1}^{\infty} f_n(z)$ with each $f_n(z)$ analytic in the region G. Given $\epsilon > 0$, there is a positive number K such that

$$\left| f(z) - \sum_{n=1}^{k} f_n(z) \right| < \frac{\epsilon}{3},$$

for $k > K$ and all z in G. For any z_0 in G and fixed $k \; (> K)$, there is a δ such that

$$\left| \sum_{n=1}^{k} f_n(z) - \sum_{n=1}^{k} f_n(z_0) \right| < \frac{\epsilon}{3}$$

whenever z is in G and $|z - z_0| < \delta$, since the partial sum is continuous. Thus, by the triangle inequality,

$$|f(z) - f(z_0)| \leqslant \left| f(z) - \sum_{n=1}^{k} f_n(z) \right|$$

$$+ \left| \sum_{n=1}^{k} f_n(z) - \sum_{n=1}^{k} f_n(z_0) \right| + \left| \sum_{n=1}^{k} f_n(z_0) - f(z_0) \right|$$

$$< \epsilon$$

whenever z lies in G and $|z - z_0| < \delta$. Hence, f is continuous in G. By Cauchy's theorem, for any pws closed curve γ lying in a disk contained in G,

$$\int_{\gamma} f_n(z)\, dz = 0, \; n = 1, 2, \ldots, k,$$

so that

$$\left| \int_{\gamma} f(z)\, dz \right| = \left| \int_{\gamma} f(z)\, dz - \sum_{n=1}^{k} \int_{\gamma} f_n(z)\, dz \right|$$

$$\leqslant \int_{\gamma} \left| f(z) - \sum_{n=1}^{k} f_n(z) \right| |dz| < \frac{\epsilon L}{3},$$

where L is the length of γ. Since ϵ can be made arbitrarily close to zero, the extension of Morera's theorem (Exercise 23, Section 2.3) holds and $f(z)$ is analytic in G. (If G is simply connected, it follows directly from Morera's theorem.) In particular, the above shows that on any pws arc γ in G,

$$\int_{\gamma} f(z)\, dz = \sum_{n=1}^{\infty} \int_{\gamma} f_n(z)\, dz.$$

Furthermore, by Cauchy's formula for derivatives,

$$\left| f'(z) - \sum_{n=1}^{k} f_n'(z) \right| = \left| \frac{1}{2\pi i} \int_{|\zeta - z_0| = r} \frac{f(\zeta) - \sum_{n=1}^{k} f_n(\zeta)}{(\zeta - z)^2} \, d\zeta \right| < \frac{4\epsilon}{3r},$$

for all z satisfying $|z - z_0| < r/2$, $k > K$, and $|\zeta - z_0| \leqslant r$ contained in G (see Figure 3.2). Thus, the series $\Sigma f_n'(z)$ converges uniformly to $f'(z)$ on $|z - z_0| < r/2$, and the proof is complete. ∎

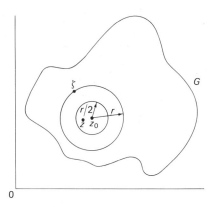

FIGURE 3.2. $|\zeta - z| > r/2$

Weierstrass's theorem can be applied to the power series $\Sigma_1^\infty a_n (z - z_0)^n$, since each term in the series is an entire function. Before proceeding in this direction, it is useful to make some comments about power series. Note that the substitution $\zeta = z - z_0$ transforms the series above into the power series $\Sigma_1^\infty a_n \zeta^n$, so we shall only consider power series of this last form.

From elementary calculus, recall the concept of the **radius of convergence** R of a power series $\Sigma_1^\infty r_n x^n$, where the coefficients r_n are real. The number $0 \leqslant R \leqslant \infty$ has the property that the series converges absolutely for $|x| < R$ and diverges for $|x| > R$, and R can be calculated by the formula

$$R = \lim_{n \to \infty} \left| \frac{r_n}{r_{n+1}} \right|,$$

provided the limit exists. Unfortunately for series like

$$2 + x + 2x^2 + x^3 + \ldots + 2x^{2k} + x^{2k+1} + \ldots,$$

the ratio of the coefficients $|r_n/r_{n+1}|$ is alternately $\frac{1}{2}$ and 2, so no limit exists. We now give a formula that can always be used in calculating the radius of convergence R of a power series $\Sigma_1^\infty a_n z^n$, and prove that R behaves in the same fashion as it did for real power series.

HADAMARD'S FORMULA

The radius of convergence R of a power series $\Sigma_1^\infty a_n z^n$ is given by

$$R^{-1} = \lim_{n \to \infty} \sup \sqrt[n]{|a_n|} = \lim_{n \to \infty} [\text{lub}(|a_n|^{1/n}, |a_{n+1}|^{1/(n+1)}, \ldots)].$$

The least upper bound (lub) either decreases or remains constant as n increases, so this limit always exists (with infinity a permissible value). Since $2^{1/(2n)} \to 1$ as $n \to \infty$, the series

$$2 + x + 2x^2 + x^3 + \ldots$$

has radius of convergence $R = 1$.

ABEL'S THEOREM

Let R be the radius of convergence of the power series $\Sigma_1^\infty a_n z^n$. Then:

   (i) The series converges absolutely in $|z| < R$ and uniformly in $|z| \leqslant \rho, \rho < R$.
   (ii) The series diverges in $|z| > R$.
   (iii) The sum of the series is analytic in $|z| < R$, and its derivative, obtained by termwise differentiation, has the same radius of convergence.

PROOF: (i) Let $|z| < r < R$. Then $r^{-1} > R^{-1}$, so the definition of lim sup implies the existence of an integer N such that $|a_n| < r^{-n}$ for all $n \geqslant N$. Then

$$\sum_{n=N}^\infty |a_n| |z|^n < \sum_{n=N}^\infty \left|\frac{z}{r}\right|^n = \frac{\left|\dfrac{z}{r}\right|^N}{1 - \left|\dfrac{z}{r}\right|},$$

by the geometric series (Example 2, Section 3.1), since $|z/r| < 1$. Thus, the series converges absolutely in $|z| < r$ for any $r < R$, hence in $|z| < R$. To show uniform convergence, we select $|z| \leqslant \rho < r < R$, then by what was done above and the triangle inequality,

$$\left| \sum_{n=k}^{\infty} a_n z^n \right| \leqslant \sum_{n=N}^{\infty} |a_n| \, |z|^n < \frac{\left(\dfrac{\rho}{r}\right)^N}{1 - \left(\dfrac{\rho}{r}\right)},$$

for all $k \geqslant N$ and $|z| \leqslant \rho$.

(ii) If $|z| > r > R$, then $r^{-1} < R^{-1}$ and the definition of lim sup yields the existence of infinitely many integers n for which $r^{-n} < |a_n|$. Therefore, infinitely many terms of the sequence satisfy $|a_n z^n| > |z/r|^n$ and thus are unbounded.

(iii) That the sum is analytic in $|z| < R$ and its derivative can be obtained by termwise differentiation follows from Weierstrass's theorem. Finally, set $\sqrt[n]{n} = 1 + c_n$; then by the binomial theorem,

$$n = (1 + c_n)^n > 1 + \frac{n(n-1)}{2} c_n^2,$$

implying that $c_n^2 < 2/n$, so $c_n \to 0$ as $n \to \infty$. We compute the radius of convergence of the derivative $\sum_1^{\infty} n a_n z^{n-1}$ by noting that

$$\lim_{n \to \infty} \sup \sqrt[n]{|a_n|} \leqslant \lim_{n \to \infty} \sup \sqrt[n]{n \, |a_n|}$$

$$\leqslant \lim_{n \to \infty} \sqrt[n]{n} \cdot \lim_{n \to \infty} \sup \sqrt[n]{|a_n|} = \lim_{n \to \infty} \sup \sqrt[n]{|a_n|}.$$

EXAMPLE 1.

Consider the series

$$1 - z^2 + z^4 - z^6 + \ldots .$$

Then $|a_n|$ vanishes for odd n and equals 1 for even n. So $R = 1$; thus, the series converges absolutely in $|z| < 1$, uniformly in $|z| \leqslant r < 1$, and diverges in $|z| > 1$. Moreover, it represents some analytic function in $|z| < 1$. Nothing is said about $|z| = 1$; however, observe that it diverges everywhere on $|z| = 1$, since its general term does not tend to zero. Applying Example 2, Section 3.1, we find that

$$\frac{1}{1 + z^2} = 1 - z^2 + z^4 - z^6 + \ldots, \qquad |z| < 1.$$

Note that the series is analytic only in $|z| < 1$, whereas the function $(1 + z^2)^{-1}$ is analytic everywhere in $\mathcal{C}$ except at $z = \pm i$. We can integrate the series termwise on any path inside the unit circle, obtaining

$$\int_0^z \frac{dz}{1 + z^2} = z - \frac{z^3}{3} + \frac{z^5}{5} - \frac{z^7}{7} + \ldots, \qquad |z| < 1.$$

In particular, the function

$$f(z) = \frac{1}{z} \int_0^z \frac{dz}{1 + z^2} = 1 - \frac{z^2}{3} + \frac{z^4}{5} - \ldots, \qquad 0 < |z| < 1,$$

$$f(0) = 1$$

is analytic in $|z| < 1$, illustrating a useful way to show analyticity.

EXAMPLE 2.

Find the radius of convergence of the power series

$$\text{(a)} \ \sum_{n=1}^{\infty} \frac{z^n}{n}, \qquad \text{(b)} \ \sum_{n=0}^{\infty} \frac{z^n}{n!}, \qquad \text{(c)} \ \sum_{n=1}^{\infty} 2^n z^{n!}.$$

SOLUTION: (a) $(1/n)^{1/n} = 1/\sqrt[n]{n}$ tends to 1 as $n \to \infty$, by the proof of Abel's theorem. Thus, $R = 1$ for the series (a). Observe that this result could have been obtained using the ratio formula for the radius of convergence.

For (b) it is easier to use the ratio formula

$$\left| \frac{r_n}{r_{n+1}} \right| = \frac{(n + 1)!}{n!} = n + 1 \to \infty, \qquad n \to \infty,$$

and $R = \infty$.

The series in (c) is one in which the ratio formula cannot be used, as it contains infinitely many zero coefficients. Here $R = 1$, since we have

$$(2^n)^{1/n!} = e^{\ln 2/(n-1)!} \to e^0 = 1, \qquad n \to \infty,$$

and all the other terms vanish.

EXAMPLE 3.

Find the analytic solution of the differential equation
$$f''(z) - 2zf'(z) - 2f(z) = 0$$
with initial conditions $f(0) = 1, f'(0) = 0$.

SOLUTION: Differentiate the series
$$f(z) = a_0 + a_1 z + a_2 z^2 + \ldots + a_n z^n + \ldots, \qquad a_0 = 1,$$
twice, obtaining
$$f'(z) = a_1 + 2a_2 z + 3a_3 z^2 + \ldots + na_n z^{n-1} + \ldots, \qquad a_1 = 0,$$
$$f''(z) = 2a_2 + 6a_3 z + \ldots + (n + 2)(n + 1)a_{n+2} z^n + \ldots.$$
Then, gathering terms with like powers,
$$f''(z) - 2zf'(z) - 2f(z) = \sum_{n=0}^{\infty} [(n + 2)(n + 1)a_{n+2} - 2(n + 1)a_n] z^n = 0,$$

so $(n + 1)[(n + 2)a_{n+2} - 2a_n] = 0$, for $n = 0, 1, 2, \ldots$. The recursion equation $a_{n+2} = 2a_n/(n + 2)$ implies that the general analytic solution of the differential equation is

$$f(z) = a_0 \left(1 + z^2 + \frac{z^4}{2!} + \frac{z^6}{3!} + \ldots \right)$$

$$+ a_1 z \left(1 + \frac{1}{3!} (2z)^2 + \frac{2}{5!} (2z)^4 + \frac{3}{7!} (2z)^6 + \ldots \right).$$

Since $a_0 = 1$ and $a_1 = 0$, we obtain the entire function

$$f(z) = 1 + z^2 + \frac{z^4}{2!} + \frac{z^6}{3!} + \ldots = e^{z^2}$$

as the analytic solution to the initial value problem.

EXERCISES

Find the radius of convergence of the series given in Exercises 1-6.

1. $\sum_{n=1}^{\infty} (nz)^n$

2. $\sum_{n=1}^{\infty} \frac{n! z^n}{n^n}$

3. $\sum_{n=0}^{\infty} z^{2n}$

4. $\sum_{n=1}^{\infty} z^{n!}$

5. $\sum_{n=0}^{\infty} [2 + (-1)^n]^n z^n$

6. $\sum_{n=0}^{\infty} (\cos in) z^n$

If the radius of convergence of the series $\sum_0^{\infty} a_n z^n$ is $R(0 < R < \infty)$, find the radius of convergence of the series given in Exercises 7–12.

7. $\sum_{n=0}^{\infty} n^k a_n z^n$

8. $\sum_{n=1}^{\infty} n^{-n} a_n z^n$

9. $\sum_{n=1}^{\infty} a_n^k z^n$

10. $\sum_{n=0}^{\infty} a_n z^{n+k}$

11. $\sum_{n=0}^{\infty} a_n^n z^{n^2}$

12. $\sum_{n=0}^{\infty} a_n z^{n^2}$

In Exercises 13–16, expand the functions in power series centered at 0 and find their radius of convergence, without using Taylor's theorem.

13. $\dfrac{2}{(1-z)^3}$

14. $\text{Log}(1+z)$

15. $\displaystyle\int_0^z \dfrac{\sin z}{z}\, dz$

16. $f(z) = \begin{cases} \dfrac{e^{az}-1}{z}, & z \neq 0 \\ a, & z = 0 \end{cases}$

17. Find the most general power series (involving two arbitrary constants) satisfying the differential equation $f''(z) + f(z) = 0$. Express the sum in terms of elementary functions.

18. Find a Maclaurin series satisfying the differential equation $f'(z) = 1 + zf(z)$ with the initial condition $f(0) = 0$. What is its radius of convergence?

19. Determine the general Maclaurin series solution of the differential equation $zf''(z) + f'(z) + zf(z) = 0$ and show that it is entire. (*Hint:* Prove $\sqrt[n]{n!} \to \infty$ as $n \to \infty$, since e^z is entire.)

20. Find the general Maclaurin series solution of the differential equation

$$(1 - z^2) f''(z) - 2zf'(z) + n(n+1)f(z) = 0.$$

21. Suppose $f(z)$ and $g(z)$ are analytic in a neighborhood of z_0 and $f(z_0) = g(z_0) = 0$ while $g'(z_0) \neq 0$. Prove **L'Hospital's theorem**

$$\lim_{z \to z_0} \frac{f(z)}{g(z)} = \frac{f'(z_0)}{g'(z_0)}.$$

22. Solve Exercise 2, Section 2.4, by the method in this section.
23. Find the sum in $|z| < 1$ of the series

$$\sin \frac{2\pi}{3} + \sin \left(\frac{4\pi}{3} \right) z + \sin \left(\frac{6\pi}{3} \right) z^2 + \ldots$$

24. Prove the **ratio test**: If

$$\lim_{n \to \infty} \left| \frac{a_n}{a_{n+1}} \right| = R,$$

then R is the radius of convergence of the series $\Sigma \, a_n z^n$.

For Exercises 25–27, let $g(t)$ be a continuous complex-valued function on $0 \le t \le 1$ and define

$$f(z) = \int_0^1 g(t) e^{zt} \, dt.$$

25. Show that for fixed z, the series

$$g(t) e^{zt} = \sum_{n=0}^{\infty} \frac{z^n t^n g(t)}{n!}$$

converges uniformly on $0 \le t \le 1$.
26. Prove that $f(z)$ is entire. (*Hint:* Use Exercise 25 to interchange sum and integral.)
27. Prove that $f'(z) = \int_0^1 t g(t) e^{zt} \, dt$.
28. Let $g(t)$ be continuous on $0 \le t \le 1$ and define

$$f(z) = \int_0^1 g(t) \sin(zt) \, dt.$$

Prove that $f(z)$ is entire and find its derivative.
29. Let $g(t)$ be continuous on $0 \le t \le 1$ and define

$$f(z) = \int_0^1 \frac{g(t)}{1 - zt} \, dt, \qquad |z| < 1.$$

Prove that $f(z)$ is analytic in $|z| < 1$ and find its derivative.
30. Use a Maclaurin series to solve the functional equation

$$f(z^2) = z + f(z).$$

Where does the series converge?

3.3 LAURENT SERIES

A series of the form

$$a_0 + \frac{a_1}{z} + \frac{a_2}{z^2} + \ldots + \frac{a_n}{z^n} + \ldots$$

can be considered a power series in the variable $1/z$. If R is its radius of convergence, the series will converge absolutely whenever $|1/z| < R$ or $|z| > 1/R$. The convergence is uniform in every region $|z| \geqslant \rho$, $\rho > 1/R$, and the series diverges for $|z| < 1/R$. Thus the series represents an analytic function in $|z| > 1/R$. Combining a series of the above type with an ordinary power series, we obtain a series of the form

$$\sum_{n=-\infty}^{\infty} a_n z^n.$$

Suppose the ordinary part converges in a disk $|z| < R$ and the other part converges in a region $|z| > r$. If $r < R$, then there is an open annulus where both series converge: $r < |z| < R$ (see Figure 3.3).

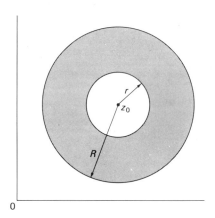

FIGURE 3.3. Region of convergence of a Laurent series about z_0

The series represents an analytic function in this annulus. Similarly,

$$\sum_{n=-\infty}^{\infty} a_n (z - z_0)^n$$

represents an analytic function in the annulus $r < |z - z_0| < R$. We call an expansion of this type a **Laurent series**. Conversely, we now prove that a function analytic in an annulus $r < |z - z_0| < R$ can be expanded into a Laurent series.

LAURENT'S THEOREM

If $f(z)$ is analytic in the annulus $r < |z - z_0| < R$, then it can be uniquely expanded into a Laurent series

$$f(z) = \sum_{n=-\infty}^{\infty} a_n (z - z_0)^n,$$

where

$$a_n = \frac{1}{2\pi i} \int_{|\zeta - z_0| = \rho} \frac{f(\zeta)\, d\zeta}{(\zeta - z_0)^{n+1}}, \qquad n = 0, \pm 1, \pm 2, \ldots, \quad r < \rho < R.$$

PROOF: Let γ_1 and γ_2 denote the circles $|z - z_0| = r + \epsilon$ and $|z - z_0| = R - \epsilon$, respectively, with $0 < \epsilon < (R - r)/2$ (see Figure 3.4). By the Cauchy integral formula,

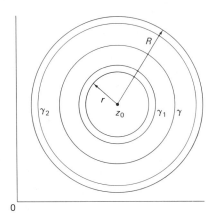

FIGURE 3.4.

$$f(z) = \frac{1}{2\pi i} \int_{\gamma_2} \frac{f(\zeta)\, d\zeta}{\zeta - z} - \frac{1}{2\pi i} \int_{\gamma_1} \frac{f(\zeta)\, d\zeta}{\zeta - z},$$

for all z satisfying $r + \epsilon < |z - z_0| < R - \epsilon$. Riemann's theorem (or Cauchy's theorem and Cauchy's theorem for derivatives) yields the analyticity of each integral in the complements of the curves γ_1 and

γ_2. Proceeding exactly as we did in proving Taylor's theorem, the first integral becomes

$$\frac{1}{2\pi i} \int_{\gamma_2} \frac{f(\zeta) \, d\zeta}{\zeta - z} = \sum_{n=0}^{\infty} a_n (z - z_0)^n$$

with

$$a_n = \frac{1}{2\pi i} \int_{\gamma_2} \frac{f(\zeta) \, d\zeta}{(\zeta - z_0)^{n+1}}, \qquad n = 0, 1, 2, \ldots .$$

For the second integral observe that by the geometric series (Example 2, Section 3.1),

$$\frac{-1}{\zeta - z} = \frac{1}{(z - z_0) - (\zeta - z_0)} = \frac{1}{z - z_0} \sum_{n=0}^{\infty} \left(\frac{\zeta - z_0}{z - z_0} \right)^n,$$

since $|\zeta - z_0| < |z - z_0|$ on γ_1. Furthermore, the series converges uniformly in ζ for ζ in γ_1. By Weierstrass's theorem, we may integrate termwise, obtaining

$$- \frac{1}{2\pi i} \int_{\gamma_1} \frac{f(\zeta)}{\zeta - z} \, d\zeta = \sum_{n=0}^{\infty} \left\{ (z - z_0)^{-(n+1)} \cdot \frac{1}{2\pi i} \int_{\gamma_1} f(\zeta)(\zeta - z_0)^n \, d\zeta \right\}.$$

Thus we have

$$- \frac{1}{2\pi i} \int_{\gamma_1} \frac{f(\zeta)}{\zeta - z} \, d\zeta = \sum_{n=0}^{\infty} a_{-n-1} (z - z_0)^{-n-1},$$

with

$$a_{-n-1} = \frac{1}{2\pi i} \int_{\gamma_1} \frac{f(\zeta) \, d\zeta}{(\zeta - z_0)^{-n}}.$$

Finally, since $f(\zeta)/(\zeta - z_0)^{n+1}$ is analytic inside and on $\gamma - \gamma_1$ or $\gamma_2 - \gamma$, where γ is the circle $|\zeta - z_0| = \rho$, $r < \rho < R$, Cauchy's theorem implies we may replace γ_1 or γ_2 by γ in the calculation of the coefficients a_n. Noticing that ϵ can be chosen arbitrarily close to zero yields the desired representation on the annulus $r < |z - z_0| < R$. ∎

The Laurent series representation for a given function is unique, for if $f(z)$ had representations

$$f(z) = \sum_{n=-\infty}^{\infty} a_n (z - z_0)^n, \qquad f(z) = \sum_{n=-\infty}^{\infty} b_n (z - z_0)^n,$$

then multiplying by $(z - z_0)^k$, for any integer k, and integrating along $|z - z_0| = \rho$ would yield by uniform convergence

$$\sum_{n=-\infty}^{\infty} a_n \cdot \int_\gamma (z - z_0)^{n+k} \, dz = \sum_{n=-\infty}^{\infty} b_n \cdot \int_\gamma (z - z_0)^{n+k} \, dz.$$

Since all powers of $z - z_0$ except $(z - z_0)^{-1}$ have analytic anti-derivatives in $r < |z - z_0| < R$, their integrals vanish by the fundamental theorem. Thus $2\pi i a_{-k-1} = 2\pi i b_{-k-1}$, implying $a_k = b_k$ for all integers k.

The coefficients a_n are not often obtained by use of their integral formulas. We give examples of the techniques employed in avoiding this computation.

EXAMPLE 1.

Using the Maclaurin series of $\cos z$ and e^z, we have

$$\frac{\cos z}{z^2} = \sum_{n=0}^{\infty} (-1)^n \frac{z^{2n-2}}{(2n)!}$$

$$= \sum_{n=-1}^{\infty} (-1)^{n+1} \frac{z^{2n}}{(2n+2)!}, \qquad 0 < |z| < \infty,$$

$$e^{1/z^2} = \sum_{n=0}^{\infty} \frac{z^{-2n}}{n!} = \sum_{n=-\infty}^{0} \frac{z^{2n}}{(-n)!}, \qquad 0 < |z|.$$

EXAMPLE 2.

Consider the function $(z^2 - 3z + 2)^{-1}$. It is analytic everywhere except at $z = 1, 2$. Find its Laurent series on the regions (a) $1 < |z| < 2$, (b) $|z| < 1$, (c) $|z| > 2$, and (d) $0 < |z - 1| < 1$.

SOLUTION: (a) In the annulus $1 < |z| < 2$, writing

$$\frac{1}{(z - 1)(z - 2)} = \frac{1}{z - 2} - \frac{1}{z - 1},$$

we may expand the fractions in the form

$$\frac{1}{z-2} - \frac{1}{z-1} = \frac{-\dfrac{1}{2}}{1 - \dfrac{z}{2}} - \frac{\dfrac{1}{z}}{1 - \dfrac{1}{z}}$$

$$= -\frac{1}{2} \sum_{n=0}^{\infty} \left(\frac{z}{2}\right)^n - \frac{1}{z} \sum_{n=0}^{\infty} \left(\frac{1}{z}\right)^n,$$

since $|1/z| < 1$ and $|z/2| < 1$. Thus,

$$\frac{1}{(z-1)(z-2)} = -\sum_{n=-\infty}^{-1} z^n - \frac{1}{2} \sum_{n=0}^{\infty} \left(\frac{z}{2}\right)^n.$$

(b) For $|z| < 1$, we expand the expression as

$$\frac{1}{z-2} - \frac{1}{z-1} = \frac{-\dfrac{1}{2}}{1 - \dfrac{z}{2}} + \frac{1}{1-z}$$

$$= -\frac{1}{2} \sum_{n=0}^{\infty} \left(\frac{z}{2}\right)^n + \sum_{n=0}^{\infty} z^n.$$

Hence,

$$\frac{1}{(z-1)(z-2)} = \sum_{n=0}^{\infty} \left(1 - \frac{1}{2^{n+1}}\right) z^n, \qquad |z| < 1.$$

(c) Here

$$\frac{1}{z-2} - \frac{1}{z-1} = \frac{\dfrac{1}{z}}{1 - \dfrac{2}{z}} - \frac{\dfrac{1}{z}}{1 - \dfrac{1}{z}}$$

$$= \frac{1}{z} \sum_{n=0}^{\infty} \left(\frac{2}{z}\right)^n - \frac{1}{z} \sum_{n=0}^{\infty} \left(\frac{1}{z}\right)^n$$

$$= \sum_{n=-\infty}^{-1} \left(\frac{1}{2^{n+1}} - 1\right) z^n, \qquad 2 < |z|.$$

(d) In $0 < |z - 1| < 1$, we obtain

$$\frac{1}{z - 2} - \frac{1}{z - 1} = -\frac{1}{z - 1} - \frac{1}{1 - (z - 1)}$$

$$= -\frac{1}{z - 1} - \sum_{n=0}^{\infty} (z - 1)^n.$$

Hence, in $0 < |z - 1| < 1$,

$$\frac{1}{(z - 1)(z - 2)} = -\frac{1}{z - 1} - \frac{1}{1 - (z - 1)} = -\sum_{n=-1}^{\infty} (z - 1)^n.$$

EXERCISES

Find the Laurent series of the function $(z^2 + z)^{-1}$ in the regions given in Exercises 1-3.

1. $0 < |z| < 1$ 2. $0 < |z - 1| < 1$ 3. $1 < |z - 1| < 2$

Represent the function $(z^3 - z)^{-1}$ as a Laurent series in the regions given in Exercises 4-7.

4. $0 < |z| < 1$ 5. $1 < |z|$
6. $0 < |z - 1| < 1$ 7. $1 < |z - 1| < 2$

Find the Laurent series of the function $z/(z^2 + z - 2)$ in the regions given in Exercises 8-13.

8. $|z| < 1$ 9. $0 < |z - 1| < 3$
10. $0 < |z + 2| < 3$ 11. $1 < |z| < 2$
12. $|z| > 2$ 13. $|z + 2| > 3$

Represent the functions given in Exercises 14-17 as Laurent series in the region $0 < |z| < \infty$.

14. $ze^{1/z}$ 15. $e^{z+1/z}$

16. $\sin z \sin \dfrac{1}{z}$ 17. $\sin \left(z + \dfrac{1}{z} \right)$

Find the Laurent series of the functions given in Exercises 18-21 in the region $0 < |z - 1| < 1$.

18. $\dfrac{1}{z - 1} \sin \dfrac{1}{z}$ 19. $\dfrac{1}{z} \sin \dfrac{1}{z - 1}$

20. $\sin \dfrac{1}{z(z-1)}$ **21.** $z \sin \dfrac{1}{z-1}$

22. Suppose $f(z)$ is analytic and bounded by M in $r < |z - z_0| < R$. Show that the coefficients of its Laurent series satisfy

$$|a_n| \leqslant MR^{-n}, \qquad |a_{-n}| \leqslant Mr^n, \qquad n = 0, 1, 2, \ldots.$$

Suppose $r = 0$. Can $f(z_0)$ be defined in such a way that $f(z)$ is analytic in $|z - z_0| < R$?

23. Bessel's function $J_n(z)$ is defined as the nth coefficient $(n \geqslant 0)$ of the Laurent series of the function

$$e^{(z/2)(\zeta - 1/\zeta)} = \sum_{n=-\infty}^{\infty} J_n(z)\zeta^n.$$

Show that

$$J_n(z) = \frac{1}{\pi} \int_0^\pi \cos(n\theta - z \sin \theta) \, d\theta.$$

24. Using the coefficient of the Laurent series of $e^{1/z}$ in $|z| > 0$, show that

$$\frac{1}{2\pi} \int_{-\pi}^\pi e^{\cos \theta} \cos(\sin \theta - n\theta) d\theta = \frac{1}{n!}, \qquad n = 0, 1, 2, \ldots.$$

(*Hint:* Integrate on $|z| = 1$.)

25. Evaluate the integral

$$\int_{-\pi}^\pi (\cos \theta)^m \cos n\theta \, d\theta, \qquad m, n \text{ integers},$$

by comparing the coefficients of the Laurent series of $(z + 1/z)^m$ with its binomial expansion.

26. Find the Laurent series of $\csc z$ in $\pi < |z| < 2\pi$.

3.4 ISOLATED SINGULARITIES

A function $f(z)$ analytic in a region $0 < |z - z_0| < R$, but not analytic or even necessarily defined at z_0, is said to have an **isolated singularity** at z_0. These singularities are classified into three categories.

(i) **Removable singularities** are those where it is possible to assign a complex number to $f(z_0)$ in such a way that $f(z)$ becomes analytic in $|z - z_0| < R$. In this case it is necessary that $f(z) \to A \ (\neq \infty)$ as $z \to z_0$. But if so, $f(z)$ is analytic on $0 < |z - z_0| < R/2$ and continuous on $|z - z_0| \leqslant R/2$; hence the maximum principle implies that $f(z)$ is bounded on $|z - z_0| \leqslant R/2$. By Cauchy's estimate or by Exercise 22, Section 3.3, the Laurent series of $f(z)$ is a convergent Taylor series, thus an analytic function in $|z - z_0| < R$. Therefore, the existence of a limit is necessary and sufficient to guarantee that the singularity is removable.

(ii) **Poles** occur whenever $f(z) \to \infty$ as $z \to z_0$. In this case observe that the function $g(z) = 1/f(z)$ has a removable singularity at z_0 with $g(z_0) = 0$, and z_0 is an isolated zero of $g(z)$, since $f(z)$ is nonzero and finite in a set $0 < |z - z_0| < r \leqslant R$. If n is the order of the zero of $g(z)$ at z_0, $g(z) = (z - z_0)^n g_n(z)$, then $f(z) = (z - z_0)^{-n} f_n(z)$, where $f_n(z) = 1/g_n(z)$ is analytic in $|z - z_0| < r$. We call n the **order of the pole of f at z_0**. Moreover, the Laurent series of $f(z)$ centered at z_0 is the product of $(z - z_0)^{-n}$ with the Taylor series of $f_n(z)$ at z_0, thus is of the form

$$\sum_{k=-n}^{\infty} a_k (z - z_0)^k, \qquad a_{-n} \neq 0.$$

Since $g(z)$ is a nonconstant analytic function in $|z - z_0| < r$, the order of the zero of g at z_0 must be finite. Thus, the order of the pole of $f(z)$ at z_0 must also be finite.

(iii) **Essential singularities** are all isolated singularities that are not removable or poles. In this case $f(z)$ does not have a limit as $z \to z_0$, and infinitely many of the coefficients a_{-n}, $n = 1, 2, 3, \ldots$, of its Laurent series about z_0 do not vanish, since otherwise z_0 is a pole or a removable singularity.

We give some examples to amplify our definitions. Observe that

$$\frac{\sin z}{z} = 1 - \frac{z^2}{3!} + \frac{z^4}{5!} - \cdots$$

has a removable singularity at $z = 0$. Thus,

$$f(z) = \begin{cases} \dfrac{\sin z}{z}, & z \neq 0, \\ 1, & z = 0, \end{cases}$$

is entire. On the other hand,

$$\frac{\cos z}{z^2} = \frac{1}{z^2} - \frac{1}{2!} + \frac{z^2}{4!} - \dots$$

has a pole of order 2 at $z = 0$. Finally,

$$e^{1/z} = 1 + \frac{1}{z} + \frac{1}{2!z^2} + \dots$$

has an essential singularity at $z = 0$.

The concept of isolated singularities also applies to (single-valued) functions $f(z)$ analytic in a neighborhood $R < |z| < \infty$ of ∞. By convention, we classify an isolated singularity at ∞ according to whether $g(z) = f(1/z)$ has a removable singularity, pole, or essential singularity at $z = 0$.

Singularities need not be isolated. For example, the function

$$f(z) = \left(\sin \frac{1}{z}\right)^{-1}$$

has singularities at $z = (\pi n)^{-1}$ for all positive integers n. Thus, $z = 0$ is not an isolated singularity.

DEFINITION

A function analytic in a region G, except for poles, is said to be **meromorphic** in G.

If $f(z)$ and $g(z)$ are analytic in G, and $g(z)$ is not identically zero, then the singularities of the quotient $f(z)/g(z)$ agree with the zeros of $g(z)$. They are poles whenever $f(z)$ is nonzero or has a zero of order less than that of $g(z)$; otherwise they are removable singularities. Extending $f(z)/g(z)$ by continuity over the removable singularities, we obtain a meromorphic function in G. For example, $f(z) = \tan z = \sin z/\cos z$ is meromorphic in $\mathbb{C}$ with poles at $z = (k + \frac{1}{2})\pi$, $k = 0, \pm 1, \pm 2, \dots$, and $z = \infty$ is an accumulation point of poles.

The behavior of a function in an ϵ-neighborhood of an essential singularity is very complicated, as the following result demonstrates.

WEIERSTRASS–CASORATI THEOREM

An analytic function approaches any given value arbitrarily closely in any ϵ-neighborhood of an essential singularity.

PROOF: If the theorem is false, we can find a complex number A and a $\delta > 0$ such that $|f(z) - A| > \delta$ in every neighborhood $0 < |z - z_0| < \epsilon$ of the essential singularity z_0. Then

$$\left| \frac{f(z) - A}{z - z_0} \right| > \frac{\delta}{|z - z_0|} \to \infty \qquad \text{as } z \to z_0,$$

implying that $g(z) = [f(z) - A]/(z - z_0)$ has a pole at z_0. Thus, $g(z)$ is meromorphic in $|z - z_0| < \epsilon$. But then so is $f(z) = A + (z - z_0)g(z)$, contradicting the hypothesis that z_0 is an essential singularity. ∎

In fact, more can be shown, although the proof is difficult and will not be given here:

PICARD'S THEOREM

An analytic function assumes every complex number, with possibly one exception, infinitely often in any ϵ-neighborhood of an essential singularity.

EXAMPLE 1.

Find and classify the singularities of the functions

(a) $f(z) = \dfrac{z}{z^2 + z}$, (b) $g(z) = e^{-1/z^2}$, (c) $h(z) = \csc z$.

SOLUTION: (a) The singularities occur at the zeros of the denominator: $z = 0, -1$. Since these are simple zeros and the numerator has a simple zero at $z = 0$, $f(z)$ has a removable singularity at $z = 0$ and a simple pole at $z = -1$.

(b) Notice that $g(z) \to 1$, since $1/z^2 \to 0$ as $z \to \infty$, so $g(z)$ has a removable singularity at $z = \infty$. But

$$g(z) = 1 - \frac{1}{z^2} + \frac{1}{2!} \frac{1}{z^4} - \cdots$$

is the Laurent series of $g(z)$ centered at $z = 0$; hence $g(z)$ has an essential singularity at $z = 0$.

(c) Since

$$\sin z = (-1)^k \sin(z - \pi k) = (-1)^k \left[(z - \pi k) - \frac{(z - \pi k)^3}{3!} + \cdots \right],$$

$h(z)$ has simple poles at $z = \pi k$, $k = 0, \pm 1, \pm 2, \ldots$, and an accumulation point of poles at $z = \infty$.

EXAMPLE 2.

Prove that $\sin z$ assumes all values of $\mathcal{C}$ in any neighborhood of ∞.

SOLUTION: Every strip $(2n - 1)\pi/2 \leqslant x \leqslant (2n + 1)\pi/2$ of $\mathcal{C}$ is mapped onto $\mathcal{C}$ by $w = \sin z$. Since infinitely many such strips can always be found in $|z| > R$ for every real R, $\sin z$ assumes all values of $\mathcal{C}$ in every neighborhood of ∞.

EXERCISES

For each of the functions in Exercises 1–6, find and classify the singularities.

1. $\dfrac{z}{z^3 + z}$

2. $\dfrac{e^z}{1 + z^2}$

3. $ze^{1/z}$

4. $e^{z - 1/z}$

5. $\sin \dfrac{1}{z} + \dfrac{1}{z^2}$

6. $e^{\tan 1/z}$

7. Construct a function having a removable singularity at $z = -1$, a pole of order 3 at $z = 0$, and an essential singularity at $z = 1$. Then find its Laurent series in $0 < |z| < 1$.

Show that each of the integrands in Exercises 8–11 has a removable singularity at $z = 0$. Remove the singularity and obtain the Maclaurin series of each integral.

8. $\text{Si}(z) = \displaystyle\int_0^z \dfrac{\sin \zeta}{\zeta} \, d\zeta$

9. $C(z) = \displaystyle\int_0^z \dfrac{\cos \zeta - 1}{\zeta} \, d\zeta$

10. $E(z) = \displaystyle\int_0^z \dfrac{e^\zeta - 1}{\zeta} \, d\zeta$

11. $L(z) = \displaystyle\int_0^z \dfrac{\text{Log}(1 + \zeta)}{\zeta} \cdot d\zeta$

12. Show that the function $f(z) = e^{1/z}$ assumes every value except 0 infinitely often in any ϵ-neighborhood of $z = 0$.

13. Prove that an entire function having a nonessential singularity at

∞ must be a polynomial. What kind of singularity do e^z, $\sin z$, and $\cos z$ have at ∞?

14. Show that a function meromorphic in $\mathfrak{M}$ must be the quotient of two polynomials.

15. Prove that an entire function that omits the values 0 and 1 is constant. (*Hint:* Use Picard's theorem.)

3.5 ANALYTIC CONTINUATION (Optional)

It often happens that the expression $f_0(z)$, such as an infinite series or an integral, defining an analytic function is meaningful only in some limited region G_0 in the plane. The question arises as to whether or not there is any way of extending the definition of the function so that it becomes analytic on a larger region. In particular, is it possible to find an expression $f_1(z)$ analytic on a region G_1 meeting G_0 such that $f_0(z) = f_1(z)$, for all z in $G_0 \cap G_1$? If so, we can extend our function to the region $G_0 \cup G_1$, and we say that the **elements** (f_0, G_0) and (f_1, G_1) form a **direct analytic continuation** of each other. Any direct analytic continuation of (f_0, G_0) to G_1 is necessarily unique, for two functions analytic on G_1 and agreeing on $G_0 \cap G_1$ must coincide on G_1 (see Exercise 20, Section 3.1).

A procedure for obtaining analytic continuations begins by expanding the given expression into a Taylor series

$$f_0(z) = \sum_{n=0}^{\infty} a_n (z - z_0)^n$$

converging in a disk $|z - z_0| < R_0$ centered at a point z_0 in G_0. If z_1 satisfies $|z_1 - z_0| < R_0$, we can expand f_0 in a power series

$$f_1(z) = \sum_{n=0}^{\infty} b_n (z - z_1)^n, \qquad b_n = \frac{f_0^{(n)}(z_1)}{n!},$$

which converges in a disk $|z - z_1| < R_1$. Certainly, $R_1 \geqslant R_0 - |z_1 - z_0|$. If equality holds, the contact point of the circles $|z - z_0| = R_0$ and $|z - z_1| = R_1$ must be a singularity of the function, since Taylor's theorem implies the existence of a singularity on each circle of convergence. Otherwise, a part of $|z - z_1| < R_1$ lies outside $|z - z_0| < R_0$ and $(f_1, \{|z - z_1| < R_1\})$ is a direct analytic continuation of $(f_0, \{|z - z_0| < R_0\})$, as both series agree on the overlap (see Figure 3.5).

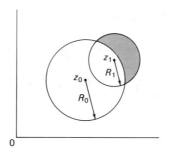

FIGURE 3.5. Direct analytic continuation

EXAMPLE 1.

The power series

$$f_0(z) = \sum_{n=0}^{\infty} (z - \tfrac{1}{2})^n$$

has radius of convergence $R = 1$, so the region of convergence G_0 is the disk $|z - \tfrac{1}{2}| < 1$. We can continue (f_0, G_0) to a disk centered at 0 by computing

$$f_0(0) = \sum_{n=0}^{\infty} (-\tfrac{1}{2})^n, \qquad f_0'(0) = \sum_{n=1}^{\infty} n(-\tfrac{1}{2})^{n-1}, \ldots,$$

but it is easier to notice that $f_0(z) = (3/2 - z)^{-1}$ in G_0. Then, by Example 2, Section 3.1, we have

$$f_1(z) = \frac{2}{3} \cdot \frac{1}{1 - \left(\dfrac{2z}{3}\right)} = \frac{2}{3} \sum_{n=0}^{\infty} \left(\frac{2z}{3}\right)^n, \qquad |z| < \frac{3}{2},$$

implying G_1 is the disk $|z| < 3/2$, and $z = 3/2$ is a singularity of the function.

This procedure can be continued, but care must be exercised, as a sequence of disks might return to overlap the first, and they might not coincide on the overlap. This occurs when the function is multivalued and the disks have taken us around a branch point of the function and onto a different branch of its Riemann surface (see Figure 3.6). Thus, even if (f_2, G_2) is a direct analytic continuation of

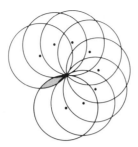

FIGURE 3.6. An analytic continuation

(f_1, G_1), it need not be one for (f_0, G_0), and only a multivalued function will serve to define the extension.

EXAMPLE 2.

Consider the function $f(z) = 1/\sqrt{z}$ at the points $z = e^{\pi i/4}, e^{7\pi i/4}$. Using the binomial theorem, we can obtain the Taylor series expansion about each of these two points:

$$\frac{1}{\sqrt{z}} = e^{-\pi i/8} \frac{1}{\sqrt{1 - (1 - ze^{-\pi i/4})}}$$

$$= e^{-\pi i/8} \sum_{n=0}^{\infty} \frac{(2n)!}{2^{2n}(n!)^2} (1 - ze^{-\pi i/4})^n, |z - e^{\pi i/4}| < 1,$$

and

$$\frac{1}{\sqrt{z}} = e^{-7\pi i/8} \frac{1}{\sqrt{1 - (1 - ze^{-7\pi i/4})}}$$

$$= e^{-7\pi i/8} \sum_{n=0}^{\infty} \frac{(2n)!}{2^{2n}(n!)^2} (1 - ze^{-7\pi i/4})^n, |z - e^{7\pi i/4}| < 1.$$

Evaluating the first expression at e^0 and the second at $e^{2\pi i}$, we obtain $e^0 = 1$ and $e^{-\pi i} = -1$, respectively. Note that in the Riemann surface $[\,\mathbb{C} - \{0\}\,]^2$, for $f(z) = 1/\sqrt{z}$, the point e^0 does not belong to the disk $|z - e^{7\pi i/4}| < 1$.

Each element of a chain of elements (f_0, G_0), (f_1, G_1), . . ., (f_n, G_n), where (f_j, G_j) is a direct analytic continuation of (f_{j-1}, G_{j-1}), is called an **analytic continuation** of the others. Thus, the above procedure can be used to construct analytic continuations, the selection of the centers $z_1, z_2, \ldots, z_n$ determining the values of the function. In particular, if γ is a curve joining z_0 to a point z' not in the disk $|z - z_0| < R_0$, we can construct an analytic continuation consisting of disks of convergence $|z - z_j| < R_j$ of series representations of the function such that z_j follows z_{j-1} in the parametrization of γ. If z' can be reached by a finite chain of such disks, we say we have an **analytic continuation of the function along the curve γ** (see Figure 3.7). Otherwise, we have infinitely many disks whose centers z_j converge to a point z^* on γ, and hence their radii tend to zero.

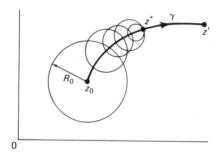

FIGURE 3.7. Analytic continuation along γ

Moreover, a singularity of the function must lie on the boundary of each of these disks, and these singularities also tend to z^*. Since every ϵ-neighborhood of z^* contains a singularity, the function cannot be analytic at z^*. We have proved the following theorem.

THEOREM

The power series $\sum_0^\infty a_n (z - z_0)^n$ can be continued analytically along any curve γ that starts in its disk $|z - z_0| < R_0$ of convergence until one of its singularities is encountered.

Intuitively, if γ and γ' are arcs disjoint except at their two end-points z_0 and z' such that no singularities lie on or inside the closed curve $\gamma - \gamma'$, then *the result of the analytic continuation is the same for each path*, for the inside could be covered by disks overlapping those in the analytic continuation along the two arcs (see Figure 3.8). This result is called the **monodromy theorem**; its proof is complicated and will not be given.

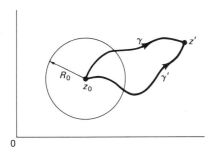

FIGURE 3.8.

DEFINITION

A **global analytic function** is a collection $\mathcal{F}$ of elements (f, G), any two of which are analytic continuations of each other by a chain of elements in $\mathcal{F}$.

EXAMPLE 3.

Let G_k be the region consisting of all points z satisfying $|\arg z - (k\pi/2)| < \pi/2$, for all integers k, and let $f_k(z) = \log z$, for all z in G_k. Then the collection (f_0, G_0), (f_1, G_1), . . ., (f_n, G_n), . . . is a global analytic function, as is the collection of elements (f_j, G_j) for all integers j.

Two elements (f_0, G_0), (f_1, G_1) are said to determine the same **branch** of a global analytic function at a point z_0 in $G_0 \cap G_1$ if $f_0(z) = f_1(z)$ in an ϵ-neighborhood of z_0. Note, however, that it is not necessary that the function elements be direct analytic continuations of each other.

EXAMPLE 4.

If G_k consists of all points z satisfying $|\arg z - (k\pi/2)| < 3\pi/4$, and $f_k(z) = \log z$, for all z in G_k and integers k, then $e^{i\pi/2}$ lies in $G_0 \cap G_2$ and $f_0(z) = f_2(z)$, for all z with $|\arg z - (\pi/2)| < \pi/4$, although (f_0, G_0) and (f_2, G_2) are not direct analytic continuations of each other (see Figure 3.9).

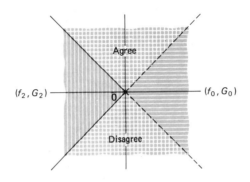

FIGURE 3.9.

The points on the boundary of the domain of definition of a global analytic function fall into two classes: (i) points on which the function can be continued analytically (**regular points**) and (ii) **singularities.**

Singularities may or may not be isolated. If isolated, a singularity is called a **branch point of order** $n - 1$ if all points in an ϵ-neighborhood of the singularity have n distinct branches. If $n = \infty$, we call it a **logarithmic branch point.**

EXERCISES

In Exercises 1–3, first find an analytic function agreeing with the given series on its disk of convergence.

1. Expand $\sum_1^{\infty} z^n/n$ in a neighborhood of $z = \frac{1}{2}$, and determine its radius of convergence.

2. Expand $\sum_{n=1}^{\infty} z^n$ into a Taylor series in a neighborhood of $z = a$, $|a| < 1$. What is the new series's radius of convergence?

3. Show that the series

$$\sum_{1}^{\infty} \frac{z^n}{n} \qquad \text{and} \qquad i\pi + \sum_{1}^{\infty} (-1)^n \frac{(z-2)^n}{n}$$

have no common region of convergence yet are analytic continuations of each other.

For Exercises 4–7, find the Taylor series of each of the functions in the disk $|z - 1| < 1$ of their principal branch. Then analytically continue each function along $\gamma: z(t) = e^{it}$, $0 \le t \le 2\pi$. Do the values at $z(2\pi)$ coincide with those at $z(0)$?

4. $\log z$ 5. $z^{1/2}$
6. $\sin (z^{1/2})\pi/2$ 7. $(\sin z\pi/2)^{1/2}$
8. Show that the function

$$\sum_{n=1}^{\infty} z^{2^n}$$

is analytic in $|z| < 1$, yet cannot be continued outside this set. We call $|z| = 1$ its **natural boundary**. (*Hint:* Since $f(z) = z^2 + z^4 + \ldots + z^{2^n} + f(z^{2^n})$, show that the points $\zeta = e^{\pi i/2^k}$ satisfy $f(t\zeta) \to \infty$ as $t \to 1^-$.)

9. Show that $|z| = 1$ is a natural boundary for

$$\sum_{n=0}^{\infty} z^{n!}.$$

10. Show that the imaginary axis is a natural boundary for the function

$$\sum_{n=0}^{\infty} e^{-n!z}.$$

Where is the function analytic?

11. Find a series representation centered at $z = 1$ for the function

$$f(z) = \int_{0}^{\infty} t^2 e^{-zt} \, dt, \qquad 0 < t < \infty,$$

analytic in Re $z > 0$. What is its analytic continuation to the whole plane?

12. The **gamma function** is defined in the right half plane by means of the integral

$$\Gamma(z) = \int_0^\infty e^{-t} t^{z-1} \, dt, \qquad 0 < t < \infty.$$

Prove that it satisfies the functional equation $\Gamma(z + 1) = z\Gamma(z)$ and is analytic in Re $z > 0$. Show that it has an analytic continuation to the whole plane as a meromorphic function with simple poles at $0, -1, -2, \ldots$.

13. **Schwarz Reflection Principle.** Let $f = u + iv$ be analytic in a region $G+$ lying in the upper half plane and having a segment γ of the real axis as part of its boundary. If f is continuous and real-valued on γ, then f has a unique analytic continuation across γ into the region $G-$, the reflection of $G+$ across the real axis. (*Hint:* Show that $\overline{f(\bar{z})}$ is analytic on $G-$ and apply Morera's theorem or the Cauchy–Riemann equations to $G+ \cup \gamma \cup G-$.)

14. Prove **Pringsheim's theorem**: A power series

$$f(z) = \sum_{n=0}^\infty a_n z^n$$

with radius of convergence 1 and nonnegative real coefficients a_n has a singularity at $z = 1$.

15. Show that even though $\sum_{n=1}^\infty z^n/n^2$ converges at each point of $|z| = 1$, it is not analytic at $z = 1$.

NOTES

Important theorems due to Mittag–Leffler and Weierstrass concerning infinite series and products representations for meromorphic functions have been omitted. The reader is urged to study these topics, a development of which may be found in [A, pp. 185–196].

SECTION 3.4

Two different proofs of Picard's theorem may be found in [A, p. 297] and [V, p. 144].

SECTION 3.5

The method indicated for constructing direct analytic continuations, though fine in theory, is rarely useful in practice. The problem lies in computing the coefficients b_n, which, unless additional information is known, are sums of infinite series.

A proof of the monodromy theorem may be found in [A, p. 285].

4 CONTOUR INTEGRATION

4.1 THE RESIDUE THEOREM

We have shown in Section 3.3 that a function $f(z)$, analytic in a region $0 < |z - z_0| < R$, can be expanded in a Laurent series about z_0. The coefficient

$$a_{-1} = \frac{1}{2\pi i} \int_{|\zeta - z_0| = \rho} f(\zeta) \, d\zeta, \qquad 0 < \rho < R,$$

of this Laurent series is called the **residue** of the function $f(z)$ at z_0 and is denoted by $\text{Res}_{z_0} f(z)$. Note that knowing the residue of f at z_0 provides an alternative method for evaluating the integral

$$\int_\gamma f(\zeta) \, d\zeta = 2\pi i \, \text{Res}_{z_0} f(z),$$

where γ is any pws Jordan curve with z_0 inside γ.

The next theorem is of fundamental importance in complex analysis and is the core concept in the development of the techniques of this chapter.

RESIDUE THEOREM

Let $f(z)$ be analytic in a region G containing the set of all points inside and on a pws Jordan curve γ except for a finite number of singularities $z_1, \ldots, z_k$ inside γ. Then

$$\int_\gamma f(z)\, dz = 2\pi i \sum_{n=1}^{k} \mathrm{Res}_{z_n} f(z).$$

PROOF: We can draw circles $|z - z_n| = r_n \;(> 0), n = 1, \ldots, k$, inside γ such that the disks $|z - z_n| \leqslant r_n$ are disjoint from each other. By the generalization of Cauchy's theorem to multiply connected regions (see Section 2.3),

$$\int_\gamma f(z)\, dz = \sum_{n=1}^{k} \int_{|z - z_n| = r_n/2} f(z)\, dz,$$

and in each region $0 < |z - z_n| < r_n$, the Laurent series development of $f(z)$ about z_n yields

$$\mathrm{Res}_{z_n} f(z) = a_{-1} = \frac{1}{2\pi i} \int_{|z - z_n| = r_n/2} f(z)\, dz, \qquad n = 1, \ldots, k.$$

Combining these two identities, we obtain the desired result. ■

For this theorem to be useful we need to obtain simple methods for evaluating residues. In particular, we wish to avoid the integration process whenever possible. If the Laurent series is known explicitly, then the residue equals a_{-1}. For nonessential singularities, we note that a_{-1} vanishes at removable singularities, and if z_0 is a pole of order k, then

$$(z - z_0)^k f(z) = \sum_{n=-k}^{\infty} a_n (z - z_0)^{n+k},$$

so that for $k = 1$,

$$\lim_{z \to z_0} (z - z_0) f(z) = a_{-1},$$

whereas for $k > 1$,

$$\lim_{z \to z_0} \frac{d^{k-1}}{dz^{k-1}} [(z - z_0)^k f(z)] = (k - 1)! a_{-1}.$$

EXAMPLE 1.

Find the residue at all singularities in $\mathcal{C}$ of the functions

(a) $f(z) = z^2 \sin \dfrac{1}{z}$,   (b) $g(z) = \dfrac{e^z}{z^3 - z^2}$,   (c) $h(z) = \dfrac{z}{\sin z}$.

SOLUTION: (a) We know the Laurent series of $f(z)$ centered on $z = 0$,

$$f(z) = z^2 \left(\frac{1}{z} - \frac{1}{3!z^3} + \frac{1}{5!z^5} - \cdots \right)$$

$$= z - \frac{1}{3!z} + \frac{1}{5!z^3} - \cdots, \qquad 0 < |z| < \infty,$$

which implies that $\operatorname{Res}_0 f(z) = -\frac{1}{6}$.

(b) Observe that $g(z)$ has a simple pole at $z = 1$ and a pole of order 2 at $z = 0$. Thus, we have

$$\operatorname{Res}_1 g(z) = \lim_{z \to 1} (z - 1) g(z) = e$$

and

$$\operatorname{Res}_0 g(z) = \lim_{z \to 0} [z^2 g(z)]' = \lim_{z \to 0} \frac{e^z (z - 2)}{(z - 1)^2} = -2.$$

(c) Note that $h(z)$ has a removable singularity at $z = 0$ and simple poles at $z = \pi k$, $k = \pm 1, \pm 2, \ldots$ (see Example 1c, Section 3.4). Since $\sin(z - \pi k) = (-1)^k \sin z$, the complete solution is given by

$$\operatorname{Res}_{\pi k} h(z) = \lim_{z \to \pi k} \frac{(z - \pi k) z}{\sin z} = (-1)^k \pi k, \qquad k = 0, \pm 1, \pm 2, \ldots.$$

EXAMPLE 2.

Evaluate the integral

$$\int_{|z - \frac{1}{2}| = 1} \frac{e^z}{z^3 - z} \, dz.$$

SOLUTION: The singularities of the integrand occur at $z = 0, \pm 1$. Hence, only the residues at the simple poles at 0 and 1 need to be calculated:

$$\operatorname{Res}_0 \frac{e^z}{z(z^2 - 1)} = \lim_{z \to 0} \frac{e^z}{z^2 - 1} = -1$$

and

$$\text{Res}_1 \; \frac{e^z}{z(z^2 - 1)} = \lim_{z \to 1} \; \frac{e^z}{z(z + 1)} = \frac{e}{2}.$$

Thus,

$$\int_{|z - \frac{1}{2}| = 1} \frac{e^z}{z^3 - z} \, dz = \pi i \, (e - 2).$$

Although the residue theorem is stated in terms of the residues at the set S of singularities of the integrand inside the pws Jordan curve γ, the residues at the set S^* of singularities of the integrand *outside* of γ can sometimes be used in evaluating integrals.

INSIDE-OUTSIDE THEOREM

Let

$$F(z) = \frac{a_n z^n + a_{n-1} z^{n-1} + \ldots + a_1 z + a_0}{b_m z^m + b_{m-1} z^{m-1} + \ldots + b_1 z + b_0},$$

where $m \geqslant n + 2$. Then

$$\int_\gamma F(z) \, dz = \begin{cases} 2\pi i \; \Sigma_S \; \text{Res} \; F(z), \\ -2\pi i \; \Sigma_{S^*} \; \text{Res} \; F(z). \end{cases}$$

PROOF: The top equality is simple a restatement of the residue theorem. To obtain the bottom equality, we choose R sufficiently large so that γ and all poles of $F(z)$ lie inside the circle $C : |z| = R$. Let γ' be a pws arc from γ to the circle C that does not pass through any pole of $F(z)$ (see Figure 4.1). Then, by the residue theorem,

$$2\pi i \; \Sigma_{S^*} \; \text{Res} \; F(z) = \int_{-\gamma + \gamma' + C - \gamma'} F(z) \, dz,$$

where γ is traversed in a clockwise direction. The integral on γ' cancels that on $-\gamma'$ (see part (iii) of the first theorem in Section 2.3), so that

$$\int_\gamma F(z) \, dz = -2\pi i \; \Sigma_{S^*} \; \text{Res} \; F(z) + \int_{|z| = R} F(z) \, dz.$$

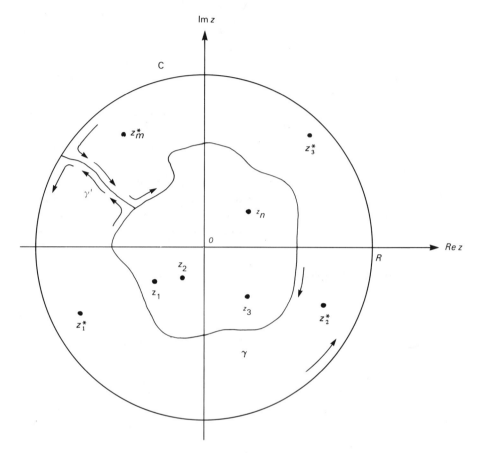

FIGURE 4.1. $S = \overset{n}{\underset{k=1}{\cup}} z_k$: poles inside γ; $S^* = \overset{m}{\underset{k=1}{\cup}} z_k^*$: poles outside γ

As R is arbitrary, the proof will be complete if we show that

$$\lim_{R \to \infty} \int_{|z|=R} F(z) \, dz = 0.$$

By hypothesis, $|z^2 F(z)|$ is bounded by a constant $0 \leqslant M < \infty$ as z tends to ∞, because

$$z^2 F(z) = \frac{a_n z^{n+2-m} + \ldots + a_0 z^{2-m}}{b_m + b_{m-1} z^{-1} + \ldots + b_0 z^{-m}}, \qquad m \geqslant n + 2.$$

Hence,

$$\left| \int_{|z|=R} F(z)\, dz \right| \leqslant \int_{|z|=R} \frac{|z^2 F(z)|}{|z|^2}\, |dz| \leqslant \frac{2\pi RM}{R^2},$$

which tends to 0 as $R \to \infty$.

EXAMPLE 3.

Evaluate the integral

$$\int_{|z|=1} \frac{z+a}{z^n\,(z+b)}\, dz, \qquad |b| > 1.$$

SOLUTION: The integrand has a pole of order n at 0 and a simple pole at $-b$. Even if n were given, the calculation of the residue at 0 would be messy for $n > 1$, as $n-1$ differentiations of $(z+a)/(z+b)$ would be required. This difficulty can be avoided by using the inside–outside theorem as

$$\int_{|z|=1} \frac{z+a}{z^n\,(z+b)}\, dz = -2\pi i\ \mathrm{Res}_{-b}\ \frac{z+a}{z^n\,(z+b)}$$

$$= -2\pi i\ \lim_{z \to -b} \frac{z+a}{z^n} = \frac{2\pi i(a-b)}{(-1)^{n+1}\,b^n}.$$

Note that the integral would be zero if $|b| < 1$.

EXERCISES

Find the residue at all singularities in $\mathbb{C}$ for the functions given in Exercises 1–12.

1. $f(z) = \dfrac{z^3}{z^2+1}$

2. $f(z) = \dfrac{e^z}{z^3 - z}$

3. $f(z) = \dfrac{z^2+1}{z^3}$

4. $f(z) = \dfrac{z}{(z^2+1)^2}$

5. $f(z) = ze^{1/z}$

6. $f(z) = z\cos\dfrac{1}{z}$

7. $f(z) = (z-1)e^{1/z}$

8. $f(z) = (z-1)^2 \sin\dfrac{1}{z}$

9. $f(z) = \dfrac{z}{\sinh z}$ 10. $f(z) = \tan z$

11. $f(z) = \cot z$ 12. $f(z) = \left(z + \dfrac{\pi}{2}\right) \sec z$

Evaluate the integrals in Exercises 13–24. In Exercises 15–18, n is a nonnegative integer.

13. $\displaystyle\int_{|z|=2} \dfrac{z^3}{z^2 + 1}\, dz$ 14. $\displaystyle\int_{|z|=2} \dfrac{e^z}{z^3 + z}\, dz$

15. $\displaystyle\int_{|z-1|=2} \dfrac{dz}{z^n\,(z^2 + 1)}$ 16. $\displaystyle\int_{|z-1|=\sqrt{5}/2} \dfrac{dz}{z^n\,(z^2 + 1)}$

17. $\displaystyle\int_{|z-i|=3/2} \dfrac{dz}{z^n\,(z^2 + 1)}$ 18. $\displaystyle\int_{|z-i|=1/2} \dfrac{dz}{z^n\,(z^2 + 1)}$

19. $\displaystyle\int_{|z-1/2|=1} \dfrac{\sin z}{z^3 + z}\, dz$ 20. $\displaystyle\int_{|z|=2} \dfrac{\sin z}{(z^3 + z)^2}\, dz$

21. $\displaystyle\int_{|z|=1} z e^{1/z}$ 22. $\displaystyle\int_{|z|=1} \tan z\, dz$

23. $\displaystyle\int_{|z|=2} \tan z\, dz$ 24. $\displaystyle\int_{|z|=5} \tan z\, dz$

25. Suppose $P(z)$, $Q(z)$ are polynomials. Show that all the residues of the function $[P(z)/Q(z)]'$ are zero.

4.2 EVALUATION OF DEFINITE REAL INTEGRALS

We now present, in this and the next three sections, a number of useful techniques for applying the residue theorem in the evaluation of definite integrals.

Integrals of the form

$$\int_0^{2\pi} F(\cos\theta,\, \sin\theta)\, d\theta,$$

where $F(s,\, t)$ is the quotient of two polynomial functions in s and t, may be transformed into a line integral by the substitution $z = e^{i\theta}$, $0 \leqslant \theta \leqslant 2\pi$, since

$$\cos \theta = \frac{1}{2} \left(e^{i\theta} + e^{-i\theta} \right) = \frac{1}{2} \left(z + \frac{1}{z} \right),$$

$$\sin \theta = \frac{1}{2i} \left(e^{i\theta} - e^{-i\theta} \right) = \frac{1}{2i} \left(z - \frac{1}{z} \right),$$

and

$$\frac{dz}{d\theta} = ie^{i\theta} = iz.$$

Thus, we have proved the following theorem.

THEOREM

$$\int_0^{2\pi} F(\cos \theta, \sin \theta) \, d\theta = \int_{|z|=1} F\left[\frac{1}{2} \left(z + \frac{1}{z} \right), \frac{1}{2i} \left(z - \frac{1}{z} \right) \right] \frac{dz}{iz}.$$

EXAMPLE 1.

Show that

$$\int_0^\pi \frac{d\theta}{a + b \cos \theta} = \frac{\pi}{\sqrt{a^2 - b^2}}, \qquad a > b > 0.$$

SOLUTION: Since $\cos \theta$ takes on the same values on $[\pi, 2\pi]$ as it does on $[0, \pi]$, the integral above equals

$$\frac{1}{2} \int_0^{2\pi} \frac{d\theta}{a + b \cos \theta} = \frac{1}{i} \int_{|z|=1} \frac{dz}{bz^2 + 2az + b}.$$

Factoring the denominator into $b(z - p)(z - q)$, where

$$p = \frac{-a + \sqrt{a^2 - b^2}}{b}, \qquad q = \frac{-a - \sqrt{a^2 - b^2}}{b},$$

and observing that $pq = 1$ and $|q| > a/b > 1$, we see that the only singularity of the integrand on the unit disk is at p. Furthermore, it is a pole of order 1, so the residue of the integrand at p equals

$$\lim_{z \to p} \frac{1}{b(z-q)} = \frac{1}{b(p-q)} = \frac{1}{2\sqrt{a^2-b^2}}.$$

The answer now follows by the residue theorem.

EXAMPLE 2.

Prove that

$$\int_0^\pi \frac{d\theta}{(a+b\cos\theta)^2} = \frac{\pi a}{\sqrt{(a^2-b^2)^3}}, \qquad a > b > 0.$$

SOLUTION: Again, the integral equals

$$\frac{2}{i}\int_{|z|=1} \frac{z\,dz}{(bz^2+2az+b)^2} = \frac{2}{ib^2}\int_{|z|=1} \frac{z\,dz}{(z-p)^2(z-q)^2},$$

with a pole of order 2 at p as the only singularity. The residue at p equals

$$\lim_{z \to p}\left[\frac{z}{(z-q)^2}\right]' = \lim_{z \to p}\frac{-(z+q)}{(z-q)^3} = \frac{-(p+q)}{(p-q)^3} = \frac{ab^2}{4\sqrt{(a^2-b^2)^3}},$$

and the result is now immediate.

EXERCISES

Evaluate the integrals in Exercises 1-9 by the method shown in this section. In Exercises 6-8, n is a nonnegative integer.

1. $\displaystyle\int_0^{\pi/2} \frac{d\theta}{a+\sin^2\theta} = \frac{\pi}{2\sqrt{a^2+a}}, \qquad a > 0$

2. $\displaystyle\int_0^{\pi/2} \frac{d\theta}{(a+\sin^2\theta)^2} = \frac{\pi(2a+1)}{4\sqrt{(a^2+a)^3}}, \qquad a > 0$

3. $\displaystyle\int_0^{2\pi} \frac{d\theta}{a^2\cos^2\theta + b^2\sin^2\theta} = \frac{2\pi}{ab}, \qquad a, b > 0$

4. $\displaystyle\int_0^{2\pi} \frac{d\theta}{(a^2\cos^2\theta + b^2\sin^2\theta)^2} = \frac{\pi(a^2+b^2)}{a^3 b^3}, \qquad a, b > 0$

5. $\displaystyle\int_0^{2\pi} \frac{d\theta}{1 - 2a\cos\theta + a^2} = \begin{cases} \dfrac{2\pi}{1 - a^2}, & \text{if } |a| < 1 \\[2ex] \dfrac{2\pi}{a^2 - 1}, & \text{if } |a| > 1 \end{cases}$

6. $\displaystyle\int_0^{2\pi} \cos^n\theta \, d\theta = \begin{cases} \dfrac{n!\,\pi}{2^{n-1}\left[\left(\dfrac{n}{2}\right)!\right]^2}, & \text{if } n \text{ is even} \\[3ex] 0, & \text{if } n \text{ is odd} \end{cases}$

7. $\displaystyle\int_0^{2\pi} (a\cos\theta + b\sin\theta)^n \, d\theta = \begin{cases} \dfrac{n!\,\pi}{2^{n-1}\left[\left(\dfrac{n}{2}\right)!\right]^2} \cdot \sqrt{(a^2+b^2)^n}, & n \text{ even} \\[3ex] 0, & n \text{ odd} \end{cases}$

a, b real

8. $\displaystyle\int_0^{2\pi} e^{\cos\theta} \cos(n\theta - \sin\theta) \, d\theta = \frac{2\pi}{n!}$

9. $\displaystyle\int_0^{2\pi} \cot(\theta + ib) \, d\theta = -2\pi i \operatorname{sign} b, \qquad b \text{ real and nonzero}$

4.3 EVALUATION OF IMPROPER REAL INTEGRALS

In the theorem given in Section 4.2, the interval of integration was automatically transformed into a closed curve, allowing us to apply the residue theorem. In the next application this is not possible, so instead we replace the given curve by a closed curve such that in the limit the values of the integrals agree.

THEOREM

Suppose $F(z)$ is the quotient of two polynomials in z such that

(i) $F(z)$ has no poles on the real axis, and

(ii) $F(1/z)$ has a zero of order at least 2 at $z = 0$, that is, the degree of the denominator exceeds the degree of the numerator by at least 2.

Then

$$\int_{-\infty}^{\infty} F(x) \begin{Bmatrix} \cos ax \\ \sin ax \end{Bmatrix} dx = \begin{Bmatrix} \mathrm{Re} \\ \mathrm{Im} \end{Bmatrix} 2\pi i \sum_{y>0} \mathrm{Res}\, F(z) e^{iaz}, \qquad a \geqslant 0,$$

the sum being taken only on the poles of $F(z)$ in the upper half plane.

PROOF: Let γ be the closed curve obtained by taking the line segment $(-R, R)$ on the real axis followed by the semicircle $z = Re^{i\theta}$, $0 \leqslant \theta \leqslant \pi$. Since $F(z)$ is the quotient of polynomials, its poles, and hence those of $F(z)e^{iaz}$, occur only at zeros of the denominator and thus are finite in number. If R is chosen sufficiently large, all poles of $F(z)$ in the upper half plane will lie inside γ (see Figure 4.2). Then the residue theorem implies

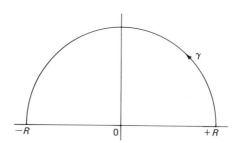

FIGURE 4.2.

$$2\pi i \sum_{y>0} \mathrm{Res}\, F(z) e^{iaz} = \int_{\gamma} F(z) e^{iaz}\, dz$$

$$= \int_{-R}^{R} F(x) e^{iax}\, dx + \int_{0}^{\pi} F(Re^{i\theta}) e^{iaRe^{i\theta}} iRe^{i\theta}\, d\theta.$$

By (ii), $|z^2 F(z)|$ is bounded by a constant M at all points of the upper half plane not inside γ. Thus,

$$\left| \int_{0}^{\pi} F(Re^{i\theta}) e^{iaRe^{i\theta}} iRe^{i\theta}\, d\theta \right| \leqslant \frac{M}{R} \int_{0}^{\pi} e^{-aR \sin \theta}\, d\theta \leqslant \frac{M\pi}{R},$$

since $e^{-aR \sin \theta} \leqslant 1$. By (ii) and the comparison theorem of improper integrals in calculus it follows that

$$\int_{-\infty}^{\infty} F(x) \cos ax \, dx, \qquad \int_{-\infty}^{\infty} F(x) \sin ax \, dx, \qquad a \geqslant 0,$$

both converge. Letting $R \to \infty$, we have

$$\int_{-\infty}^{\infty} F(x) e^{iax} \, dx = 2\pi i \sum_{y>0} \text{Res } F(z) e^{iaz}, \qquad a \geqslant 0,$$

from which the result follows by taking the real and imaginary parts of both sides. ∎

REMARK: If $a > 0$, condition (ii) can be replaced by

(ii)′ $F(1/z)$ has a zero of order 1 at $z = 0$.

In this case we cannot use the comparison theorem to obtain the convergence of the integral

$$\int_{-\infty}^{\infty} F(x) e^{iax} \, dx, \qquad a > 0.$$

In fact, we must prove that

$$\int_{-X_1}^{X_2} F(x) e^{iax} \, dx, \qquad a > 0,$$

has a limit as X_1 and X_2 tend independently to ∞. Let γ be the boundary of the rectangle with vertices at the points $-X_1$, X_2, $X_2 + iY$, $-X_1 + iY$, the constants X_1, X_2, Y chosen large enough that the poles of $F(z)$ in the upper half plane lie inside γ (see Figure 4.3). Condition (ii)′ now shows $|zF(z)|$ bounded by M at all points in $y > 0$ not inside γ. The integral

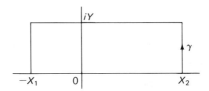

FIGURE 4.3.

$$\left| \int_{X_2}^{X_2+iY} F(z)e^{iaz}\,dz \right| \leqslant M \int_0^Y \frac{e^{-ay}}{|X_2+iy|}\,dy$$

$$\leqslant \frac{M}{X_2} \int_0^Y e^{-ay}\,dy < \frac{M}{aX_2}.$$

Similarly, the integral on the line segment joining $-X_1 + iY$ to $-X_1$ is bounded by M/aX_1 and

$$\left| \int_{X_2+iY}^{-X_1+iY} F(z)e^{iaz}\,dz \right| \leqslant \frac{Me^{-aY}}{Y} \int_{-X_1}^{X_2} dx = \frac{Me^{-aY}}{Y}(X_1+X_2).$$

Using the residue theorem and the triangle inequality,

$$\left| \int_{-X_1}^{X_2} F(x)e^{iax}\,dx - 2\pi i \sum_{y>0} \text{Res}\, F(z)e^{iaz} \right|$$

$$< M \left[\frac{1}{aX_1} + \frac{1}{aX_2} + \frac{e^{-aY}}{Y}(X_1+X_2) \right]$$

First letting $Y \to \infty$ and then letting X_1 and X_2 tend indpendently to ∞ yield the result.

EXAMPLE 1.

Prove that

$$\int_0^\infty \frac{\cos ax}{x^2+b^2}\,dx = \frac{\pi e^{-ab}}{2b}, \qquad a \geqslant 0, \quad b > 0.$$

SOLUTION: Here $F(z)$ equals $(z^2+b^2)^{-1}$ with poles at $\pm ib$, and $F(1/z) = z^2/(1+b^2z^2)$ has a zero of order 2 at 0. As the hypotheses of the theorem are satisfied, we have

$$\int_{-\infty}^\infty \frac{\cos ax}{x^2+b^2}\,dx = \text{Re}\left[2\pi i\, \text{Res}_{ib}\, \frac{e^{iaz}}{z^2+b^2} \right] = \text{Re}\left[\frac{\pi}{b} e^{-ab} \right],$$

from which the result follows, since the integrand is an even function. Note that

$$\int_{-\infty}^\infty \frac{\sin ax}{x^2+b^2}\,dx = 0, \qquad a \geqslant 0, \quad b > 0.$$

EXAMPLE 2.

Show that

$$\int_0^\infty \frac{x \sin ax}{x^2 + b^2} \, dx = \frac{\pi}{2} e^{-ab}, \qquad a > 0, \quad b > 0.$$

SOLUTION: Conditions (i) and (ii)′ apply to $F(z) = z/(z^2 + b^2)$, so that

$$\int_{-\infty}^\infty \frac{x \sin ax}{x^2 + b^2} \, dx = \mathrm{Im}\left[2\pi i \, \mathrm{Res}_{ib} \frac{z e^{iaz}}{z^2 + b^2}\right] = \pi e^{-ab},$$

the integrand again being an even function.

EXERCISES

Evaluate the integrals below by the method given in this section.

1. $\displaystyle\int_{-\infty}^\infty \frac{x \, dx}{(x^2 + 2x + 2)^2} = -\frac{\pi}{2}$

2. $\displaystyle\int_{-\infty}^\infty \frac{x^2 \, dx}{(x^2 + 2x + 2)^2} = \pi$

3. $\displaystyle\int_0^\infty \frac{x^2 \, dx}{(x^2 + a^2)^2} = \frac{\pi}{4a}, \qquad a > 0$

4. $\displaystyle\int_{-\infty}^\infty \frac{dx}{(x^2 + a^2)(x^2 + b^2)} = \frac{\pi}{ab(a + b)}, \qquad a, b > 0$

5. $\displaystyle\int_{-\infty}^\infty \frac{dx}{(x^2 + 1)^{n+1}} = \frac{(2n)!\pi}{2^{2n}(n!)^2}, \qquad n \text{ a nonnegative integer}$

6. $\displaystyle\int_{-\infty}^\infty \frac{\cos ax \, dx}{(x^2 + b^2)^2} = \frac{\pi(1 + ab)e^{-ab}}{2b^3}, \qquad a \geq 0, \quad b > 0$

7. $\displaystyle\int_{-\infty}^\infty \frac{x^3 \sin ax \, dx}{(x^2 + b^2)^2} = \frac{\pi}{2}(2 - ab)e^{-ab}, \qquad a, b > 0$

8. $\displaystyle\int_0^\infty \frac{\cos ax}{x^4 + b^4} \, dx = \frac{\pi}{2b^3} e^{-(ab)/\sqrt{2}} \sin\left(\frac{ab}{\sqrt{2}} + \frac{\pi}{4}\right), \qquad a \geq 0, \quad b > 0$

9. $\displaystyle\int_0^\infty \frac{x \sin ax}{x^4 + b^4}\, dx = \frac{\pi}{2b^2} e^{-(ab)/\sqrt{2}} \sin \frac{ab}{\sqrt{2}}, \qquad a \geqslant 0, \quad b > 0$

10. $\displaystyle\int_0^\infty \frac{x^3 \sin ax}{x^4 + b^4}\, dx = \frac{\pi}{2} e^{-(ab)/\sqrt{2}} \cos \frac{ab}{\sqrt{2}}, \qquad a, b > 0$

4.4 INTEGRALS WITH POLES ON THE REAL AXIS

Throughout the discussion in Section 4.3, we assumed the condition that $F(z)$ had no poles on the real axis, since otherwise the integral

$$\int_{-\infty}^{\infty} F(x) e^{iax}\, dx, \qquad a > 0,$$

diverges. However, the real or imaginary part of the integral above may converge if $F(z)$ has poles of order 1 coinciding with zeros of $\cos ax$ or $\sin ax$.

Suppose $F(z)$ has a pole of order 1 at $z = 0$ and no other poles on the real axis. Then

$$\int_{-\infty}^{\infty} F(x) \sin ax\, dx, \qquad a > 0,$$

converges. The technique for integration consists of using the boundary γ of the rectangle with vertices at $-X_1$, X_2, $X_2 + iY$, $-X_1 + iY$, except that the origin is avoided by following a small semicircle E of radius r in the lower half plane (see Figure 4.4). Assume X_1, X_2, Y,

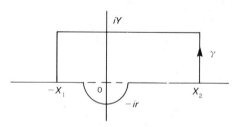

FIGURE 4.4.

$1/r$ are chosen sufficiently large that all poles of $F(z)$ not in the lower half plane lie inside γ. Then $F(z)e^{iaz} = (a_{-1}/z) + f(z)$ with $a_{-1} = \mathrm{Res}_0\, F(z)e^{iaz}$ and $f(z)$ analytic in a closed ϵ-neighborhood of $z = 0$.

Now on the semicircle E with $r < \epsilon$,

$$\int_E F(z)e^{iaz}\, dz = i \int_{-\pi}^{0} [a_{-1} + f(re^{i\theta})re^{i\theta}]\, d\theta$$

$$= \pi i a_{-1} + ir \int_{-\pi}^{0} f(re^{i\theta})e^{i\theta}\, d\theta.$$

Since $f(z)$ is bounded in $|z| \leqslant \epsilon$ by a constant N,

$$\left| ir \int_{-\pi}^{0} f(re^{i\theta})e^{i\theta}\, d\theta \right| \leqslant rN\pi,$$

and the second term vanishes as r tends to 0. By the residue theorem and the inequalities developed in Section 4.3, we get

$$\int_{-X_1}^{-r} + \int_E + \int_r^{X_2} F(x)e^{iax}\, dx - 2\pi i \sum_{y \geqslant 0} \text{Res } F(z)e^{iaz} \to 0$$

by letting $Y \to \infty$, and then X_1 and X_2 tend independently to ∞. Now letting $r \to 0$, we find that

$$\lim_{r \to 0} \int_{-\infty}^{-r} + \int_r^{\infty} F(x)e^{iax}\, dx = 2\pi i \left[\sum_{y > 0} \text{Res } F(z)e^{iaz} + \frac{a_{-1}}{2} \right].$$

The limit on the left-hand side of the expression is referred to as the **Cauchy principal value** of the integral and written

$$\text{PV} \int_{-\infty}^{\infty} F(x)e^{iax}\, dx = 2\pi i \left[\sum_{y > 0} \text{Res } F(z)e^{iaz} + \frac{a_{-1}}{2} \right].$$

Observe that only half the residue at 0 has been included on the right.

We digress briefly to make some comments about Cauchy principal values. Let $f(x)$ be defined on the real line and consider the limits

$$\lim_{R \to \infty} \int_{-R}^{R} f(x)\, dx, \tag{1}$$

$$\lim_{R_1 \to \infty} \int_{-R_1}^{0} f(x)\, dx + \lim_{R_2 \to \infty} \int_0^{R_2} f(x)\, dx. \tag{2}$$

If the limit in (1) exists, the improper integral of f is said to **converge in the sense of Cauchy** and we write

$$\text{PV} \int_{-\infty}^{\infty} f(x)\, dx = \lim_{R \to \infty} \int_{-R}^{R} f(x)\, dx.$$

If the limits in (2) exist, we say the improper integral **converges** and we set

$$\int_{-\infty}^{\infty} f(x)\ dx = \lim_{R_1 \to \infty} \int_{-R_1}^{0} f(x)\ dx + \lim_{R_2 \to \infty} \int_{0}^{R_2} f(x)\ dx.$$

Note that convergence of the integral implies convergence (to the same value) in the sense of Cauchy, but that an integral may have a principal value without being convergent. For example,

$$\text{PV} \int_{-\infty}^{\infty} x\ dx = \lim_{R \to \infty} \left(\frac{x^2}{2} \Big|_{-R}^{R} \right) = 0,$$

yet neither limit in (2) exists. In some situations, such as net charge on an infinite plate, the limit in (1) would be used; in other situations, such as total charge on the plate, the limit in (2) would be employed. We choose the tool to fit the problem.

A similar development arises when $f(x)$ is defined in an interval $a \leqslant x \leqslant b$ but is unbounded in every neighborhood of a point c, $a < c < b$. The improper integral converges provided the right side of the equation

$$\int_{a}^{b} f(x)\ dx = \lim_{\epsilon \to 0} \int_{a}^{c-\epsilon} f(x)\ dx = \lim_{\epsilon \to 0} \int_{c+\eta}^{b} f(x)\ dx, \qquad \epsilon > 0, \quad \eta > 0, \quad (3)$$

exists. Even if these limits fail to exist, the Cauchy principal value of the integral

$$\text{PV} \int_{a}^{b} f(x)\ dx = \lim_{\epsilon \to 0} \left(\int_{a}^{c-\epsilon} f(x)\ dx + \int_{c+\epsilon}^{b} f(x)\ dx \right), \epsilon > 0, \quad (4)$$

may exist. For example,

$$\text{PV} \int_{-1}^{1} \frac{dx}{x} = \lim_{\epsilon \to 0} \left(\log \epsilon + \log \frac{1}{\epsilon} \right) = 0, \qquad \epsilon > 0,$$

but neither of the limits in (3) exists.

As before, convergence implies convergence in the sense of Cauchy. Moreover, an improper integral of mixed type may have a Cauchy principal value even though the integral diverges:

$$\text{PV} \int_{-\infty}^{\infty} \frac{dx}{x} = \text{PV} \left(\int_{-\infty}^{-1} + \int_{1}^{\infty} \right) \frac{dx}{x} + \text{PV} \int_{-1}^{1} \frac{dx}{x}$$

$$= \lim_{R \to \infty} \left(\int_{-R}^{-1} + \int_{1}^{R} \right) \frac{dx}{x} = 0.$$

If $F(z)$ has several poles of order 1 on the real axis coinciding with the zeros of either $\cos ax$ or $\sin ax$, then including as many semicircles as there are poles to γ and treating them as the semicircle E was treated above yield the following general result.

THEOREM

Suppose $F(z)$ is the quotient of two polynomials in z such that

(i)′ all poles of $F(z)$ lying on the real axis are of order 1 and coincide with zeros of either $\cos ax$ or $\sin ax$, $a > 0$, and

(ii)′ $F(1/z)$ has a zero of order at least 1 at $z = 0$.

Then

$$\text{PV} \int_{-\infty}^{\infty} F(x)e^{iax}\, dx = 2\pi i \left[\sum_{y>0} \text{Res } F(z)e^{iaz} + \tfrac{1}{2} \sum_{y=0} \text{Res } F(z)e^{iaz} \right].$$

EXAMPLE 1.

Prove that

$$\int_0^{\infty} \frac{\sin x}{x}\, dx = \frac{\pi}{2}.$$

SOLUTION: Since $F(z) = 1/z$, it is clear that (i)′ and (ii)′ hold, so we have

$$\text{PV} \int_{-\infty}^{\infty} \frac{e^{ix}}{x}\, dx = \pi i \, \text{Res}_0 \frac{e^{iz}}{z} = \pi i.$$

Equating the imaginary parts yields the desired result, since the integrand is an even function and $x = 0$ is a removable singularity of $(\sin x)/x$.

Integrals containing powers of $\cos ax$ or $\sin ax$ may be evaluated by the same technique.

EXAMPLE 2.

Show that

$$\int_0^\infty \frac{\sin^2 x}{x^2}\, dx = \frac{\pi}{2}.$$

SOLUTION: Using the double-angle formula $2\sin^2 x = 1 - \cos 2x$, we obtain

$$\int_{-\infty}^\infty \frac{1 - \cos 2x}{4x^2}\, dx,$$

which converges by the comparison theorem of calculus. Integrating $(1 - e^{2iz})/4z^2$ along the curve γ shown in Figure 4.5, we have

$$\int_\gamma \frac{1 - e^{2iz}}{4z^2}\, dz = 2\pi i \operatorname{Res}_0 \frac{1 - e^{2iz}}{4z^2} = \pi.$$

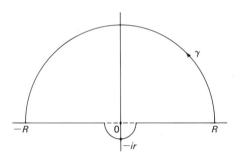

FIGURE 4.5.

The absolute value of the integral along the arc $|z| = R$, $0 \leqslant \arg z \leqslant \pi$, is bounded by

$$\frac{1}{4R} \int_0^\pi |1 - e^{2iR\, e^{i\theta}}|\, d\theta \leqslant \frac{\pi}{2R},$$

which vanishes as $R \to \infty$. Since

$$\frac{1 - e^{2iz}}{4z^2} = \frac{-i}{2z} + f(z)$$

with $f(z)$ analytic in a closed disk centered at 0 containing the semicircle E, we have

$$\left| \int_E \frac{1 - e^{2iz}}{4z^2} \, dz - \frac{\pi}{2} \right| = \left| ir \int_{-\pi}^0 f(re^{i\theta})e^{i\theta} \, d\theta \right| \leqslant rN\pi,$$

and this bound vanishes as $r \to 0$. Thus,

$$\text{PV} \int_{-\infty}^{\infty} \frac{1 - e^{2ix}}{4x^2} \, dx = \frac{\pi}{2},$$

and the solution is complete.

EXERCISES

Evaluate the integrals in Exercises 1-9 by the method shown in this section.

1. $\displaystyle\int_{-\infty}^{\infty} \frac{\cos \pi x}{4x^2 - 1} \, dx = \frac{-\pi}{2}$

2. $\displaystyle\int_{-\infty}^{\infty} \frac{\sin \pi x}{x^5 - x} \, dx = \frac{\pi}{2} \left(e^{-\pi} - 3\right)$

3. $\displaystyle\int_{-\infty}^{\infty} \frac{\sin \pi x \cos \pi x}{2x^2 - x} \, dx = -\pi$

4. $\displaystyle\int_{-\infty}^{\infty} \frac{\sin x}{x} \frac{x^2 + a^2}{x^2 + b^2} \, dx = \frac{\pi}{b^2} \left[a^2 + e^{-b}(b^2 - a^2)\right], \qquad a, b > 0$

5. $\displaystyle\int_0^{\infty} \frac{\sin x}{x(x^2 + b^2)} \, dx = \frac{\pi}{2b^2} \left(1 - e^{-b}\right), \qquad b > 0$

6. $\displaystyle\int_0^{\infty} \frac{\sin ax}{x(x^2 + b^2)^2} \, dx = \frac{\pi}{2b^4} \left[1 - \frac{e^{-ab}}{2}(ab + 2)\right], \qquad a, b > 0$

7. $\displaystyle\int_0^{\infty} \frac{\cos ax - \cos bx}{x^2} \, dx = \frac{b - a}{2} \pi, \qquad a, b \geqslant 0$

8. $\displaystyle\int_0^{\infty} \frac{\sin^3 x}{x^3} \, dx = \frac{3\pi}{8}$

9. $\displaystyle\int_{-\infty}^{\infty} \frac{\sin m(x - a)}{x - a} \frac{\sin n(x - b)}{x - b} \, dx = \pi \frac{\sin n(a - b)}{a - b},$
$$m \geqslant n \geqslant 0, \quad a, b \text{ real}, \quad a \neq b$$

10. Prove the identity

$$\text{PV } \frac{1}{2\pi i} \int_{-\infty}^{\infty} \frac{e^{itx} \, dx}{x} = \begin{cases} \frac{1}{2}, & t > 0, \\ 0, & t = 0, \\ -\frac{1}{2}, & t < 0, \end{cases}$$

using the method shown in this section. If we add $\frac{1}{2}$ to this function, we obtain the "impulse function," often found in engineering books, representing a sudden switch-in of current into an open circuited electric line.

4.5 INTEGRATION OF MULTIVALUED FUNCTIONS (Optional)

When we work with integrals involving multivalued functions, we must take into account the branch points and branch cuts of the integrand, in addition to its isolated singularities. We do this because to use the residue theorem, we must select a region in which the integrand is single-valued.

THEOREM

Let $F(z)$ be the quotient of two polynomials in z satisfying

(i) $F(z)$ has no poles on the positive real axis, and
(ii) $z^{a+1} F(z)$ vanishes as z tends to 0 or ∞, where a is real but not an integer.

Then

$$\int_0^{\infty} x^a F(x) \, dx = \frac{2\pi i}{1 - e^{2\pi i a}} \sum_{z \neq 0} \text{Res}(z^a F(z)),$$

the sum being taken over all nonzero poles of $F(z)$.

PROOF: Since $F(z)$ has only finitely many poles in $\mathcal{C}$, there are numbers $0 < r < R$ such that all the nonzero poles will be inside the annulus $r < |z| < R$. For the function z^a, we select the branch of $\mathcal{R}$ whose argument lies between 0 and 2π, with branch points 0 and ∞. Let $\gamma = \gamma_1 + \gamma_2 + \gamma_3 + \gamma_4$ consist of the boundary of the region obtained by cutting $r < |z| < R$ along the linear segment $r < x < R$, labeled as shown in Figure 4.6.

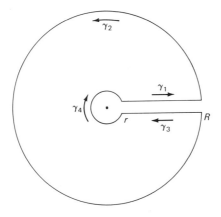

FIGURE 4.6.

Strictly speaking, we cannot apply the residue theorem directly to γ, since $z^a F(z)$ is multivalued on the branch cut. However, we can apply the residue theorem to the boundaries of the regions D_1, D_2 indicated in Figure 4.7, with the integrals along the arcs γ_5 and γ_6 canceling out, thus extending the residue theorem to γ.

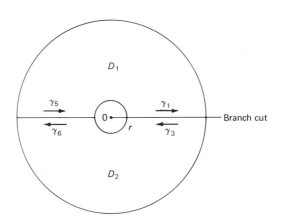

FIGURE 4.7.

Note that the integrand has different values on γ_1 and γ_3. By the residue theorem,

$$\int_\gamma z^a F(z)\, dz = 2\pi i \sum_{z \neq 0} \text{Res}(z^a F(z)),$$

but

$$\left| \int_{\gamma_j} z^a F(z)\, dz \right| \leqslant \int_0^{2\pi} |z^{a+1} F(z)|\, d\theta, \qquad j = 2, 4,$$

which vanishes by (ii) as $R \to \infty$ or $r \to 0$. Now

$$z^a F(z) = \begin{cases} x^a F(x) & \text{on } \gamma_1, \\ x^a e^{2\pi i a} F(x) & \text{on } \gamma_3, \end{cases}$$

so

$$\int_{\gamma_1 + \gamma_3} z^a F(z)\, dz = (1 - e^{2\pi i a}) \int_r^R x^a F(x)\, dx$$

yields the required formula by letting $R \to \infty$ and $r \to 0$. ∎

EXAMPLE 1.

Show that

$$\int_0^\infty \frac{x^a\, dx}{x + b} = \frac{-\pi b^a}{\sin \pi a}, \qquad 0 > a > -1, \quad b > 0.$$

SOLUTION: Here $0 < a + 1 < 1$, so it is clear that (i) and (ii) hold. Selecting the branch of $\mathfrak{R}$ whose argument lies between 0 and 2π, we have

$$\frac{2\pi i}{1 - e^{2\pi i a}} \, \text{Res}_{-b} \, \frac{z^a}{(z + b)} = \frac{2\pi i b^a}{e^{-\pi i a} - e^{\pi i a}},$$

since on this branch

$$(-b)^a = b^a e^{\pi i a}.$$

The same type of procedure can be applied to other multivalued functions. We illustrate this in the next two examples.

EXAMPLE 2.

Prove that

$$\int_0^\infty \frac{\log x}{x^2 + b^2}\, dx = \frac{\pi}{2b} \log b, \qquad b > 0.$$

SOLUTION: Here we use the curve γ shown in Figure 4.8. Then

$$\int_\gamma \frac{\log z}{z^2 + b^2}\, dz = 2\pi i\, \mathrm{Res}_{bi}\, \frac{\log z}{z^2 + b^2} = \frac{\pi}{b}\left[\log b + \frac{i\pi}{2} \right].$$

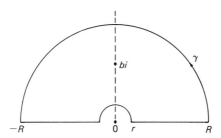

FIGURE 4.8.

But

$$\left| iR \int_0^\pi \frac{\log R + i\theta}{(Re^{i\theta})^2 + b^2}\, e^{i\theta}\, d\theta \right| \leqslant \frac{R(\,|\log R| + \pi)}{|R^2 - b^2|}\, \pi,$$

which vanishes as $R \to \infty$ or 0 by L'Hospital's theorem. Since the integral is convergent,

$$\frac{\pi}{b}\left[\log b + \frac{i\pi}{2} \right] = \int_{-\infty}^\infty \frac{\log x\, dx}{x^2 + b^2}$$

$$= \int_{-\infty}^\infty \frac{\log |x|}{x^2 + b^2}\, dx + i\pi \int_{-\infty}^0 \frac{dx}{x^2 + b^2},$$

from which the result follows as the first integrand is even.

EXAMPLE 3.†

Prove that

$$\int_0^\infty \frac{\sinh ax}{\sinh \pi x} \, dx = \frac{1}{2} \tan \frac{a}{2}, \qquad -\pi < a < \pi.$$

SOLUTION: The integral of the function $e^{az}/\sinh \pi z$ vanishes over the curve γ shown in Figure 4.9, as no singularities lie inside γ. But

$$|\sinh \pi(R + iy)| \geqslant |\sinh \pi R|$$

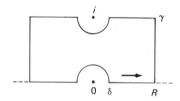

FIGURE 4.9.

(see Exercise 27, Section 1.8), implying that

$$\left| \int_R^{R+i} \frac{e^{az}}{\sinh \pi z} \, dz \right| \leqslant \frac{e^{aR}}{|\sinh \pi R|} \to 0$$

as $R \to \pm\infty$. Since $i \sinh z = \sin iz$, $1/\sinh \pi z$ has poles of order 1 at all integral multiples of i; thus,

$$\text{Res}_0 \; \frac{e^{az}}{\sinh \pi z} = \lim_{z \to 0} \frac{z e^{az}}{\sinh \pi z} = \frac{1}{\pi}$$

and

$$\text{Res}_i \; \frac{e^{az}}{\sinh \pi z} = \lim_{z \to i} \frac{(z - i)e^{az}}{\sinh \pi z} = \frac{-e^{ai}}{\pi}$$

by L'Hospital's theorem. Integrating over the two semicircles yields

$$-\pi i \left(\frac{1}{\pi} - \frac{e^{ai}}{\pi} \right)$$

plus an integral that vanishes as $\delta \to 0$. But

$$\sinh \pi(x + i) = -\sinh \pi x,$$

so we obtain

$$\text{PV}(1 + e^{ai}) \int_{-\infty}^{\infty} \frac{e^{ax}}{\sinh \pi x} \, dx = i(1 - e^{ai}),$$

or

$$\text{PV} \int_{-\infty}^{\infty} \frac{e^{ax}}{\sinh \pi x} \, dx = \tan \frac{a}{2},$$

from which the result follows, since the integrand of the original equation is even.

EXERCISES

Evaluate the integrals in Exercises 1–16, using the method of this section.

1. $\displaystyle\int_{0}^{\infty} \frac{x^a}{(x + b)^2} \, dx = \frac{\pi a b^{a-1}}{\sin \pi a}$, $1 > a > -1$, $b > 0$

2. $\displaystyle\int_{-\infty}^{\infty} \frac{e^{ay} \, dy}{1 + be^{-y}} = \frac{-\pi b^a}{\sin \pi a}$, $0 > a > -1$, $b > 0$

3. $\displaystyle\int_{0}^{\infty} \frac{x^a \, dx}{x^2 + b^2} = \frac{\pi b^{a-1}}{2 \cos \dfrac{\pi a}{2}}$, $1 > a > -1$, $b > 0$

4. $\displaystyle\int_{0}^{\infty} \frac{x^a \, dx}{x^2 + 2x \cos \theta + 1} = \frac{\pi}{\sin \pi a} \frac{\sin \theta a}{\sin \theta}$,

   $1 > a > -1$, $\pi > \theta > -\pi$

5. $\displaystyle\int_{0}^{\infty} \frac{x^a \, dx}{(x^2 + b^2)^2} = \frac{\pi b^{a-3}(1 - a)}{4 \cos \dfrac{\pi a}{2}}$, $3 > a > -1$, $b > 0$

6. $\displaystyle\int_{0}^{\infty} \frac{x^a \, dx}{x^3 + b^3} = \frac{2\pi b^{a-2}}{3 \sin \pi a} \left[\cos \frac{\pi}{3}(1 - 2a) - \frac{1}{2} \right]$,

   $2 > a > -1$, $b > 0$

7. $\displaystyle\int_0^\infty \frac{\log x}{(x^2 + b^2)^2}\, dx = \frac{\pi}{4b^3}\,(\log b - 1), \qquad b > 0$

8. $\displaystyle\int_0^\infty \frac{x^a \log x}{x + b}\, dx = \frac{\pi b^a}{\sin^2 \pi a}\,(\pi \cos \pi a - \sin \pi a \cdot \log b),$

$0 > a > -1, \quad b > 0$

9. $\displaystyle\int_0^\infty \frac{x^a \log x}{x^2 + b^2}\, dx = \frac{\pi b^{a-1}}{2 \cos^2 \dfrac{\pi a}{2}}\left[\frac{\pi}{2}\,\sin\frac{\pi a}{2} + \log b \cdot \cos\frac{\pi a}{2}\right],$

$1 > a > -1, \quad b > 0$

10. $\displaystyle\int_0^\infty \frac{x}{\sinh x}\, dx = \frac{\pi^2}{4}$

11. $\displaystyle\int_0^\infty \frac{\sin ax}{\sinh x}\, dx = \frac{\pi}{2}\,\tanh\frac{a\pi}{2}, \qquad a \text{ real}$

12. $\displaystyle\int_0^\infty \frac{x \cos ax}{\sinh x}\, dx = \frac{\pi^2}{4}\,\text{sech}^2\frac{a\pi}{2}, \qquad a \text{ real}$

13. $\displaystyle\int_0^\infty \frac{\cosh ax}{\cosh \pi x}\, dx = \frac{1}{2}\,\sec\frac{a}{2}, \qquad -\pi < a < \pi$

14. $\displaystyle\int_0^\infty x^{a-1}\begin{Bmatrix}\cos bx\\ \sin bx\end{Bmatrix} dx = \begin{Bmatrix}\cos\dfrac{\pi a}{2}\\[2mm] \sin\dfrac{\pi a}{2}\end{Bmatrix}\cdot \frac{\Gamma(a)}{b^a}, 1 > a > 0,\ b > 0,$

where $\Gamma(a) = \int_0^\infty e^{-x} x^{a-1}\, dx$ is the *gamma function*. (*Hint:* Integrate $z^{a-1} e^{-bz}$ around a suitable contour and use the inequality $\cos \theta \geqslant 1 - (2/\pi)\theta,\ 0 \leqslant \theta \leqslant \pi/2$.)

15. $\displaystyle\int_0^\infty \frac{\sin x^a}{x^a}\, dx = \frac{\Gamma\left(\dfrac{1}{a}\right)\cos\dfrac{\pi}{2a}}{a - 1}, \qquad 1 > a > \frac{1}{2}$

(*Hint:* Show $x\Gamma(x) = \Gamma(x + 1)$ by differentiating by parts.)

16. $\displaystyle\int_0^\infty \frac{\cos x}{\sqrt{x}}\, dx = \int_0^\infty \frac{\sin x}{\sqrt{x}}\, dx = \int_0^\infty \frac{e^{-x}}{\sqrt{2x}}\, dx.$

17. Prove that

$$\int_0^{\pi/2} (\tan \theta)^a \, d\theta = \frac{\pi}{2 \cos \dfrac{\pi a}{2}}, \qquad 1 > a > -1,$$

and

$$\int_0^{\pi/2} \log \tan \theta \, d\theta = 0.$$

(*Hint:* Use Exercise 3.)

4.6 THE ARGUMENT PRINCIPLE

Another application of the residue theorem, useful in detemining the number of zeros and poles of a meromorphic function, is the following result.

THE ARGUMENT PRINCIPLE

Let $w = f(z)$ be meromorphic in the simply connected region G and γ be a pws Jordan curve in G that avoids the zeros and poles of $f(z)$. Then

$$\frac{1}{2\pi i} \int_{f(\gamma)} \frac{dw}{w} = \frac{1}{2\pi i} \int_{\gamma} \frac{f'(z) \, dz}{f(z)} = Z - P,$$

where Z and P are, respectively, the number of zeros and poles, including multiplicities, of $f(z)$ lying inside γ.

PROOF: Note that the first integral equals the number of times the closed curve $f(\gamma)$ winds around 0; in other words, it measures the **variation of the argument** of $f(z)$ as z traverses the curve γ, leading to the name of the theorem. (See Example 4 of Section 2.1.)

If a is a zero of order k of $f(z)$, then we write $f(z) = (z - a)^k f_0(z)$, with $f_0(z)$ analytic and nonzero in an ϵ-neighborhood of a. Thus,

$$\frac{f'(z)}{f(z)} = \frac{k}{z - a} + \frac{f_0'(z)}{f_0(z)},$$

and, since f_0'/f_0 is analytic in an ϵ-neighborhood of a, we see that f'/f has a pole of order 1 with residue k at $z = a$.

On the other hand, if a is a pole of order h of $f(z)$, then $f(z) = f_0(z)/(z-a)^h$ with $f_0(z)$ again nonzero and analytic in an ϵ-neighborhood of a. So

$$\frac{f'(z)}{f(z)} = \frac{-h}{z-a} + \frac{f_0'(z)}{f_0(z)}$$

has a pole of order 1 with residue $(-h)$ at $z = a$. By the residue theorem, it follows that

$$\frac{1}{2\pi i} \int_\gamma \frac{f'(z)}{f(z)}\, dz = Z - P,$$

where Z is the sum of all the orders k of the zeros of $f(z)$, and P is the sum of all the orders h of the poles of $f(z)$, lying inside γ. ■

A most useful application of the argument principle is the following result.

ROUCHÉ'S THEOREM

Let $f(z)$ and $g(z)$ be analytic in a simply connected region G. If $|f(z)| > |g(z) - f(z)|$ at all points of the pws Jordan curve γ lying in G, then $f(z)$ and $g(z)$ have the same number of zeros inside γ.

PROOF: The hypothesis $|f(z)| > |g(z) - f(z)|$ forces both functions to be nonzero on γ; thus γ avoids the poles and zeros of $F(z) = g(z)/f(z)$. However, for all z on γ,

$$\left| \frac{g(z)}{f(z)} - 1 \right| < 1.$$

Thus $F(\gamma)$ does not wind around 0, so the argument principle implies that $F(z)$ has the same number of zeros as it has poles inside γ. But these correspond to the zeros of $g(z)$ and $f(z)$, respectively, so the proof is complete. ■

EXAMPLE 1.

Find the number of roots of the equation

$$z^4 + 5z + 1 = 0$$

lying inside the circle $|z| = 1$.

SOLUTION: Let $f(z) = 5z$ and $g(z) = z^4 + 5z + 1$. Then, by the triangle inequality,

$$|g(z) - f(z)| \leqslant |z|^4 + 1 < |5z| = |f(z)|$$

on $|z| = 1$. Since $f(z)$ has one zero inside $|z| = 1$, so does $g(z)$. On the other hand, letting $f(z) = z^4$, we have

$$|5z + 1| \leqslant 11 < 16 = |z|^4$$

on $|z| = 2$. Thus, $g(z)$ has four zeros inside $|z| = 2$, three of which lie in the annulus $1 < |z| < 2$, since no zeros lie on $|z| = 1$.

EXAMPLE 2.

Show that $z - e^z + a = 0$, $a > 1$, has one root in the left half plane.

SOLUTION: Let $f(z) = z + a$ and $g(z) = z - e^z + a$. For $z = iy$ or $|z| = R > 2a$, $x < 0$, we have

$$|g(z) - f(z)| = e^{\operatorname{Re} z} \leqslant 1 < a < |f(z)|,$$

and $f(z)$ has only one root (at $z = -a$), so the proof is complete.

EXAMPLE 3.

Find the number of roots of the equation

$$z^4 + iz^3 + 3z^2 + 2iz + 2 = 0$$

lying in the upper half plane.

SOLUTION: Let $f(z) = z^4 + 3z^2 + 2 = (z^2 + 2)(z^2 + 1)$ and $g(z) = z^4 + iz^3 + 3z^2 + 2iz + 2$. For $z = x$ or $|z| = R \geqslant 2$, we have

$$|g(z) - f(z)| = |z| |z^2 + 2| < |z^2 + 1| |z^2 + 2| = |f(z)|,$$

so $g(z)$ has two roots in the upper half plane.

EXAMPLE 4.

Find the number of roots of the equation
$$7z^3 - 5z^2 + 4z - 2 = 0$$
in the disk $|z| \leqslant 1$.

SOLUTION: If we multiply the equation by $z + 1$, we obtain
$$7z^4 + 2z^3 - z^2 + 2z - 2 = 0.$$
Letting $f(z) = 7z^4$ and $g(z) = 7z^4 + 2z^3 - z^2 + 2z - 2$, we find by the triangle inequality that
$$|g(z) - f(z)| \leqslant 2\,|z|^3 + |z|^2 + 2\,|z| + 2 < 7\,|z|^4 = |f(z)|,$$
whenever $|z| = 1 + \epsilon$, $\epsilon > 0$. Hence, $g(z)$ has four roots in $|z| \leqslant 1$, implying that the original equation had three roots in the closed unit disk.

EXERCISES

For Exercises 1-4, find the number of roots of the given equations inside the circle $|z| = 1$.

1. $z^5 + 8z + 10 = 0$
2. $z^8 - 2z^5 + z^3 - 8z^2 + 3 = 0$
3. $z^6 + 3z^5 - 2z^2 + 2z - 9 = 0$
4. $z^7 - 7z^6 + 4z^3 - 1 = 0$
5. How many of the roots of the equations given in Exercises 1-4 lie inside $|z| = 2$?
6. How many roots of the equation
$$3z^4 - 6iz^3 + 7z^2 - 2iz + 2 = 0$$
lie in the upper half plane?
7. How many roots of the equation
$$z^6 + z^5 - 6z^4 - 5z^3 + 10z^2 + 5z - 5 = 0$$
lie in the right half plane?
8. Find the number of roots of the equation
$$9z^4 + 7z^3 + 5z^2 + z + 1 = 0$$
lying in the disk $|z| \leqslant 1$.

9. How many roots of the equation

$$z^4 + 2z^3 - 3z^2 - 3z + 6 = 0$$

lie in the disk $|z| \leqslant 1$?

10. Show that the function

$$f(z) = \frac{z - a}{1 - \bar{a}z}, \qquad |a| < 1,$$

assumes in $|z| < 1$ every value c satisfying $|c| < 1$ exactly once, and no value c for which $|c| > 1$. Thus, $f(z)$ maps $|z| < 1$ one-to-one and onto itself. (*Hint:* Show $|f(z)| = 1$ on $|z| = 1$ and apply Rouché's theorem to $f(z) - c$.)

11. Assuming the hypothesis of the argument principle, show that the number of times $f(\gamma)$ winds around the point a equals $Z_a - P$, where Z_a is the number of a values of $f(z)$ including multiplicities.

12. Let $f(z)$ be analytic in a region G, a in G, and suppose $f(z) - f(a)$ has a zero of order n at $z = a$. Prove that for sufficiently small $\epsilon > 0$, there is a $\delta > 0$ such that for all ζ in $|\zeta - f(a)| < \delta$ the equation $f(z) - \zeta = 0$ has exactly n roots in $|z - a| < \epsilon$.

13. Use the result in Exercise 12 to prove that nonconstant analytic functions map open sets onto open sets, and use this fact to get an immediate proof of the maximum principle. (*Hint:* Show that interior points are mapped onto interior points.)

14. Use Rouché's theorem to prove the fundamental theorem of algebra.

NOTES

SECTION 4.1

The results in this section are easily extended to arbitrary pws closed curves γ in $\mathcal{C}$. However, in most applications, γ is a Jordan curve. Consequently, this extra hypothesis was incorporated to simplify the statements of the theorems. For the more general statements, see [A, pp. 147–151].

SECTIONS 4.2–4.5

The reader may have noticed that a number of the integrals, which depended on one or more arbitrary parameters, could be obtained by differentiation or integration of other integrals with respect to these parameters. For example: Exercises 1 and 2 of Section 4.2, Examples 1 and 2 of Section 4.3, Exercises 8 and 9 of Section 4.3, Exercises 5 and 6 of Section 4.4, Exercises 3 and 5 of Section 4.5. A sufficient condition for the validity of these procedures is the uniform convergence of the integrals on the interval of definition of the parameters. The relevant theorems and proofs may be found in most advanced calculus books. For example, see [B, pp. 204–212]. This technique is usually easier than evaluating the integrals by the residue method.

SECTION 4.6

These results may also be extended to arbitrary pws closed curves γ (see [A, pp. 151–153]).

5 CONFORMAL MAPPINGS

5.1 GEOMETRIC CONSIDERATIONS

Let us investigate the change in the slope of a smooth arc passing through the point z_0 under the mapping $w = f(z)$, when f is analytic at z_0 and $f'(z_0) \neq 0$.

If $\gamma \colon z = z(t)$, $z(0) = z_0$, is such an arc, its tangent at z_0 has slope

$$\frac{dy}{dx} = \frac{y'(0)}{x'(0)} = \tan \arg z'(0),$$

and its image $f(\gamma) \colon w = f(z(t))$ has a tangent at $f(z_0)$ with slope $\tan \arg w'(0)$. But by the chain rule,

$$\arg w'(0) = \arg[f'(z_0)z'(0)] = \arg f'(z_0) + \arg z'(0).$$

Hence, the change in direction is equal to the constant $\arg f'(z_0)$ *independent of the arc chosen*. Thus, the angle formed by the tangents of two smooth arcs intersecting at z_0 is preserved under the mapping $w = f(z)$, as both directions are changed by the same amount (see Figure 5.1). Mappings that preserve angle size and orientation are said to be **conformal**. We have proved the following theorem.

THEOREM

If $f(z)$ is analytic in a region G, then $w = f(z)$ is conformal at all points z_0 in G for which $f'(z_0) \neq 0$.

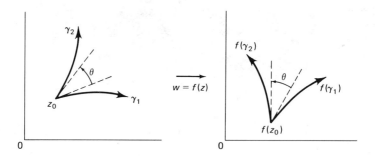

FIGURE 5.1. Mapping conformal at z_0

EXAMPLE 1.

The mapping $w = e^z$ is conformal at all points in $\mathbb{C}$, since its derivative is nonzero. This function maps the real axis in the z-plane onto the positive reals in the w-plane. The imaginary axis in the z-plane is mapped repeatedly on the unit circle in the w-plane because $|e^{iy}| = 1$. Thus, the right angle between the coordinate axes in the first quadrant of the z-plane is transformed into the right angle between the positive real axis and the unit circle in the first quadrant of the w-plane (see Figure 5.2).

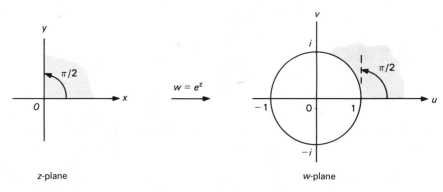

z-plane   w-plane

FIGURE 5.2. The conformal mapping $w = e^z$

Let $w = f(z)$ be conformal in a region G containing the point z_0. Consider the effect of this mapping on a disk centered at z_0 lying in

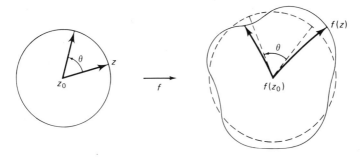

FIGURE 5.3. Mapping a disk centered at z_0

G (see Figure 5.3). The angles between radial lines are preserved, although their lengths are not. However, since

$$|f'(z_0)| = \lim_{z \to z_0} \frac{|f(z) - f(z_0)|}{|z - z_0|},$$

the radial lines are subject to approximately the same change of scale $|f'(z_0)|$ when the radius is small. Roughly, *"small" circles about z_0 are mapped onto "small" circles about $f(z_0)$ with change of scale $|f'(z_0)|$.* Moreover, this indicates that the mapping is *locally* one-to-one, although it is clear nothing can be said about its *global* behavior. For example, $f(z) = e^z$ is locally one-to-one, since $f'(z) = e^z \neq 0$, but $f(0) = f(2\pi i)$, so that it is not one-to-one in $\mathbb{C}$.

Angles are magnified at all points where the derivative is zero. For example, $f(z) = z^2$ has a derivative that vanishes at the origin. Since $f(1) = f(-1) = 1$ and $f(i) = f(-i) = -1$, the right angles between the axes are mapped into $180°$ angles. This doubling of angles causes circles around the origin to be mapped to circular curves that wind around the origin twice. This motivates the following theorem.

THEOREM

Let $f(z)$ be analytic in a region G containing the point z_0 at which $f'(z)$ has a zero of order k. Then all angles at z_0 are magnified by the factor $k + 1$.

PROOF: We can write $f'(z) = (z - z_0)^k g(z)$, with g analytic and non-zero in an ϵ-neighborhood of z_0. Thus, the terms $f'(z_0), f''(z_0), \ldots,$

$f^{(k)}(z_0)$ all vanish in the Taylor series for $f'(z)$. Hence, the Taylor series of $f(z)$ is

$$f(z) = f(z_0) + \frac{f^{(k+1)}(z_0)}{(k+1)!}(z-z_0)^{k+1} + \ldots,$$

implying that

$$\arg[f(z) - f(z_0)] = (k+1)\arg(z-z_0) + \arg\left[\frac{f^{(k+1)}(z_0)}{(k+1)!} + \ldots\right].$$

The first two arguments compare the angles between the horizontal direction and the vector pointing from $f(z_0)$ to $f(z)$ and from z_0 to z. If z tends to z_0 along a fixed vector making an angle θ with the horizontal direction, the angle of the vector from $f(z_0)$ to $f(z)$ with the horizontal tends to

$$(k+1)\theta + \arg\left[\frac{f^{(k+1)}(z_0)}{(k+1)!}\right],$$

the last argument being independent of θ. Thus, the angle between the tangents of two smooth arcs intersecting at z_0 is magnified by the factor $k+1$. ∎

EXAMPLE 2.

The mapping $w = 1 - \cos z$ is entire and conformal except at the zeros $0, \pm\pi, \pm 2\pi, \ldots$ of the derivative $(1 - \cos z)' = \sin z$. To examine the behavior of this mapping at $z = 0$, note that $\sin z$ has a zero of order 1 at $z = 0$ and

$$\sin z = z - \frac{z^3}{3!} + \frac{z^5}{5!} - \ldots = z\left(1 - \frac{z^2}{3!} + \ldots\right).$$

The preceding theorem implies that $w = 1 - \cos z$ will magnify all angles at $z = 0$ by 2. Observe in Figure 5.4 that the right angle

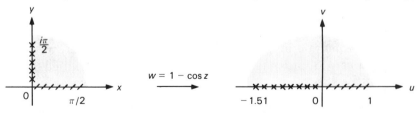

FIGURE 5.4.   Local behavior of $w = 1 - \cos z$ at $z = 0$

between the coordinate axes in the first quadrant is transformed into a 180° angle, since $1 - \cos x > 0$ for $0 < x < \pi/2$ and

$$1 - \cos iy = 1 - \cosh y < 0 \text{ for } 0 < y < \pi/2.$$

Turning to global properties, it is reasonable to ask when a given region G can be mapped conformally onto a given region H. The next result, whose proof is beyond the scope of this book, is the fundamental result in this direction.

RIEMANN MAPPING THEOREM

Let z_0 be a point in a simply connected region G ($\neq \mathbb{C}$). Then there is a unique analytic function $w = f(z)$ mapping G one-to-one onto the disk $|w| < 1$ such that $f(z_0) = 0$ and $f'(z_0) > 0$.

Now suppose G and H are two simply connected regions different from $\mathbb{C}$. The theorem yields the existence of analytic functions f, g mapping G, H onto the unit disk. Thus $g^{-1}f$ is a one-to-one mapping of G onto H. If we can show that g^{-1}, and thus the composition, is analytic, we then have a conformal mapping of G onto H, proving that *any two simply connected regions different from the plane can be mapped conformally onto each other.* Since g is conformal (it is one-to-one and analytic), so is g^{-1}. The inverse function theorem of calculus (see [B, p. 278]) shows that g^{-1} has continuous first partial derivatives, and these satisfy the Cauchy–Riemann equations, since

$$x_u = \frac{1}{u_x} = \frac{1}{v_y} = y_v \qquad y_u = \frac{1}{u_y} = \frac{-1}{v_x} = -x_v.$$

Hence, g^{-1} is analytic.

The conditions $f(z_0) = 0$ and $f'(z_0) > 0$ imply that the image of any smooth arc γ through z_0 will have the same slope at 0 as the arc γ does at z_0, since $\arg f'(z_0) = 0$. This is not a limitation; instead, it is a normalization indicating that there are three "degrees of freedom" in choosing the mapping: the x- and y-coordinates of the point z_0 and the change of direction of angles. Should we wish to change the direction by the angle θ, we need only multiply the mapping by the constant of unit length $e^{i\theta}$.

Although the Riemann mapping theorem asserts the existence of a function mapping a given region conformally onto a disk, it does not show how to find it. Construction of the function can be a matter of great difficulty. The rest of this chapter is devoted to the construction of *specific* conformal mappings and their application in fluid flow, heat flow, and electrostatics.

EXERCISES

In Exercises 1–4, indicate where the mappings are conformal.

1. $w = e^z$ 2. $w = \sin z$

3. $w = \dfrac{1}{z}$ 4. $w = z^2 - z$

Describe what each of the mappings in Exercises 5–8 does to the right angle between the coordinate axes in the first quadrant.

5. $w = z^3 \sin z$ 6. $w = z - \sin z$
7. $w = e^z - z$ 8. $w = e^{z^2} - \cos z$

9. Show that the image under the mapping $w = z^2$ of the circle $|z - r| = r, r > 0$, is the cardioid with polar equation

$$\rho = 2r^2 (1 + \cos \theta).$$

10. Show that the mapping $w = z + 1/z$ maps circles $|z| = r$ onto ellipses

$$\frac{x^2}{\left(r + \dfrac{1}{r}\right)^2} + \frac{y^2}{\left(r - \dfrac{1}{r}\right)^2} = 1.$$

11. If $w = f(z)$ is an analytic function, show that its Jacobian satisfies

$$\frac{\partial(u, v)}{\partial(x, y)} = |f'(z)|^2.$$

12. Let $f(z) = u(x, y) + iv(x, y)$ be conformal and have continuous first partial derivatives u_x, u_y, v_x, v_y in a region G. Show that $f(z)$ is analytic in G. (*Hint:* Show that the Cauchy–Riemann equations hold.)

13. Why does the Riemann mapping theorem state that it is not possible to map the simply connected complex plane $\mathcal{C}$ conformally onto the unit disk?

14. A region G is *convex* if the line segment between any two points in G lies in G. Prove the *Noshiro–Warshawski theorem:* Suppose $w = f(z)$ is analytic in a convex region G. If Re $f'(z) > 0$ for all z in G, then f is one-to-one in G. (*Hint:* Express $f(z_1) - f(z_2)$ as an integral.)

15. Use Exercise 14 to prove that if f is analytic at z_0 and $f'(z_0) \neq 0$, then there is a neighborhood of z_0 on which f is one-to-one.

5.2 LINEAR FRACTIONAL TRANSFORMATIONS

A simple, but important, type of conformal mapping is given by the expression

$$w = w(z) = \frac{az + b}{cz + d}, \qquad ad - bc \neq 0,$$

where a, b, c, d are complex constants. Such a mapping is called a **linear fractional transformation**. The condition $ad - bc \neq 0$ prevents its derivative

$$w' = \frac{ad - bc}{(cz + d)^2}$$

from vanishing, as otherwise the function is constant. We can solve for z, obtaining

$$z = \frac{-dw + b}{cw - a},$$

and using the convention that $w(-d/c) = \infty$ and $w(\infty) = a/c$, it follows that w maps $\mathfrak{M}$ (the Riemann sphere) one-to-one onto itself. Moreover, the mapping is conformal except at $z = \infty$, $-d/c$, because at these points $w' = 0$ or ∞.

EXAMPLE 1.

Consider the linear fractional transformation

$$w = \frac{z - 1}{z + 1}.$$

Find the images of the points i, $-2i$, and ∞. What points are mapped onto the points 0, 1, and ∞?

SOLUTION: We have

$$i \to \frac{i - 1}{i + 1} \cdot \frac{1 - i}{1 - i} = \frac{2i}{2} = i \quad \text{and} \quad -2i \to \frac{-2i - 1}{-2i + 1} = \frac{3 - 4i}{5}.$$

Writing

$$w = \frac{1 - (1/z)}{1 + (1/z)},$$

we see that ∞ is mapped onto 1. To find what point is mapped onto 0, we observe that $z = 1$ causes the numerator of the right side of the linear fractional transformation to vanish. Hence, 1 is mapped onto 0. Similarly, -1 causes the denominator to vanish, so -1 is mapped onto ∞.

A composition of two linear fractional transformations is again a linear fractional transformation, since

$$\frac{a\left(\dfrac{\alpha z + \beta}{\gamma z + \delta}\right) + b}{c\left(\dfrac{\alpha z + \beta}{\gamma z + \delta}\right) + d} = \frac{(a\alpha + b\gamma)z + (a\beta + b\delta)}{(c\alpha + d\gamma)z + (c\beta + d\delta)}$$

with

$$(a\alpha + b\gamma)(c\beta + d\delta) - (a\beta + b\delta)(c\alpha + d\gamma) = (ad - bc)(\alpha\delta - \beta\gamma) \neq 0.$$

Any linear fractional transformation is a composition of four special types of such transformations:

(i) *Translation:* $w = z + \alpha$, α complex,
(ii) *Rotation:* $w = e^{i\theta} z$, θ real,
(iii) *Magnification:* $w = kz$, $k > 0$,
(iv) *Inversion:* $w = 1/z$.

If $c \neq 0$, we can write

$$\frac{az + b}{cz + d} = \frac{bc - ad}{c^2 \left(z + \dfrac{d}{c}\right)} + \frac{a}{c},$$

showing that the transformation may be decomposed into a translation by d/c, followed by a rotation by $e^{2i \arg c}$, a magnification $|c|^2$, an inversion, a rotation, a magnification, and a translation.

If $c = 0$,

$$\frac{az + b}{d} = \frac{a}{d}\left(z + \frac{b}{a}\right),$$

proving that the decomposition consists of a translation, a rotation, and a magnification.

EXAMPLE 2.

Find a linear fractional transformation that maps the circle $|z - i| = 1$ onto the circle $|w + 1| = 2$.

SOLUTION: Consider the sequence of linear fractional transformations shown in Figure 5.5: a translation $\zeta = z - i$, followed

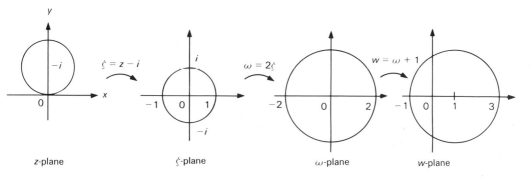

FIGURE 5.5.

by a magnification $\omega = 2\zeta$, and followed by another translation $w = \omega + 1$. The composition of these three mappings is

$$w = \omega + 1 = 2\zeta + 1 = 2(z - i) + 1$$

or

$$w = 2z + (1 - 2i),$$

and this linear fractional transformation maps $|z - i| = 1$ onto $|w + 1| = 2$.

The fundamental property of linear fractional transformations is that *they map circles onto circles in* $\mathfrak{M}$. A circle in $\mathfrak{M}$ corresponds to a circle or a straight line in $\mathfrak{C}$, as lines in the plane correspond to circles through ∞ on the Riemann sphere (see Section 1.3). Geometrically, it is clear that translations and rotations carry circles onto circles. Before considering the other two transformations, observe that the line $y = \tan\theta \cdot x + b$ can be written in the form

$$\operatorname{Re}(-ie^{-i\theta}z) = y\cos\theta - x\sin\theta = b\cos\theta, \qquad |\theta| < \frac{\pi}{2}.$$

The magnification $w = kz$, $k > 0$, maps (by substitution) circles $|z - z_0| = r$ onto circles $|w - kz_0| = kr$, and lines $\operatorname{Re}(\alpha z) = c$ onto lines $\operatorname{Re}(\alpha w) = ck$, with $|\alpha| = 1$, c real. Under inversion, the circle $|z - z_0| = r \, (> 0)$ satisfies

$$0 = |z - z_0|^2 - r^2 = |z|^2 + |z_0|^2 - 2\operatorname{Re}\bar{z}z_0 - r^2$$

$$= \frac{1}{|w|^2} + (|z_0|^2 - r^2) - \frac{2}{|w|^2}\operatorname{Re}z_0 w. \qquad (1)$$

If $|z_0| = r$, indicating the circle passes through the origin, we obtain the equation

$$0 = \frac{1 - 2\operatorname{Re}z_0\,w}{|w|^2}, \qquad (2)$$

yielding the line $\operatorname{Re}(z_0 w) = \frac{1}{2}$ through ∞. If $|z_0| \neq r$, the origin does not lie on the circle, so multiplying equation (1) by the nonzero quantity $|w|^2/(|z_0|^2 - r^2)$, we have

$$0 = \frac{1}{|z_0|^2 - r^2} + |w|^2 - \frac{2}{|z_0|^2 - r^2}\operatorname{Re}z_0\,w$$

$$= \left|w - \frac{\bar{z}_0}{|z_0|^2 - r^2}\right|^2 - \frac{r^2}{(|z_0|^2 - r^2)^2},$$

a circle. That lines map onto circles through the origin follows by reversing the steps leading to equation (2).

Since any linear fractional transformation is a composition of these special transformations, we have proved the following theorem.

THEOREM

Linear fractional transformations map circles onto circles in $\mathfrak{M}$.

EXAMPLE 3.

Map the intersection of the disks $|z - 1| < 1$ and $|z - i| < 1$ conformally onto the first quadrant.

SOLUTION: Since the circles $|z - 1| = 1$, $|z - i| = 1$ intersect at the points 0 and $1 + i$, we employ the mapping

$$\zeta = \frac{z}{z - (1 + i)}$$

that sends 0 to 0 and $1 + i$ to ∞. The circles map to lines perpendicular to each other at the origin, since the mapping is conformal and the tangent lines to the circles are perpendicular at $z = 0$. Since $\zeta(2) = 1 + i$ and $\zeta((1 + i)/2) = -1$, the lines have slope ± 1 in the ζ-plane and the overlap corresponds to the set $|\arg \zeta - \pi| < \pi/4$ (see Figure 5.6). The rotation

$$w = e^{-3\pi i/4} \zeta = \frac{e^{-3\pi i/4} z}{z - (1 + i)}$$

yields the desired mapping.

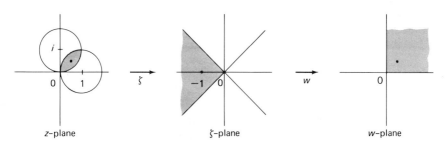

z-plane   ζ-plane   w-plane

FIGURE 5.6.

EXAMPLE 4.

Map the right half plane onto the unit disk $|z| < 1$ so that the point 1 is mapped onto the origin.

SOLUTION: Observe that the mapping of Example 1

$$w = \frac{z-1}{z+1} \qquad (3)$$

sends 1 to 0, 0 to -1, and ∞ to 1. Furthermore, $\pm i$ are mapped onto themselves (such points are called **fixed points** of the mapping (3)). Since three points determine a circle, it follows that the imaginary axis is mapped onto the unit circle (see Figure 5.7).

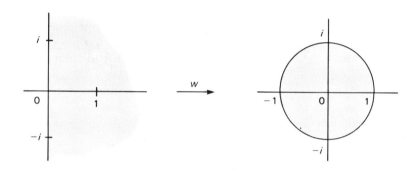

FIGURE 5.7.

EXAMPLE 5.

Find the number of roots of the equation

$$p(z) = 11z^4 - 10z^3 - 4z^2 + 10z + 9 = 0$$

lying in the right half plane.

SOLUTION: Since transformation (3) maps the right half plane onto the disk, by substituting the inverse mapping

$$z = \frac{1+w}{1-w},$$

we obtain the equivalent problem of finding the number of roots of the equation

$$g(w) = w^4 + 3w^3 + 8w^2 - 2w + 1 = 0$$

lying in $|w| < 1$. Letting $f(w) = 8w^2$, we find that

$$|g(w) - f(w)| \le 7 < 8 \, |w|^2 = |f(w)|$$

on $|w| = 1$, implying by Rouché's theorem that $p(z)$ has two roots in the right half plane.

EXERCISES

In Exercises 1-4, describe the image of the region indicated under the given mapping.

1. The disk $|z| < 1$; $\quad w = i\, \dfrac{z - 1}{z + 1}$

2. The quadrant $x > 0$, $\quad y > 0$; $\quad w = \dfrac{z - i}{z + i}$

3. The angular sector $|\arg z| < \dfrac{\pi}{4}$; $\quad w = \dfrac{z}{z - 1}$

4. The strip $0 < x < 1$; $\quad w = \dfrac{z}{z - 1}$

5. Find the number of roots of the equation
$$11z^4 - 20z^3 + 6z^2 + 20z - 1 = 0$$
   lying in the right half plane.

6. How many roots of the equation
$$17z^4 + 26z^3 + 56z^2 + 38z + 7 = 0$$
   lie in the first quadrant?

7. Using the exponential function, map the region lying inside $|z| = 2$ and outside $|z - 1| = 1$ onto the upper half plane.

8. Map the region $|z - 1| < 1$, $\operatorname{Im} z < 0$, onto the upper half plane.

9. Map the sector $|\arg z| < \pi/4$ onto the set $|\operatorname{Re} w| < 1$, $\operatorname{Im} w > 0$. (*Hint:* Use the sine function.)

5.3 THE SYMMETRY PRINCIPLE

Given three distinct points z_1, z_2, z_3 in $\mathfrak{M}$, there is a linear fractional transformation carrying them onto 0, 1, ∞, respectively. If none of the points is ∞, it is given by the *cross ratio*

$$w = \frac{(z - z_1)(z_2 - z_3)}{(z - z_3)(z_2 - z_1)},$$

and becomes

$$\frac{z_2 - z_3}{z - z_3}, \qquad \frac{z - z_1}{z - z_3}, \qquad \frac{z - z_1}{z_2 - z_1},$$

if z_1, z_2, or $z_3 = \infty$. If w^* is another linear fractional transformation with the same property, then the composition w^*w^{-1} keeps the points 0, 1, ∞ fixed. Thus, we have a linear fractional transformation

$$\zeta = \frac{az + b}{cz + d}, \qquad ad - bc \neq 0,$$

satisfying the equations

$$0 = \frac{b}{d}, \qquad 1 = \frac{a + b}{c + d}, \qquad \infty = \frac{a}{c},$$

But then $b = c = 0$ and $a = d$, implying $w^*w^{-1} = I$, the identity mapping, and hence $w^* = w$. Therefore, w is the only linear fractional transformation mapping the points z_1, z_2, z_3 onto 0, 1, ∞, respectively.

Since a circle is determined by three of its points, we can now easily determine a linear fractional transformation carrying a given circle in the z-plane onto a given circle in the w-plane. We select distinct points z_1, z_2, z_3 on the first circle and w_1, w_2, w_3 on the second circle. Then

$$\frac{(w - w_1)(w_2 - w_3)}{(w - w_3)(w_2 - w_1)} = \frac{(z - z_1)(z_2 - z_3)}{(z - z_3)(z_2 - z_1)}$$

maps z_1, z_2, z_3 onto w_1, w_2, w_3, as the right-hand side of the equation maps z_1, z_2, z_3 onto 0, 1, ∞, and the inverse of the left-hand side maps 0, 1, ∞ onto w_1, w_2, w_3.

EXAMPLE 1.

Find the linear fractional transformation mapping the points 1, i, -1 onto the points 2, 3, 4, respectively.

SOLUTION: Solve the equation

$$\frac{(w - 2)(3 - 4)}{(w - 4)(3 - 2)} = \frac{(z - 1)(i + 1)}{(z + 1)(i - 1)}$$

for w, obtaining

$$w = \frac{(2 - 4i)z + (2 + 4i)}{(1 - i)z + (1 + i)}.$$

The points w and $\bar{w}$ are symmetric with respect to the real axis. We can generalize this concept to any circle C in $\mathfrak{M}$.

DEFINITION

The points z and z^* are *symmetric with respect to the circle C*, in the extended z-plane, if there is a linear fractional transformation w mapping C onto the real axis and satisfying $\overline{w(z)} = w(z^*)$.

At first glance it might appear that symmetry with respect to C depends on the transformation w, but if w^* is a linear fractional transformation also mapping C onto the real axis, then $\zeta = w^* w^{-1}$ maps the real axis onto itself. Therefore, it is of the form

$$\frac{(\zeta - b_1)(b_2 - b_3)}{(\zeta - b_3)(b_2 - b_1)} = \frac{(w - a_1)(a_2 - a_3)}{(w - a_3)(a_2 - a_1)},$$

with $a_j, b_j, j = 1, 2, 3$ real. Solving for ζ, we obtain

$$\zeta = \frac{\alpha w + \beta}{\gamma w + \delta},$$

with $\alpha, \beta, \gamma, \delta$ real; thus,

$$\overline{w^*(z)} = \overline{\zeta(w(z))} = \zeta(\overline{w(z)}) = \zeta(w(z^*)) = w^*(z^*),$$

and symmetry is independent of the transformation employed. Moreover, *symmetry is preserved under linear fractional transformations*, for if z and z^* are symmetric with respect to the circle C and w^* is any such transformation, then $w^*(z)$ and $w^*(z^*)$ are symmetric with respect to $w^*(C)$ under the mapping ww^{*-1}. This fact is called the **symmetry principle**.

EXAMPLE 2.

Find the point that is symmetric to the point i with respect to the circle $|z + 1| = 1$.

SOLUTION: First we need to find the linear fractional trans-
formation of the circle $|z + 1| = 1$ onto the real axis. Selecting
$0, -1 + i, -2$ to map onto $0, 1, \infty$ yields the mapping

$$w = \frac{-iz}{z + 2},$$

which transforms i onto $w = (2 - i)/5$. Thus, $\bar{w} = (2 + i)/5$ and
the inverse mapping

$$z = \frac{-2w}{w + i}$$

sends w onto $\frac{1}{2}(-1 + i)$. Thus, $z* = -\frac{1}{2} + \frac{1}{2}i$.

EXAMPLE 3.

Find the number $a(< 1)$ for which there is a linear fractional
transformation mapping the right half plane with the disk
$|z - 1| \leqslant a$ deleted onto the ring $1 < |w| < 2$.

SOLUTION: We begin by finding points z_0, z_0^* symmetric with
respect to both the imaginary axis and the circle $|z - 1| = a$.
Rotating the imaginary axis by $90°$ we see that

$$iz_0^* = \overline{iz_0} = -i\bar{z}_0,$$

so that $z_0^* = -\bar{z}_0$. The linear fractional transformation mapping
$1 + a, 1 + ia, 1 - a$ onto $0, 1, \infty$ is given by

$$w = -i \left[\frac{z - (1 + a)}{z - (1 - a)} \right].$$

Thus

$$-i \overline{\left[\frac{z_0 - (1 + a)}{z_0 - (1 - a)} \right]} = -i \left[\frac{z_0^* - (1 + a)}{z_0^* - (1 - a)} \right] = -i \left[\frac{-\bar{z}_0 - (1 + a)}{-\bar{z}_0 - (1 - a)} \right]$$

so that

$$i \left[\frac{\bar{z}_0 - (1 + a)}{\bar{z}_0 - (1 - a)} \right] = -i \left[\frac{\bar{z}_0 + (1 + a)}{\bar{z}_0 + (1 - a)} \right],$$

from which we obtain $\bar{z}_0{}^2 = 1 - a^2 > 0$. Hence z_0 is real and we may assume $z_0 > 0$, since $z_0^* = -z_0$, implying that $z_0 = \sqrt{1 - a^2}$. By the symmetry principle, the mapping

$$\zeta = \frac{z - z_0}{z + z_0}$$

sends z_0 to 0 and $-z_0$ to ∞ and maps the imaginary axis and the circle $|z - 1| = a$ onto concentric circles centered at the origin. Since $\zeta(\infty) = 1$, and

$$\zeta(1 + a) = \frac{(1 + a) - z_0}{(1 + a) + z_0} \cdot \frac{(1 + a) - z_0}{(1 + a) - z_0} = \frac{1 - z_0}{a} < 1,$$

we magnify by $a/(1 - z_0) = 2$, yielding $a = \frac{4}{5}$ and

$$w = 2\zeta = 2 \left(\frac{z - \frac{3}{5}}{z + \frac{3}{5}} \right).$$

EXERCISES

Find the linear fractional transformation mapping the points -1, i, $1 + i$, respectively, onto the points given in Exercises 1-4.

1. $0, 1, \infty$ 2. $1, \infty, 0$ 3. $2, 3, 4$ 4. $0, 1, i$

5. Is $w = \bar{z}$ a linear fractional transformation?

6. Show that any four distinct points can be mapped by a linear fractional transformation to the points 1, -1, k, $-k$, where k depends on the original points.

Find the points symmetric to the point $3 + 4i$ with respect to the circles given in Exercises 7-9.

7. $|z| = 1$ 8. $|z - 1| = 1$ 9. $|z - i| = 2$

10. Map the unit circle onto itself such that the point α goes onto 0, and $\alpha/|\alpha|$ onto 1, $|\alpha| < 1$. (*Hint:* Map α^* to ∞.)

11. Find the linear fractional transformation that carries $|z| = 1$ onto $|z - 1| = 1$, the point -1 onto 0, and 0 onto $2i$.

12. Find a linear fractional transformation that carries $|z| = 1$ and $|z - 1| = 3$ onto concentric circles. What is the ratio of the radii?

13. Do Exercise 12 for $|z| = 1$ and Im $z = 2$.

14. Prove every conformal mapping of a disk onto another is given by a linear fractional transformation. Why does this imply the uniqueness of the function in the Riemann mapping theorem? (*Hint:* Use Schwarz's lemma, Exercise 3, Section 2.4.)

15. Suppose $z*$ is symmetric to the point z with respect to the circle $|z - a| = R$. Prove that $(z* - a)(\bar{z} - \bar{a}) = R^2$.

16. Use the result in Exercise 15 to verify that the construction shown in Figure 5.8 can be used to locate points symmetric with respect to a circle.

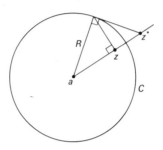

FIGURE 5.8. Geometric construction of symmetric points with respect to the circle $|z - a| = R$

5.4 COMPOSITIONS OF ELEMENTARY CONFORMAL MAPPINGS

In Section 5.1 we proved that the elementary functions e^z, $\cos z$, $\sin z$, $\log z$, and z^α are conformal in those regions of their domains of definition where their derivative is nonzero. In this section we illustrate how compositions of these functions with linear fractional transformations can be used to map certain regions conformally onto each other. The procedure we will use in analyzing the mapping is similar to that used in Examples 2 and 3 of Section 5.2.

EXAMPLE 1.

Find a conformal mapping of the infinite strip $|\operatorname{Im} z| < \pi/2$ onto the unit disk.

SOLUTION: If we apply the conformal mapping $\zeta = e^z$ to the infinite strip $|\operatorname{Im} z| < \pi/2$, we obtain the right half plane $\operatorname{Re} \zeta > 0$, because

$$e^{x \pm i\pi/2} = \pm e^x i \qquad \text{and} \qquad e^0 = 1.$$

The symmetry principle implies that the mapping

$$w = \frac{1 - \zeta}{1 + \zeta},$$

that sends $1, 0, -1$ onto $0, 1, \infty$ must map the imaginary axis onto the unit circle. Hence, the composition $w = w(\zeta(z))$,

$$w = \frac{1 - \zeta}{1 + \zeta} = \frac{1 - e^z}{1 + e^z} = -\tanh\left(\frac{z}{2}\right),$$

maps the strip $|\operatorname{Im} z| < \pi/2$ onto the unit disk (see Figure 5.9).

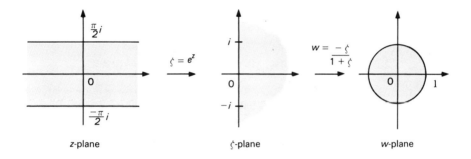

z-plane ζ-plane w-plane

FIGURE 5.9.

EXAMPLE 2.

Map the half-infinite strip $J = \{ z : |\operatorname{Re} z| < \pi/2, \operatorname{Im} z > 0 \}$ conformally onto the first quadrant.

SOLUTION: The function $\zeta = \sin z$ maps J onto the upper half plane because (see Section 1.8)

$$\sin\left(\pm \frac{\pi}{2} + iy\right) = \sin\left(\pm \frac{\pi}{2}\right)\cosh y = \pm \cosh y,$$

implying that $|\sin(\pm \pi/2 + iy)| = \cosh y \geq 1$ while $|\sin x| \leq 1$ for $|x| \leq \pi/2$. But the function $w = \sqrt{\zeta}$ maps the upper half plane onto the first quadrant, since the square root halves the argument. Thus, $w = \sqrt{\sin z}$ is the desired mapping.

EXAMPLE 3.

Map the right half plane with the line $\{z: x \geqslant 1, y = 0\}$ missing onto the upper half plane.

SOLUTION: First apply the function $\zeta = z^2$ to obtain the plane minus two slits shown in the ζ-plane in Figure 5.10. Then use

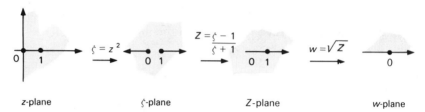

z-plane ζ-plane Z-plane w-plane

FIGURE 5.10. The mapping $w = \sqrt{\dfrac{z^2 - 1}{z^2}}$

the linear fractional transformation that takes $1, \infty, 0$ onto $0, 1, \infty$,

$$Z = \frac{\zeta - 1}{\zeta},$$

to map the two-slitted plane onto a plane with only one slit. Finally, $w = \sqrt{Z}$ produces the upper half plane, so the desired mapping is

$$w = \sqrt{\frac{\zeta - 1}{\zeta}} = \sqrt{\frac{z^2 - 1}{z^2}}.$$

EXERCISES

1. Find a conformal mapping of the unit disk onto the infinite strip $|\operatorname{Re} z| < 1$. (*Hint:* Consider the inverse mapping in Example 1.)
2. Show that

$$w = \frac{z}{\sqrt{z^2 + 1}}$$

conformally maps the upper half plane with the line $\{z: x = 0, y \geqslant 1\}$ deleted onto the upper half plane.

3. Find a mapping that carries the upper half plane onto the complement of the line segment from -1 to 1.

4. Find a conformal mapping of the square $\{z: |x| \leq 1, |y| \leq 1\}$ onto the annulus $1 < |w| < e^{2\pi}$ with the negative real axis deleted.

5. What is the image of the disk $|z - a| < a$ under the mapping $w = z^2$?

6. Show that the transformation

$$\left(\frac{w-1}{w+1}\right)^2 = i\left(\frac{z-1}{z+1}\right)$$

conformally maps the upper half of the unit disk onto the unit disk.

7. Describe the image of the hyperbola $x^2 - y^2 = \frac{1}{2}$ under the mapping $w = \sqrt{1 - z^2}$.

8. Map the complement of the line segment $\{z: y = 0, |x| \leq 1\}$ onto the unit disk.

*9. Map the outside of the parabola $y^2 = 4x$ onto the unit disk so that $0, -1$ are sent onto $1, 0$.

*10. Map the region to the left of the right-hand branch of the hyperbola $\text{Re}(z^2) = 1$ onto the unit disk. (*Hint:* Consider the mapping $w = z^3 - 3z$.)

*11. Prove that the mapping

$$w = \frac{Az^2 + Bz + C}{az^2 + bz + c}$$

can be decomposed into the three successive transformations,

$$\zeta = \frac{\alpha z + \beta}{\gamma z + \delta}, \qquad Z = \frac{1}{2}\left(\zeta + \frac{1}{\zeta}\right), \qquad w = \mu Z + \nu,$$

or into the two successive transformations

$$\zeta = \frac{\alpha z + \beta}{\gamma z + \delta} \qquad \text{and} \qquad w = \zeta^2 + \nu.$$

5.5 FLUID FLOW

In this section we will discuss a physical problem that can be analyzed with the help of analytic functions.

Since a complex function can be decomposed into two real

functions, the theory of analytic functions is very useful in solving problems involving two variables in two-dimensional space. However, since this book is not a treatise on mathematical physics, much of what follows is a heuristic outline of the physical theory.

A complete description of the motion of a fluid requires the knowledge of the velocity vector at all points of the fluid at any given time. Suppose the fluid is **incompressible** (that is, of constant density) and the flow is **steady** (independent of time) and **two-dimensional** (the same in all planes parallel to the xy-plane in three-dimensional space). Conditions of this type occur, for instance, when the fluid flows past a long cylindrical object whose axis is perpendicular to the direction of the flow. The **velocity vector** can then be given as a continuous complex-valued function of a complex variable $V = V(z)$ for all z in a region G. We also assume in this section that no **sources** or **sinks** (points at which fluid is being created or destroyed) lie in the region G.

The assumptions that the fluid is incompressible and there are no sources or sinks in G imply that a simply connected region in G always contains the same amount of fluid. Thus, the quantity of fluid per unit time passing by a length element ds on a pws Jordan curve γ, lying together with its inside in G, is $V_n\, ds$, where V_n is the component (a real number) of V in the outward normal direction to the curve (see Figure 5.11). Hence, the total outward **flow**

$$Q = \int_\gamma V_n\, ds = 0. \tag{1}$$

The line integral of the tangential component V_s of the velocity V around the curve γ,

$$\Gamma = \int_\gamma V_s\, ds, \tag{2}$$

is called the **circulation** of V along γ. If the circulation is not zero on some curve γ, then the tangential components having one sign dominate the ones having the other sign in the integral (2). Roughly, this means that the fluid rotates around γ. The flow is said to be **irrotational** if the circulation is zero along all closed curves in G. We assume the flow is irrotational so that $\Gamma = 0$.

Consider Figure 5.11, where the outward normal and tangential directions to the curve γ are indicated at a point z. Let $\alpha = \alpha(z)$ be

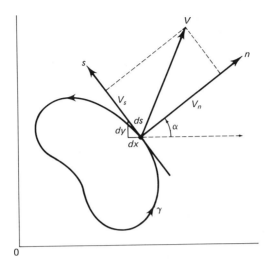

FIGURE 5.11. Components of the velocity vector

the angle between the (positive) horizontal direction and the out-ward normal to γ at z, and suppose the velocity vector V at z is as indicated.

Rotating the ns-coordinate system about the point z through an angle of $-\alpha$ yields the normal and tangential components of the velocity vector V

$$V_n = \mathrm{Re}(e^{-i\alpha} V), \qquad V_s = \mathrm{Im}(e^{-i\alpha} V).$$

In particular, we have

$$e^{-i\alpha} V = V_n + iV_s. \tag{3}$$

The length element ds is related (see Figure 5.11) to the element dx and dy by the identities

$$dx = \cos\left(\frac{\pi}{2} + \alpha\right) ds, \qquad dy = \sin\left(\frac{\pi}{2} + \alpha\right) ds,$$

implying that

$$dz = dx + i\, dy = e^{i(\pi/2+\alpha)}\, ds = ie^{i\alpha}\, ds. \tag{4}$$

Now, if γ is any pws Jordan curve contained together with its inside in G, we have, by equations (1)-(4),

$$\int_\gamma \overline{V(z)} \, dz = i \int_\gamma \overline{(e^{-i\alpha} V)} \, ds$$

$$= i \int_\gamma \overline{(V_n + i V_s)} \, ds$$

$$= \int_\gamma (V_s + i V_n) \, ds = 0,$$

implying that $\overline{V(z)}$ is analytic by Morera's theorem. If G is simply connected, the antiderivative of $\overline{V(z)}$ is an analytic function $w(z) = u(z) + iv(z)$, called the **complex potential** of the flow; u is known as the **potential function** and v as the **stream function**.* Individual particles of the fluid move along curves whose direction at each point coincides with that of the velocity vector. Such curves are called **streamlines** and are characterized by the equation $v(z) = \text{constant}$, since the tangent to such a curve has slope

$$\frac{dy}{dx} = -\frac{v_x}{v_y} = -\frac{v_x}{u_x} = -\tan \arg w' = \tan \arg V$$

by the Cauchy-Riemann equations, since $V = \overline{w'}$.

The curves $u(z) = \text{constant}$, are called **equipotential lines** and are normal to the streamlines, since

$$\frac{dy}{dx} = -\frac{u_x}{u_y} = \frac{u_x}{v_x} = \frac{-1}{\tan \arg V}.$$

Points at which $V(z) = 0$, and consequently $w'(z) = 0$, are known as **stagnation points** of the flow.

EXAMPLE 1.

Suppose we have a uniform flow of velocity A (> 0) in the positive x-direction in the upper half plane. This approximates fluid flow in extremely wide channels (see Figure 5.12).

Since $V(z) = A$, it follows that $w'(z) = A$, so the complex

---

*We avoid using the symbol ϕ for the potential function so as to emphasize the analogy later between fluid flow and heat flow and to preserve our mapping notation.

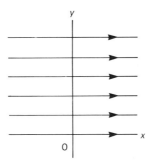

FIGURE 5.12.

potential is $w(z) = Az + c$, where $c = c_1 + ic_2$ is a complex constant. Thus, $u(z) = Ax + c_1$ and $v(z) = Ay + c_2$, so the equipotential lines are vertical and the streamlines are horizontal (neglecting the effect of viscosity on the real axis). Setting $c = 0$, the streamline $v = 0$ coincides with the real axis.

Suppose the function $\zeta = f(z)$ maps the region G conformally onto the upper half plane $f(G)$. If the complex potential $w(\zeta)$ of the fluid flow in $f(G)$ is known, then the complex potential of the flow in G is given by the analytic function $w(f(z))$. For example, if the complex potential in $f(G)$ is that given in Example 1, then the streamlines in G are those curves that are mapped by the composite function $w \circ f$ onto the straight lines $v = $ constant in the upper half plane. The determination of such composite functions is the fundamental procedure in the solution of problems in fluid dynamics.

EXAMPLE 2.

If we are interested in finding the streamlines along a right angle in a wide channel, we can approximate this situation by studying the flow in the first quadrant. The mapping $\zeta = z^2$ maps the quadrant onto the upper half plane. Thus, if we know the complex potential $w = w(\zeta)$ of the flow in the upper half plane, then $w = w(z^2)$ is the complex potential of the flow in the first quadrant. For example, if we assume that the flow is uniform and of velocity A (> 0) in the upper half of the ζ-plane (and

$c = 0$), then the complex potential in the ζ-plane is $w = A\zeta$. Thus, the complex potential in the first quadrant satisfies $w = Az^2$, the streamlines are given by the hyperbolas $2Axy =$ constant, and the velocity vector is $V(z) = 2A\bar{z}$. The origin is a stagnation point. (See Figure 5.13.)

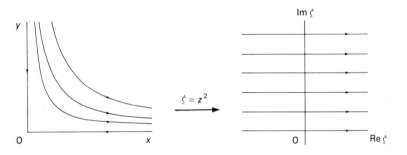

FIGURE 5.13.   Streamlines along a corner

EXAMPLE 3.

The mapping $\zeta = z + a^2/z$ has important applications in two-dimensional fluid flow. Rewriting the transformation in the form

$$\frac{(z \pm a)^2}{z} = \zeta \pm 2a,$$

we see that the image ζ of each point on the circle $|z| = b$, $b > a$, satisfies

$$|\zeta - 2a| + |\zeta + 2a| = \frac{|z - a|^2 + |z + a|^2}{b}.$$

By the law of cosines (see Figure 5.14), it follows that
$$|z - a|^2 = a^2 + b^2 - 2ab \cos \theta,$$
$$|z + a|^2 = a^2 + b^2 - 2ab \cos (\pi - \theta),$$
where $\theta = \arg z$. Hence,
$$|\zeta - 2a| + |\zeta + 2a| = \frac{2(a^2 + b^2)}{b},$$

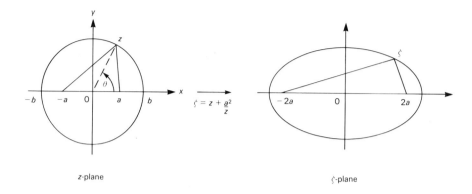

FIGURE 5.14.   The mapping $\zeta = z + \dfrac{a^2}{z}$

and since the right side is constant for fixed b, it follows that the image of the circle $|z| = b$ is an ellipse with foci at $\pm 2a$. Thus, concentric circles of radius $b > a$ centered at the origin of the z-plane map onto confocal ellipses in the ζ-plane. Moreover, the circle $|z| = a$ maps onto the straight line segment joining $-2a$ to $2a$ in the ζ-plane, since $z = ae^{i\theta}$ implies that

$$\zeta = z + \frac{a^2}{z} = ae^{i\theta} + ae^{-i\theta} = 2a \cos \theta, \qquad 0 \leqslant \theta < 2\pi.$$

Noting that $(z + a^2/z)' = 1 - (a/z)^2$, it follows that $\zeta = z + a^2/z$ conformally maps the exterior of the circle $|z| = a$ onto the exterior of the line segment joining $-2a$ to $2a$. Hence, assuming that the motion of the fluid flow in the ζ-plane is uniform and of velocity A (> 0) parallel to the real axis, we obtain the complex potential

$$w = A\left(z + \frac{a^2}{z}\right)$$

for the flow past a circular cylinder of radius a (see Figure 5.15). The stream function is obtained by setting $z = re^{i\theta}$, obtaining

$$v = A\left(r - \frac{a^2}{r}\right) \sin \theta;$$

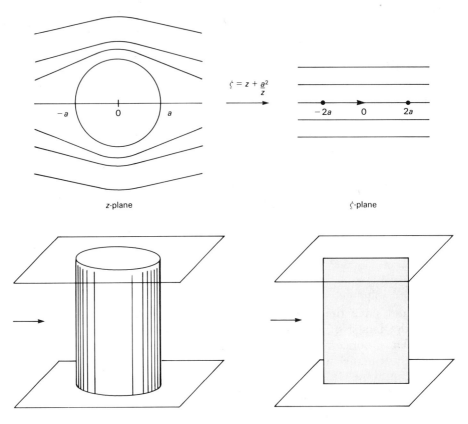

FIGURE 5.15. Flow past a cylinder

the streamline $v = 0$ consists of the circle $|z| = a$ and the real axis with $|x| \geqslant a$. The velocity of the fluid flow is

$$V = \overline{w'} = A\left[1 - \left(\frac{a}{z}\right)^2\right],$$

with stagnation points at $z = \pm a$. Note that $V \to A$ as $|z| \to \infty$, implying that although the stream is disturbed by the presence of the cylinder, this disturbance is negligible at great distances from the cylinder and the flow for large $|z|$ is essentially uniform and of velocity A parallel to the x-axis.

EXERCISES

Find the streamlines for an incompressible fluid flow without sources or sinks in each of the regions given in Exercises 1–4. Assume that the flow has velocity $A > 0$ as $z \to \infty$. Are there any stagnation points?

1. $0 < \arg z < \dfrac{3\pi}{4}$

2. $0 < \arg z < \dfrac{5\pi}{4}$

3. $0 < \arg z < \dfrac{\pi}{4}$

4. $\dfrac{-\pi}{4} < \arg z < \dfrac{5\pi}{4}$

For Exercises 5–8, calculate the **speed** $|V|$ at $z = 0$, 1, and i for the flow in the upper half plane given by each of the complex potentials. Are there any stagnation points?

5. $w = z + z^3$

6. $w = z + 2iz^2$

7. $w = 3z - iz^2$

8. $w = \sin z$

9. Find the equations of the streamlines for the complex potentials given in Exercises 5–8.

10. Find the equations of the equipotential lines for the complex potentials given in Exercises 5–8.

11. Suppose the complex potential for a flow in the z-plane is given by $w = \cosh^{-1}(z/a)$. Describe the streamlines for this flow. (*Hint:* Consider $z = a \cosh w$.)

12. Use Exercise 11 to describe a flow through an aperture bounded by the hyperbola $x^2 - y^2 = 1$.

13. Use Exercise 11 to describe a flow through an aperture of breadth $2a$ in a flat plate. Is this flow physically realizable? (*Hint:* Find the speed at the edges.)

14. Use the mapping $\zeta = \sin z$ to determine the streamlines of an incompressible fluid flow in the region shown in Figure 5.16. Assume that the flow in the ζ-plane is uniform and parallel to the imaginary axis. Is this flow *physically* realizable?

15. **Bernoulli's theorem** states that in a steady motion of an incompressible fluid, the quantity

$$\frac{p}{\rho} + \tfrac{1}{2}\,|V|^2$$

has a constant value at each point of any streamline of the flow, where p, ρ, and $|V|$ are the pressure, density, and speed, respectively. Show in Example 3, that if $A^2 > \tfrac{2}{3}p(\infty)/\rho$, then there will be points at which the pressure is negative. At these points a

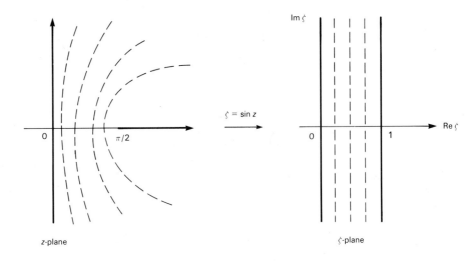

FIGURE 5.16. Vertical uniform flow in the ζ-plane

vacuum will form, causing the phenomenon of **cavitation.** Cavitation occurs, for example, near the tips of a rapidly moving propeller.

5.6 THE SCHWARZ-CHRISTOFFEL FORMULA

In the second theorem in Section 5.1 (page 187), we proved that if $f(z)$ is analytic in a region containing the point z_0 at which $f'(z)$ has a zero of order k, then all angles at z_0 are magnified by the factor $k + 1$. Consider, instead, an analytic function $f(z)$ with derivative

$$f'(z) = A(z - z_0)^{\alpha},$$

where $-1 < \alpha < 1$, $\alpha \neq 0$. The point z_0 is a branch point for f'; without loss of generality, assume that the branch cut is vertically downward from z_0. Let z be a point on the line through z_0 parallel to the real axis. Since

$$\arg f'(z) = \arg A + \alpha \arg (z - z_0),$$

the *change in direction* will be $\arg A$ if z is to the right of z_0, and $\arg A + \pi\alpha$ if z is to the left of z_0 (see Figure 5.17). Thus, the angle π at z_0 is magnified by the factor $\alpha + 1$ at $f(z_0)$.

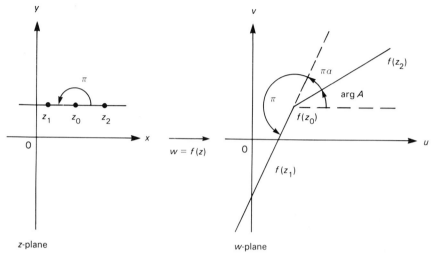

FIGURE 5.17. Effect of $f'(z) = A(z - z_0)^\alpha$

We can make use of this observation to construct a function $f(z)$ that maps the real axis onto a polygonal path. Let $x_1 < x_2 < \ldots < x_n$ be real numbers, and suppose the function $f(z)$ has the derivative

$$f'(z) = A(z - x_1)^{\alpha_1} (z - x_2)^{\alpha_2} \ldots (z - x_n)^{\alpha_n},$$

where $A \neq 0$ is a complex constant and $-1 < \alpha_k < 1$, for $k = 1, 2, \ldots, n$. Since

$$\arg f'(z) = \arg A + \alpha_1 \arg(z - x_1) + \alpha_2 \arg(z - x_2) + \ldots + \alpha_n \arg(z - x_n),$$

the images of the intervals $(-\infty, x_1)$, (x_1, x_2), . . ., (x_n, ∞) are straight line segments whose angles measured from the horizontal are given as follows.

Interval	Angle
(x_n, ∞)	$\arg A$
(x_{n-1}, x_n)	$\arg A + \pi\alpha_n$
.	.
.	.
.	.
(x_1, x_2)	$\arg A + \pi(\alpha_2 + \ldots + \alpha_n)$
$(-\infty, x_1)$	$\arg A + \pi(\alpha_1 + \ldots + \alpha_n)$

Thus, the function $w = f(z)$ maps the real axis onto a polygonal path, as shown in Figure 5.18.

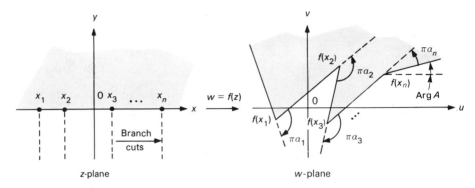

FIGURE 5.18.

By construction, $f(z)$ is analytic on the complex plane except on the branch cuts downward from each of the points $x_1, x_2, \ldots, x_n$. Thus, if z is any point in the upper half plane, we can define the conformal mapping $f(z)$ by

$$f(z) = \int_\gamma f'(\zeta) \, d\zeta$$

where γ is the straight line segment from $x_0 (\neq x_k, k = 1, 2, \ldots, n)$ to z. Any function $f(z)$ of this form is known as a **Schwarz–Christoffel transformation**. This discussion provides the converse to the following theorem, whose proof is beyond the scope of this book.

SCHWARZ–CHRISTOFFEL THEOREM

All functions mapping the upper half plane conformally onto a polygon in $\mathfrak{M}$ with exterior angles α_k, $k = 1, \ldots, n$, are of the form

$$f(z) = A + B \int_0^z (z - x_1)^{\alpha_1} (z - x_2)^{\alpha_2} \cdots \cdot (z - x_n)^{\alpha_n} \, dz,$$

where the points $x_1 < x_2 < \ldots < x_n$ are real and A, B are complex constants.

The function given by the integral equation in this theorem is called the **Schwarz–Christoffel formula** for the given polygon.

The *exterior angle* at the vertex $w_k = f(z_k)$ of the polygon is the angle $\pi \alpha_k$ required to bring the direction of the vector from w_k to w_{k+1} into coincidence with the direction of the vector from w_{k-1} to w_k. Glancing at Figure 5.19, we see that the exterior angle is measured by rotating from the next side of the polygon to the dashed line continuation of the previous side of the polygon. Note

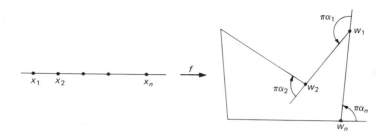

FIGURE 5.19.   Exterior angle of a polygon

that $-1 < \alpha_k < 1$, with $\alpha_k > 0$ when the rotation is counterclockwise and $\alpha_k < 0$ when the rotation is clockwise. Furthermore, if we make a complete clockwise circuit around the perimeter of the polygon, we will turn around one complete revolution, implying that

$$\pi \sum_{k=1}^{n} a_k = -2\pi \qquad \text{or} \qquad \sum_{k=1}^{n} a_k = -2.$$

The constants A and B control by translation, magnification, and rotation the location, scale, and orientation of the polygon in the w-plane, and the points x_k map to the vertices w_k of the polygon.

A linear fractional transformation of the upper half plane onto itself allows us to map three of the points x_k onto three prescribed positions on the real axis. Thus, *we are free to select the location of three of the points x_k*. Depending on the polygon, an appropriate choice of the locations of these three points can be extremely useful in obtaining a closed form solution for the integral. The location of the remaining points x_k depends on the shape of the polygon and can be very difficult to establish except in cases where the polygon is highly symmetric.

It is usually advantageous to choose $x_n = \infty$; with this choice, the term involving x_n in the Schwarz–Christoffel formula is omitted.

The following examples illustrate use of the Schwarz–Christoffel formula.

EXAMPLE 1.

Map the upper half plane conformally onto the strip $|x| < 1$, $y > 0$.

SOLUTION: Select $-1, 1, \infty$ as the points that are to be mapped onto the vertices $-1, 1, \infty$ of the strip in $\mathfrak{M}$ (see Figure 5.20). By the Schwarz–Christoffel formula,

$$w = A + B \int_0^z (z + 1)^{-1/2} (z - 1)^{-1/2} \, dx = A + B \int_0^z \frac{dz}{\sqrt{z^2 - 1}}$$

$$= A + \frac{B}{i} \int_0^z \frac{dz}{\sqrt{1 - z^2}} = A + \frac{B}{i} \sin^{-1} z.$$

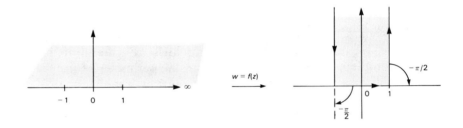

FIGURE 5.20.

Since $w(\pm 1) = \pm 1$, we have

$$A - \frac{iB\pi}{2} = 1,$$

$$A + \frac{iB\pi}{2} = -1.$$

Hence, $A = 0$ and $B = 2i/\pi$, implying that $w = (2/\pi) \sin^{-1} z$.

EXAMPLE 2.

Map the upper half plane onto the shaded region shown in the w-plane in Figure 5.21.

SOLUTION: It is easy to obtain this transformation without using the Schwarz–Christoffel formula. The composition of the mappings indicated in Figure 5.21 imply that

$$w = \sqrt{W} = \sqrt{Z - 1} = \sqrt{z^2 - 1}.$$

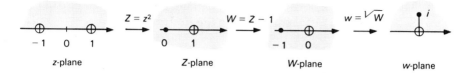

FIGURE 5.21. $w = \sqrt{z^2 - 1}$

To check that the same transformation is obtained with the Schwarz–Christoffel formula, note that we have exterior angles $-\pi/2, \pi, -\pi/2$ at $-1, 0, 1$. Hence,

$$w = A + B \int_0^z \frac{z\, dz}{\sqrt{z^2 - 1}}$$

$$= A + B\sqrt{z^2 - 1}\ \Big|_0^z = (A - Bi) + B\sqrt{z^2 - 1}.$$

Now $0 = w(1) = A - Bi$ and $i = w(0) = A$, so $B = 1$ and $w = \sqrt{z^2 - 1}$.

EXAMPLE 3.

Map the upper half plane onto the infinite strip $0 < v < \pi$.

SOLUTION: Consider the shaded triangle in Figure 5.22. Assume that the points $\infty, -1, 0$ in the z-plane are mapped onto the

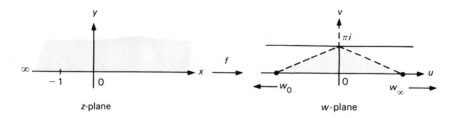

FIGURE 5.22.

points w_∞, πi, w_0 in the w-plane. If we let w_0 approach ∞ through the negative reals while w_∞ approaches ∞ through the positive reals, we obtain the infinite strip $0 < v < \pi$ in the limit.

The exterior angles at w_∞, πi, w_0 tend to $-\pi$, 0, $-\pi$, so that a Schwarz–Christoffel formula for the limiting case is

$$w = A + B \int_1^z \frac{dz}{z} = A + B \log z.$$

We choose $z = 1$ as the lower limit of integration, since $\log 0 = \infty$. Now

$$\pi i = w(-1) = A + B \log(-1) = A + B\pi i,$$

so letting $A = 0$ and $B = 1$ yields the desired transformation $w = \text{Log } z$.

EXAMPLE 4.

Map the upper half plane onto the interior of a triangle with exterior angles $-\pi\alpha$, $-\pi\beta$, $-\pi\gamma$, where $\alpha + \beta + \gamma = 2$.

SOLUTION: Selecting 0, 1, ∞ as the points we wish to map onto the vertices with exterior angles $-\pi\alpha$, $-\pi\beta$, $-\pi\gamma$, respectively, we find the function has the form

$$f(z) = A + B \int_0^z \frac{dz}{z^\alpha (z - 1)^\beta}.$$

Since A and B affect merely the position and size of the triangle, in order to find the simplest formula for the location of the vertices, we set $A = 0$, $B = e^{i\pi\beta}$, and

$$f(z) = \int_0^z \frac{dz}{z^\alpha (1 - z)^\beta}.$$

Then $f(0) = 0$ and

$$f(1) = \int_0^1 \frac{dx}{x^\alpha (1 - x)^\beta} = \frac{\Gamma(1 - \alpha)\Gamma(1 - \beta)}{\Gamma(\gamma)}.$$

The gamma function satisfies the identity $\Gamma(x)\Gamma(1 - x) = \pi (\sin \pi x)^{-1}$, so the length of this side is

$$c = \frac{1}{\pi} \sin \pi\gamma \, \Gamma(1 - \alpha)\Gamma(1 - \beta)\Gamma(1 - \gamma).$$

Using the law of sines, we find the lengths of the other two sides are

$$a = \frac{1}{\pi} \sin \pi\alpha \; \Gamma(1 - \alpha)\Gamma(1 - \beta)\Gamma(1 - \gamma),$$

$$b = \frac{1}{\pi} \sin \pi\beta \; \Gamma(1 - \alpha)\Gamma(1 - \beta)\Gamma(1 - \gamma)$$

(see Figure 5.23).

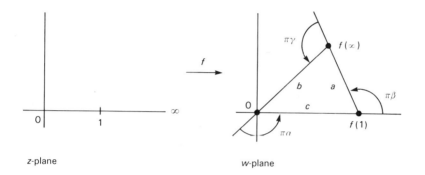

z-plane   w-plane

FIGURE 5.23.

EXAMPLE 5.

Map the upper half plane onto the interior of a rectangle.

SOLUTION: By Exercise 6, Section 5.3, any four points on the real axis may be mapped by a linear fractional transformation onto the points ± 1, $\pm k$, $k > 1$ (invert if necessary). Therefore, such a mapping is given by

$$f(z) = \int_0^z \frac{dz}{\sqrt{(1 - z^2)(k^2 - z^2)}}$$

(see Figure 5.24). From the formula it is clear that the vertices of the rectangle are symmetric with respect to the imaginary axis with

$$b = \frac{1}{k} \int_0^1 \frac{dx}{\sqrt{(1 - x^2)(1 - k^{-2}x^2)}} = \frac{1}{k} F\left(\frac{\pi}{2}, \frac{1}{k}\right)$$

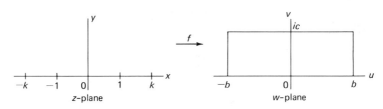

FIGURE 5.24.

(an elliptic integral of the first kind) and

$$ic = \int_1^k \frac{dx}{\sqrt{(1-x^2)(k^2-x^2)}} = \frac{i}{k}\int_1^k \frac{dx}{\sqrt{(x^2-1)(1-k^{-2}x^2)}}.$$

EXERCISES

1. Find the streamlines of an incompressible fluid flow of velocity $A \ (> 0)$ at ∞ for the shaded region in the w-plane of Figure 5.21.

2. Obtain a mapping of the upper half plane onto the shaded region in the w-plane of Figure 5.21 that sends the points $0, 1, \infty$ onto $0, i, 0$.

3. Show that the function

$$f(z) = \int_0^z \frac{dz}{\sqrt{z(z^2-1)}}$$

maps the upper half plane onto a square with sides of length

$$a = \frac{1}{2}\int_0^1 \frac{dt}{t^{3/4}\sqrt{1-t}} = \frac{\Gamma(\frac{1}{4})\Gamma(\frac{1}{2})}{2\Gamma(\frac{3}{4})} = \frac{\Gamma^2(\frac{1}{4})}{2\sqrt{2\pi}}.$$

4. Using the Schwarz–Christoffel transformation, find the mapping carrying the upper half plane onto the infinite strip $|y| < 1$.

5. Map the upper half plane conformally onto the exterior of the half strip $|x| < \pi/2, \ y > 0$.

6. Map the upper half plane conformally onto each of the regions indicated in Figure 5.25 with $0, 1, \infty \to \alpha, \beta, \gamma$, respectively.

7. Map the upper half plane conformally onto the region pictured in Figure 5.26 and show that the length (with $|B| = 1$) of the segment from α to β equals $12\pi\sqrt{2\pi}/5\Gamma^2\left(\frac{1}{4}\right)$.

*8. Map the upper half plane conformally onto the region in Figure 5.27 with $0, x, 1, \infty \to \alpha, \beta, \gamma, \delta$, respectively, and show that $x = k^2, \ 0 < k < 1$. (*Hint:* Let $s^2 = (z-1)/(z-x)$.)

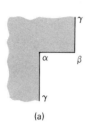

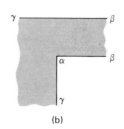

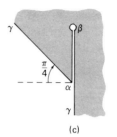

(a)   (b)   (c)

FIGURE 5.25.

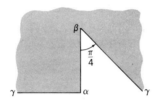

FIGURE 5.26.

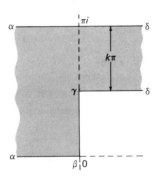

FIGURE 5.27.

*9. Show that

$$f(z) = A \left[\text{Log} \frac{\sqrt{z} + \sqrt{z - a}}{\sqrt{z} - \sqrt{z - a}} - i\sqrt{a - 1} \, \text{Log} \frac{i\sqrt{z(a - 1)} + \sqrt{z - a}}{i\sqrt{z(a - 1)} - \sqrt{z - a}} \right] + B$$

maps the upper half plane conformally onto the region in Figure 5.28 with $0, 1, a, \infty \rightarrow \alpha, \beta, \gamma, \delta$ and $a = 1 + (h^2/H^2)$.

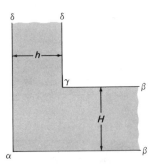

FIGURE 5.28.

5.7 **PHYSICAL APPLICATIONS IN HEAT FLOW AND ELECTROSTATICS (Optional)**

In this section we develop the basic theory for two-dimensional steady-state heat flows and electrostatic fields. It is important to note the similarities between these flows and fluid flow. In the next section we present a brief development of the theory of wakes in a fluid.

HEAT FLOW

We can approach the study of heat conduction in a solid homogenous body in much the same way that we approached fluid flow, if the solid is such that the flow is two-dimensional and the flow of heat is in steady state. Assume that no heat sources or sinks are present in the simply connected region G. Since two points may have different temperatures, there is a flow of heat, by conduction, from the hotter to the cooler parts. The **heat flow vector** $Q = Q(z)$ may be written as a continuous complex-valued function. The heat flowing out of the inside of the pws closed curve γ contained in G must satisfy

$$\int_\gamma Q_n \, ds = 0,$$

as otherwise the temperature inside will change. Since heat flows from hotter to cooler parts, it is irrotational, so we have

$$\int_\gamma Q_s \, ds = 0;$$

thus, by Morera's theorem (as in Section 5.5) $\overline{Q}$ is analytic in G. Then

$$Q(z) = -k\overline{w'(z)},$$

where k is the **coefficient of thermal conductivity** and $w = u + iv$ is the **complex potential** of the thermal field. By the Cauchy–Riemann equations, $\overline{w'(z)} = u_x + iu_y = \text{grad } u(z)$, the **gradient** of u, so $Q = -k \text{ grad } u$ (Fourier's law), implying that the heat flow is normal to the curves $u(z) = $ constant. Points on these lines must then have equal temperature; thus the curves $u(z) = $ constant are the **isotherms** and $u(z)$ is the **temperature**. The curves $v(z) = $ constant are called the **streamlines** and are orthogonal to the isotherms.

   A frequent problem in steady-state heat flow is to construct the isotherms in a region G with given boundary temperatures.

EXAMPLE 1.

   Find the isotherms of the plate G indicated in Figure 5.29 insulated along the segment $0 < x < 1$, $y = 0$, with temperature $0°$ along $z = y \geqslant 0$, and $1°$ along $z = x \geqslant 1$.

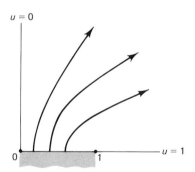

FIGURE 5.29.

SOLUTION: The function $w = (2/\pi) \sin^{-1} z$ maps G onto the strip $0 < u < 1$, $v > 0$, and is thus the complex potential. Then

$$z = \sin \frac{\pi}{2} w = \sin \frac{\pi}{2} u \, \cosh \frac{\pi}{2} v + i \cos \frac{\pi}{2} u \, \sinh \frac{\pi}{2} v.$$

Hence, we find

$$\frac{x^2}{\sin^2 \dfrac{\pi}{2} u} - \frac{y^2}{\cos^2 \dfrac{\pi}{2} u} = 1,$$

implying that the isotherms are hyperbolas.

EXAMPLE 2.

Find the isotherms of a plate G shaped as in Figure 5.25a, insulated along the segment joining $\alpha = 0$ to $\beta = 1$, with temperature $-1°$ on the ray from α to γ and $1°$ on the ray from β to γ.

SOLUTION: Since the exterior angles at 0 and 1 are $-\pi/2$ and $\pi/2$, respectively, the Schwarz-Christoffel transformation

$$z = 1 + \frac{i}{\pi} \int_1^{\zeta} \sqrt{\frac{\zeta + 1}{\zeta - 1}}\, d\zeta = 1 + \frac{i}{\pi} [\sqrt{\zeta^2 - 1} + \cosh^{-1} \zeta]$$

maps the upper half plane onto G with $-1, 1, \infty \to \alpha, \beta, \gamma$. But $\zeta = \sin(\pi w/2)$ maps the strip $|u| < 1, v > 0$, onto the upper half plane, so

$$z = 1 + \pi^{-1} \left[i \cosh^{-1} \left(\sin \frac{\pi w}{2} \right) - \cos \frac{\pi w}{2} \right]$$

maps the strip above onto G. Hence its inverse $w = w(z)$ is the complex potential. As in Example 1, the isotherms will be the images under $z = z(w)$ of the vertical lines $u = $ constant. Simplifying the first term in the parentheses, we find

$$z = \frac{w + 1}{2} - \frac{1}{\pi} \cos \frac{\pi w}{2},$$

from which the isotherms may easily be graphed.

ELECTROSTATICS

Consider a *plane* electrostatic field $E(z)$ arising from the attraction or repulsion of an arbitrary system of charges (sources and sinks) in the plane. In a simply connected region G complementary to these charges, the inside of a pws closed curve γ in G has no charge, so

$$\int_\gamma E_n \, ds = 0,$$

by Gauss's law. The **circulation** of the field is the work done by the field when a positive unit charge is taken completely around the curve γ. As no expenditure of energy is required to maintain an electrostatic field, we have

$$\int_\gamma E_s \, ds = 0.$$

Then E is said to be a **potential field**, $\overline{E}/i$ is analytic, and its anti-derivative $iw = -v + iu$ is called the **complex potential** of the field; $-v$ is the **force function** and u the **potential function**. By the Cauchy-Riemann equations, we find

$$E = -\overline{w'(z)} = -(u_x + iu_y) = -\text{grad } u.$$

The curves $v(z) = $ constant are the **lines of force**, and $u(z) = $ constant are **equipotential lines**.

We can gather all our analogies between fluid flow, heat flow, and electrostatics and present them in tabular form, as is done in Table 5.1. Similar analogies to fluid flow can be made with steady-state diffusion, static magnetic and gravitational fields, and hydromechanics.

Frequently, we wish to find the equipotential lines of a plane electrostatic field bounded by contours on which the potential is a given constant (each contour is a **conductor**).

Table 5.1   Analogies of Fluid Flow, Heat Flow, and Electrostatic Fields

	Fluid Flow	Heat Flow	Electrostatic Field
Complex potential	$w(z) = u + iv$	$w(z) = u + iv$	$iw(z) = -v + iu$
Vector field	$V = \overline{w'(z)}$	$Q = -k\overline{w'(z)}$	$E = -\overline{w'(z)}$
	$= \text{grad } u$	$= -k \text{ grad } u$	$= -\text{grad } u$
u	Potential function	Temperature	Potential function
$u(z) = $ constant	Equipotential lines	Isotherms	Equipotential lines
v	Stream function	Stream function	$-v$ is the force function
$v(z) = $ constant	Streamlines	Streamlines	Lines of force

EXAMPLE 3.

A condenser consists of two plates in the form of coplanar half planes with parallel edges separated by the distance $2a$ and with potential difference $2u_0$. Any cross section normal to the planes yields a plane field with two cuts (see Figure 5.30). Find the equipotential lines for this field.

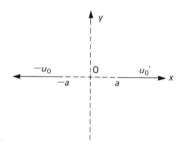

FIGURE 5.30.

SOLUTION: The function

$$w = \frac{2u_0}{\pi} \sin^{-1} \left(\frac{z}{a} \right)$$

maps the domain onto the strip $|u| < u_0$. Thus, the equipotential lines are the hyperbolas

$$\frac{x^2}{a^2 \sin^2 \dfrac{\pi u}{2u_0}} - \frac{y^2}{a^2 \cos^2 \dfrac{\pi u}{2u_0}} = 1.$$

EXAMPLE 4.

A parallel-plate capacitor consists of two parallel half planes whose edges are separated by the distance 2π and with potential difference $2u_0$. Any cross section normal to the planes yields a two-dimensional field with two cuts as shown in Figure 5.31. Find the equipotential lines for this field.

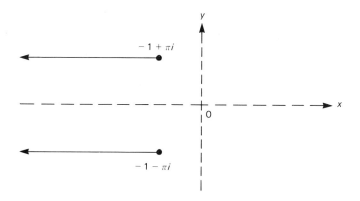

FIGURE 5.31.

SOLUTION: Look at the shaded region in Figure 5.32. Letting w_0 tend to $-\infty$ produces the region in Figure 5.31 as the limiting case. By symmetry, the points -1, 0, 1, ∞ of the boundary of the upper half plane map onto the vertices of the shaded region in Figure 5.32. Since the exterior angles at $-1 + \pi i$, w_0, $-1 - \pi i$, ∞ tend to π, $-\pi$, π, π as w_0 approaches $-\infty$, the Schwarz-Christoffel formula yields

$$w = A + B \int^z \frac{(z + 1)(z - 1)}{z}\, dz = A + B\left(\tfrac{1}{2}z^2 - \log z\right)$$

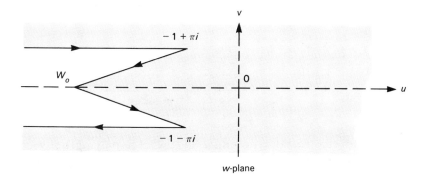

w-plane

FIGURE 5.32.

where the values of the lower limit of integration (evaluated at some point other than zero) are absorbed in the constants A and B. Evaluating this expression at $z = \pm 1$ leads to the system

$$A + B/2 \qquad = -1 - \pi i$$
$$A + B(\tfrac{1}{2} - \pi i) = -1 + \pi i,$$

with the solution $A = -\pi i$, $B = -2$, and $w = 2 \operatorname{Log} z - z^2 - \pi i$. Now observe that the upper half plane can be mapped onto the strip $|\operatorname{Im} \zeta| < \pi i$ by the mapping $\zeta = 2 \operatorname{Log} z - \pi i$. Looking at the mappings in Figure 5.33, it is evident that the function

$$w = 2 \operatorname{Log} z - \pi i - z^2 = \zeta + e^{\zeta}$$

maps the strip $|\operatorname{Im} \zeta| < \pi i$ conformally onto the shaded region in the w-plane. Since the equipotential lines are parallel to the

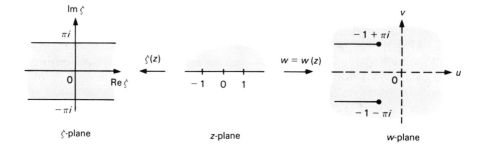

FIGURE 5.33.

real axis in the ζ-plane, the equipotential lines can be obtained in the w-plane. They are given in parametric form by

$$u = \xi + e^{\xi} \cos \eta$$
$$v = \eta + e^{\xi} \sin \eta, \qquad \eta \text{ constant},$$

where $\zeta = \xi + i\eta$.

EXERCISES

1. Find the electrostatic potential in the region lying between a solid cylinder parallel to a flat plate, where the potential on the cylinder is 1 and the potential on the plate is 0. Assume that a cross section places the plate on the imaginary axis and the cylinder is given by $|z - 2| \leqslant 1$.

2. Find the streamlines for a dam of height a if the flow is infinitely deep and has velocity $A > 0$ as $z \to \infty$. What is the velocity at 0 and ia (see Figure 5.34)?

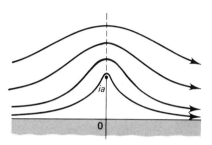

FIGURE 5.34.

3. Find the isotherms in the infinite slab $0 < y < \pi$ if the edges are insulated for $x < 0$ and the temperature satisfies $u(x) = 0°$ and $u(x + i\pi) = 1°$, for $x \geqslant 0$.

4. Find the complex potential and stagnation points for a flow of an incompressible fluid through the region in Figure 5.35, assuming the original velocity of the flow is A.

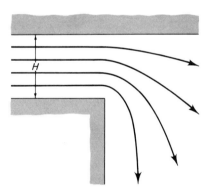

FIGURE 5.35.

5. Find the isotherms of the plate indicated in Figure 5.36 with temperature $0°$ on the horizontal side and $1°$ on the vertical sides.

6. A condenser consists of three parallel plates: the center one a half plane, the others, planes with cross section and potential as indicated in Figure 5.37. Find an expression for the equipotential lines. (*Hint:* Use the Schwarz–Christoffel mapping.)

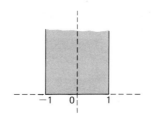

FIGURE 5.36.

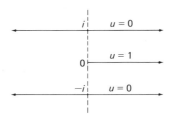

FIGURE 5.37.

7. Show that the function

$$v = \text{Im}\,[e^{-i\alpha}\, z(\cos\alpha + i\sin\alpha\sqrt{1 - (e^{i\alpha}/z)^2}\,)]$$

is the stream function for a flow around a lamina of length 2 inclined at an angle α from the horizontal, when $V(\infty) = A(> 0)$.

5.8 WAKES IN A FLUID FLOW (Optional)

Consider the direct impact of a stream of infinite width and velocity $A\ (> 0)$ on a fixed lamina of width $4a$ placed at right angles to the flow (see Figure 5.38c). The function

$$\zeta = z - a^2/z$$

maps the exterior of the circle $|z| = a$ conformally onto this region. Using the three maps shown in Figure 5.38 yields the complex potential of the flow about the fixed lamina:

$$w(\zeta) = A\omega = A(z + a^2/z) = A(2z - \zeta) = \pm A\sqrt{\zeta^2 + 4a^2},$$

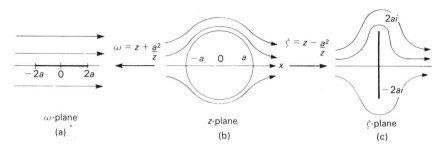

FIGURE 5.38.

since

$$0 = 4(z^2 - \zeta z - a^2) = (2z - \zeta)^2 - (\zeta^2 + 4a^2).$$

The complex potential can be used to determine the streamlines and the velocity of the flow at any point ζ:

$$V = \overline{w'} = \pm \frac{A\zeta}{\sqrt{\zeta^2 + 4a^2}}.$$

At this point a disconcerting problem arises: The velocity at $\zeta = \pm 2ai$ is infinite. Since infinite velocities are not physically possible, we seek a realistic solution to this difficulty. One hypothesis is the existence of an infinite region of water *at rest*, called the **wake**, behind the lamina. The wake will be bounded by *free streamlines* along which the speed is a finite constant (see Figure 5.39).

The existence of a wake requires a change in the analysis of the problem, since the flow is taking place in a polygon bounded by the lamina and the free streamlines. The Riemann mapping theorem guarantees that the polygon can be mapped conformally onto the upper half plane with the points $\pm 2ai$ going to ± 1. The streamlines in the upper half plane can be determined by considering the streamlines in the ω-plane in Figure 5.40c.

To determine the complex potential $w(z)$, recall that $\overline{V(z)} = dw/dz$, and consider the mapping

$$W(z) = \frac{A}{\overline{V(z)}} = \frac{A}{w'(z)}.$$

Along the lamina, the velocity is parallel to the y-axis; hence, V is pure imaginary. By symmetry, the speed will be the same at $\pm 2ai$, although the direction will be opposite. Finally, the speed will be

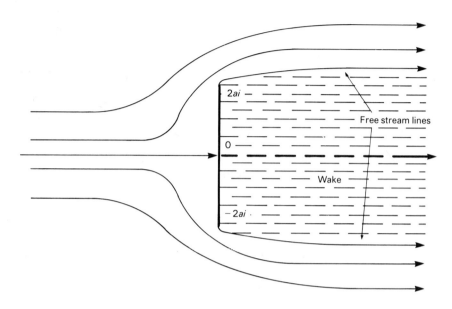

FIGURE 5.39.   Wake behind a lamina

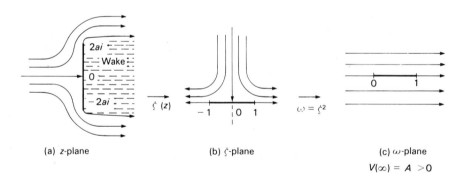

(a) z-plane   (b) ζ-plane   (c) ω-plane
                                          $V(\infty) = A > 0$

FIGURE 5.40.

constant on each free streamline. Thus, since $V(+\infty) = A$ and the free streamlines tend to ∞, every point on the two free streamlines will map to a point on the right half of the unit circle. Since $V(0) = 0$, the origin maps to ∞. Furthermore, $V(-1)$ is positive and less than A, since the flow slows down as it approaches the lamina. Thus, the flow is mapped onto the shaded region in the W-plane in Figure 5.41.

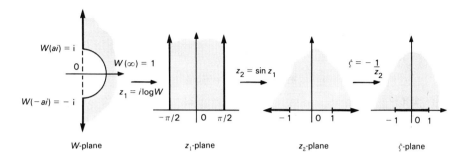

FIGURE 5.41. The hodograph

This region is called a **hodograph**. Note that this region maps onto the upper half of the ζ-plane by means of the transformation

$$\zeta = -[\sin(i \log W)]^{-1} = i\,[\sinh(\log W)]^{-1},$$

and the points $i, -i, \infty, 1$ are sent to $1, -1, 0, \infty$. Thus,

$$\sinh(\log W) = \frac{i}{\zeta},$$

or, by Exercise 24 of Section 1.9,

$$\log W = \sinh^{-1}\left(\frac{i}{\zeta}\right) = \log\left[\frac{i}{\zeta} + \sqrt{1 - \frac{1}{\zeta^2}}\right].$$

Since $W = A/w'(z)$, the equation above implies that

$$\frac{1}{A} \cdot \frac{dw}{dz} = \frac{\zeta}{i + \sqrt{\zeta^2 - 1}} = \frac{\sqrt{\zeta^2 - 1} - i}{\zeta}.$$

But

$$\frac{dw}{dz} = \frac{dw}{d\zeta} \cdot \frac{d\zeta}{dz} \qquad \text{and} \qquad \frac{dw}{d\zeta} = \frac{dw}{d\omega} \cdot \frac{d\omega}{d\zeta} = 2A\zeta,$$

so that

$$2\zeta\,\frac{d\zeta}{dz} = \frac{\sqrt{\zeta^2 - 1} - i}{\zeta}.$$

Thus,

$$z = 2 \int [\sqrt{\zeta^2 - 1} + i]\ d\zeta = \zeta\sqrt{\zeta^2 - 1} - \cosh^{-1}\zeta + 2i\ \zeta + c$$

$$= \sqrt{\omega(\omega - 1)} - \cosh^{-1}\sqrt{\omega} + 2i\sqrt{\omega} + c.$$

Since $\omega = 0$ corresponds to $z = 0$, $c = \cosh^{-1} 0 = -i\cos^{-1} 0 = -i\pi/2$. The problem is now solved, at least in implicit form, since the streamlines are given by the curves Im ω = constant.

EXERCISES

1. Consider a fluid issuing from a slit in the bottom of a large vessel (see Figure 5.42). Assume that the fluid emerges as a *jet* bounded by free streamlines along which the speed is constant, and that the flow in the jet is uniform and parallel to the imaginary axis at ∞. Find the free streamlines. (*Hint:* Use a hodograph.)

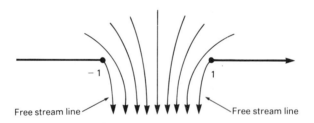

FIGURE 5.42.   Jet flow

2. Assume that a wake forms behind the dam in Exercise 2 of Section 5.7. Find the free streamlines.

NOTES

SECTION 5.1

A table of elementary conformal mappings is given in the appendix. Further mappings may be found in [Ko]. Two different proofs of the Riemann mapping theorem may be found in [V].

SECTION 5.2

Linear fractional transformations are also known as linear transformations, bilinear transformations, linear substitutions, and Möbius transformations.

SECTION 5.5

Problems involving sources will be discussed in Chapter 6. Some detailed applications may be found in [R]. For an excellent text detailing the use of complex analysis in hydrodynamics, see [MT].

SECTION 5.6

See [A, pp. 227-232] for a proof of the Schwarz–Christoffel formula. A formula for mapping the unit disk onto the *exterior* of a polygon is easily derived.

SECTION 5.8

A thorough discussion of wakes and jets may be found in [MT].

6 BOUNDARY VALUE AND INITIAL VALUE PROBLEMS

6.1 HARMONIC FUNCTIONS

Laplace's equation $u_{xx} + u_{yy} = 0$ is of fundamental importance in physics, arising in connection with heat and fluid flows as well as gravitational and electrostatic fields. For example, the temperature u in a heat flow is the real part of an analytic function $w = u + iv$. By the Cauchy–Riemann equations, we have formally

$$u_{xx} = (u_x)_x = (v_y)_x = (v_x)_y = (-u_y)_y = -u_{yy}$$

and Laplace's equation holds for u. Similarly, it holds for v.

Any real-valued function $u(x, y)$ with continuous second partial derivatives that satisfies Laplace's equation on a region G is said to be **harmonic** on G.

EXAMPLE 1.

Show that the function $u(x, y) = x^2 - y^2$ is harmonic on $\mathfrak{C}$.

SOLUTION: The function u has continuous first and second partials:

$$u_x = 2x, \qquad u_y = -2y$$

$$u_{xx} = 2, \qquad u_{xy} = u_{yx} = 0, \qquad u_{yy} = -2.$$

238

Furthermore, u satisfies Laplace's equation, since

$$u_{xx} + u_{yy} = 2 - 2 = 0.$$

Thus, u is harmonic in $\mathbb{C}$. Observe that $u = \text{Re}\,(z^2)$.

Harmonic functions are intimately connected with analytic functions.

THEOREM

(i) Let $f(z) = u(z) + iv(z)$ be analytic on the region G. Then both of the real-valued functions $u(z)$ and $v(z)$ are harmonic on G.

(ii) Let $u(z)$ be a real-valued function that is harmonic on a simply connected region G. Then the line integral

$$v(z) = \int_{\gamma} u_x \; dy - u_y \; dx,$$

where γ is any pws arc in G joining z_0 to z, is harmonic on G, and the function $f(z) = u(z) + iv(z)$ is analytic on G. We call $v(z)$ a **harmonic conjugate** of $u(z)$.

PROOF: (i) If $f = u + iv$ is analytic on G, so is

$$f'(z) = u_x + iv_x = v_y - iu_y.$$

Thus, the second partials of u and v are continuous and $u_{xy} = u_{yx}$ and $v_{xy} = v_{yx}$ from calculus. Applying the Cauchy–Riemann equations, we have

$$u_{xx} = (v_y)_x = (v_x)_y = -u_{yy} \qquad \text{and} \qquad v_{xx} = (-u_y)_x = -(u_x)_y = -v_{yy},$$

implying that u and v both satisfy Laplace's equation.

(ii) The function $F(z) = u_x - iu_y$ has continuous first partials satisfying the Cauchy–Riemann equations on G,

$$(u_x)_x = (-u_y)_y \qquad \text{and} \qquad (u_x)_y = -(-u_y)_x,$$

since u is harmonic. These are sufficient conditions for the analyticity of $F(z)$ on G. Since G is simply connected, we can use the fundamental theorem to define an analytic antiderivative of $F(z)$:

$$f(z) = \int F(z) \; dz = \int (u_x - iu_y)\,(dx + idy)$$

$$= \int (u_x \; dx + u_y \; dy) + i \int (u_x \; dy - u_y \; dx).$$

The first integrand is the exact differential of the function $u = u(x, y)$, so that

$$f(z) = \int du + i \int u_x \, dy - u_y \, dx$$

$$= u(z) + i \int u_x \, dy - u_y \, dx.$$

Thus, we have constructed an analytic function $f(z)$ having $u(z)$ as its real part. This means that the integral

$$v(z) = \int u_x \, dy - u_y \, dx,$$

defined up to an arbitrary constant, is harmonic on G. ∎

We will illustrate the use of the above integral in determining harmonic conjugates in the following examples.

EXAMPLE 2.

Find the harmonic conjugate on $\mathcal{C}$ of the function
$$u = x^2 - y^2.$$

SOLUTION: The function u was shown to be harmonic on $\mathcal{C}$ in Example 1. Since $u_x = 2x$ and $u_y = -2y$, we have

$$v(z) = \int u_x \, dy - u_y \, dx = \int 2x \, dy + 2y \, dx$$

$$= 2 \int d(xy) = 2xy + c.$$

Note that v is harmonic on $\mathcal{C}$, since $v_{xx} = v_{yy} = 0$, and that
$$f(z) = u + iv = (x^2 - y^2) + 2ixy + c = z^2 + c$$
is entire.

EXAMPLE 3.

Is there an analytic function $f = u + iv$ for which
$$u = \frac{x}{x^2 + y^2} \quad ?$$

SOLUTION: Observe that u is not defined at the origin. Thus, we shall seek a harmonic conjugate of u on some simply connected subset of $\mathbb{C} - \{0\}$. First we show that u is harmonic:

$$u_x = \frac{y^2 - x^2}{(x^2 + y^2)^2}, \qquad u_y = \frac{-2xy}{(x^2 + y^2)^2},$$

and

$$u_{xx} = \frac{2x^3 - 6xy^2}{(x^2 + y^2)^3} = -u_{yy}.$$

Hence, a harmonic conjugate is given by

$$v(z) = \int u_x \, dy - u_y \, dx = \int \frac{(y^2 - x^2) \, dy + 2xy \, dx}{(x^2 + y^2)^2}$$

$$= \int d \left(\frac{-y}{x^2 + y^2} \right) = \frac{-y}{x^2 + y^2}.$$

Note that the function

$$f(z) = u + iv = \frac{x - iy}{x^2 + y^2} = \frac{\bar{z}}{|z|^2} = \frac{1}{z}$$

is analytic for $z \neq 0$.

The correspondence between analytic and harmonic functions yields many important properties for the latter.

MAXIMUM PRINCIPLE

If $u(z)$ is harmonic and nonconstant in a simply connected region G, then $u(z)$ has no maximum or minimum in G.

PROOF: Constructing a conjugate harmonic function $v(z)$, we have that $f = u + iv$ is analytic in G. Likewise

$$F(z) = e^{f(z)} = e^{u + iv}$$

is analytic in G, and $|F(z)| = e^{u(z)}$. Since $F(z)$ is nonzero in G, applying the maximum and minimum principles for analytic functions to F, it follows that e^u has no maximum or minimum in G. Since the real function e^u is an increasing function of u, the proof is complete. ∎

MEAN VALUE THEOREM

If $u(z)$ is harmonic in $|z - \zeta| < R$, then

$$u(\zeta) = \frac{1}{2\pi} \int_0^{2\pi} u(\zeta + re^{i\theta}) \, d\theta, \qquad 0 < r < R.$$

PROOF: Construct a harmonic conjugate $v(z)$ so that $f = u + iv$ is analytic on $|z - \zeta| < R$. Gauss's mean value theorem (see Section 2.4) states that

$$f(\zeta) = \frac{1}{2\pi} \int_0^{2\pi} f(\zeta + re^{i\theta}) \, d\theta, \qquad 0 < r < R.$$

Taking the real part of both sides yields the desired equation. ∎

EXERCISES

Prove that the functions in Exercises 1–4 are harmonic in $\mathcal{C}$.

1. $\phi(x, y) = e^x \cos y$
2. $\phi(x, y) = \sin x \sinh y$
3. $\phi(x, y) = x^3 - 3xy^2$
4. $\phi(x, y) = e^{x^2 - y^2} \sin(2xy)$

Determine if analytic functions $f(z) = u + iv$ exist satisfying the conditions given in Exercises 5–7. If so, indicate the domain of definition.

5. $u = \sin x \cosh y$
6. $u = \log(x^2 + y^2)$
7. $u = e^{y/x}$

Find the conjugates of the harmonic functions given in Exercises 8–11.

8. $u = x^2 - (y - 1)^2$
9. $u = \frac{1}{2} \log(x^2 + y^2)$
10. $u = \tan^{-1} \dfrac{2xy}{x^2 - y^2}$
11. $u = \dfrac{x(x - 1) + y^2}{(x - 1)^2 + y^2}$

12. If u and v are harmonic functions, show that $au + bv$ is also harmonic, where a and b are real constants. Show that uv is harmonic if u and v are conjugate harmonic functions.
13. Show that $\log |f(z)|$ is harmonic whenever $f(z)$ is analytic and nonzero.
14. Show that the maximum principle holds for multiply connected regions.
15. Prove that $\int_0^\pi \log \sin \theta \, d\theta = -\pi \log 2$. (*Hint:* Apply the mean value theorem to $\log |1 + z|$ in $|z| \leq r < 1$, and let $r \to 1^-$.)

6.2 DIRICHLET'S PROBLEM

If we study the applications to fluid flow, heat flow, and electrostatics that were given in Chapter 5, we see that the solution in each situation is given by an analytic function called the complex potential. The real and imaginary parts of the complex potential have physical meaning as streamlines, lines of force, equal temperature, and so on. Recall from Section 6.1 that the real and imaginary parts of an analytic function are harmonic functions satisfying Laplace's equation

$$\Delta u = u_{xx} + u_{yy} = 0.$$

Hence, these applications reduce to finding a function that is harmonic in a given region G and that takes on certain preassigned *values* on the *boundary* of the region G. Any such situation is called a **boundary value problem.** More specifically, we have the following.

DIRICHLET'S PROBLEM
Given an arbitrary region G, is there a function harmonic in G having preassigned values on the boundary of G?

EXAMPLE 1.

Find a function that is harmonic in the first quadrant and that has the boundary values 0 on the real axis and 100 on the imaginary axis.

SOLUTION: The function

$$w = \frac{200}{\pi} \text{ Log } z$$

maps the first quadrant conformally onto the strip $0 \leqslant v \leqslant 100$, where $w = u + iv$ (see Figure 6.1). Note that the positive x-axis is transformed into the line $v = 0$, while the positive y-axis becomes the line $v = 100$. Since

$$u + iv = \frac{200}{\pi} (\log |z| + i \text{ Arg } z),$$

the function

$$v = \frac{200}{\pi} \text{ Arg } z$$

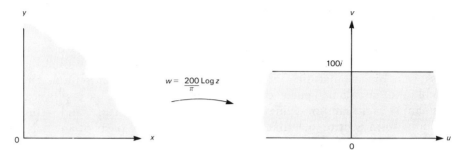

FIGURE 6.1. Graphical solution of a Dirichlet problem

is harmonic in the first quadrant and satisfies the required boundary conditions.

Not all Dirichlet problems have a solution. The existence of a solution depends on the *geometry* of the region: *a solution exists whenever no component of the complement of the region can be continuously shrunk to a point.* The proof of this assertion is beyond the scope of this book. Exercise 9 provides an example of a region for which Dirichlet's problem need not have a solution.

Dirichlet's problem always has a solution for a simply connected region G ($\neq \mathbb{C}$). To see how to obtain an explicit expression for the solution u at any point z_0 in G, let g be the function that maps G conformally onto the unit disk $|\zeta| < 1$ with $g(z_0) = 0$. (The existence of such a function is guaranteed by the Riemann mapping theorem.) For simplicity, assume g is analytic in an open set containing G and its boundary. The composite function $u \circ g^{-1}$ (see Figure 6.2) is harmonic on $|\zeta| \le 1$, so by the mean value theorem for harmonic functions (see Section 6.1), we can represent $u \circ g^{-1}(0)$ as the integral average of the values of $u \circ g^{-1}$ on $|\zeta| = 1$ by writing

$$u \circ g^{-1}(0) = \frac{1}{2\pi} \int_0^{2\pi} u \circ g^{-1}(e^{i\theta})\, d\theta.$$

Setting $\zeta = e^{i\theta}$ yields

$$u \circ g^{-1}(0) = \frac{1}{2\pi i} \int_{|\zeta|=1} u \circ g^{-1}(\zeta)\, \frac{d\zeta}{\zeta},$$

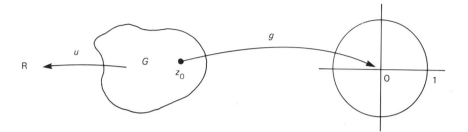

FIGURE 6.2.

and since $\zeta = g(z)$, the integral becomes

$$u(z_0) = \frac{1}{2\pi i} \int_{\partial G} u(z) \frac{g'(z)}{g(z)} \, dz. \tag{1}$$

This equation indicates that the value of a harmonic function u at points interior to the region G can be determined as an integral of the boundary values of u. Observe the similarity of this situation with the Cauchy integral formula.

The following examples illustrate the use of this integral.

EXAMPLE 2.

To solve Dirichlet's problem in $G = \{z: |z| < R\}$, note that

$$g(z) = \frac{R(z - z_0)}{R^2 - \bar{z}_0 z}$$

maps G conformally onto the open unit disk with $g(z_0) = 0$. For $|z| = R$,

$$\frac{g'(z)}{g(z)} = \frac{R^2 - |z_0|^2}{(z - z_0)(R^2 - \bar{z}_0 z)} = \frac{|z|^2 - |z_0|^2}{z|z - z_0|^2},$$

so that equation (1) becomes

$$u(z_0) = \frac{1}{2\pi i} \int_{|z|=R} u(z) \frac{|z|^2 - |z_0|^2}{|z - z_0|^2} \frac{dz}{z}.$$

Setting $z = Re^{i\phi}$ and $z_0 = re^{i\theta}$ yields **Poisson's integral formula** for the disk $|z| < R$:

$$u(re^{i\theta}) = \frac{1}{2\pi} \int_0^{2\pi} u(Re^{i\phi}) \frac{R^2 - r^2}{R^2 + r^2 - 2Rr \cos(\phi - \theta)} \, d\phi, \tag{2}$$

because

$$|z - z_0|^2 = |z|^2 + |z_0|^2 - (\bar{z}_0 z + z_0 \bar{z})$$
$$= R^2 + r^2 - Rr \left(e^{i(\phi - \theta)} + e^{-i(\phi - \theta)} \right).$$

Since $z_0 = re^{i\theta}$ is arbitrary, equation (2) is valid at all points $|z_0| < R$.

EXAMPLE 3.

Let G be the right half plane, and let z_0 be any point interior to G. Then the function

$$g(z) = (z - z_0)/(z + z_0)$$

maps G conformally onto the unit disk with $g(z_0) = 0$. Since

$$\frac{g'(z)}{g(z)} = \frac{2z_0}{z^2 - z_0^2},$$

Poisson's integral formula for the right half plane is

$$u(z_0) = \frac{z_0}{\pi i} \int_{\partial G} \frac{u(z)\, dz}{z^2 - z_0^2},$$

and setting $z = it$, we have

$$u(z_0) = \frac{-z_0}{\pi} \int_{-\infty}^{\infty} \frac{u(it)\, dt}{t^2 + z_0^2}.$$

In the discussion leading up to Poisson's integral formula in the disk $|z| < R$, the boundary values were assumed to be continuous. In many applications (such as Example 1), the boundary data are discontinuous. It is interesting to note that Poisson's integral formula yields a harmonic function in spite of discontinuities.

POISSON'S THEOREM

Let $U(\phi)$ be continuous for $0 \leqslant \phi \leqslant 2\pi$ except for a finite number of jumps. Then the function

$$u(z) = \frac{1}{2\pi} \int_0^{2\pi} U(\phi) \frac{R^2 - |z|^2}{|Re^{i\phi} - z|^2}\, d\phi$$

is harmonic in $|z| < R$ and

$$\lim_{z \to R e^{i\phi}} u(z) = U(\phi),$$

at all points of continuity of $U(\phi)$.

PROOF: Note that

$$\text{Re} \left(\frac{Re^{i\phi} + z}{Re^{i\phi} - z} \right) = \frac{R^2 - |z|^2}{|Re^{i\phi} - z|^2}. \tag{3}$$

Substituting the left side of equation (3) into the integral, we can repeatedly perform partial differentiations with respect to x and y under the integral sign, as the resulting integrand is continuous on $|z| \leqslant r < R$. Then

$$\Delta u(z) = \frac{1}{2\pi} \int_0^{2\pi} U(\phi) \, \Delta \text{Re} \left(\frac{Re^{i\phi} + z}{Re^{i\phi} - z} \right) d\phi = 0, \qquad |z| < R,$$

since the real part of the analytic function $(Re^{i\phi} + z)/(Re^{i\phi} - z)$ is harmonic. Thus, $u(z)$ is harmonic in $|z| < R$.

Letting $\zeta = Re^{i\phi}$, note that

$$\frac{1}{2\pi} \int_0^{2\pi} \text{Re} \left(\frac{Re^{i\phi} + z}{Re^{i\phi} - z} \right) d\phi = \text{Re} \left(\frac{1}{2\pi i} \int_{|\zeta| = R} \frac{\zeta + z}{\zeta - z} \frac{d\zeta}{\zeta} \right)$$

$$= \text{Re} \left[\frac{1}{2\pi i} \int_{|\zeta| = R} \left(\frac{2}{\zeta - z} - \frac{1}{\zeta} \right) d\zeta \right] = 1.$$

If $U(\phi)$ is continuous at $\phi = \alpha$, then given $\epsilon > 0$, there is a $\delta > 0$ such that $|U(\phi) - U(\alpha)| < \epsilon$ whenever $|\phi - \alpha| < \delta$ (assume that U has period 2π). Thus,

$$|u(z) - U(\alpha)| = \left| \frac{1}{2\pi} \int_0^{2\pi} (U(\phi) - U(\alpha)) \, \text{Re} \left(\frac{Re^{i\phi} + z}{Re^{i\phi} - z} \right) d\phi \right|$$

$$\leqslant \frac{1}{2\pi} \int_0^{2\pi} |U(\phi) - U(\alpha)| \, \frac{R^2 - |z|^2}{|Re^{i\phi} - z|^2} \, d\phi$$

$$\leqslant \epsilon + \frac{1}{2\pi} \int_{\delta \leqslant |\phi - \alpha| \leqslant \pi} \frac{R^2 - |z|^2}{|Re^{i\phi} - z|^2} |U(\phi) - U(\alpha)| \, d\phi.$$

Now, if $|\arg z - \alpha| < \delta/2$ and $|\phi - \alpha| \geqslant \delta$, then (see Figure 6.3)

$$|Re^{i\phi} - z| \geqslant R \sin \frac{\delta}{2},$$

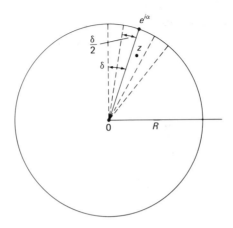

FIGURE 6.3.

so that

$$|u(z) - U(\alpha)| \leqslant \epsilon + \frac{R^2 - |z|^2}{2\pi R^2 \sin^2 \dfrac{\delta}{2}} \int_0^{2\pi} |U(\phi) - U(\alpha)| \, d\phi \to \epsilon,$$

as $z \to Re^{i\alpha}$. Since ϵ is arbitrary, the proof is complete. ∎

The following example contrasts the use of conformal mappings with Poisson's integral formula in finding solutions to Dirichlet's problem.

EXAMPLE 4.

Find a function u harmonic in the unit disk having the boundary value 1 in the right half plane and 0 in the left half plane.

SOLUTION: The linear fractional transformation

$$\zeta = i \left(\frac{i - z}{i + z} \right)$$

maps the points 0, i, $-i$, 1 onto i, 0, ∞, -1, so it maps the unit disk conformally onto the upper half plane. Following this map with $w = (1/\pi i) \operatorname{Log} \zeta$ transforms the upper half plane onto the strip $0 < u < 1$, where $w = u + iv$ (see Figure 6.4). Hence, the required harmonic function is

$$u(z) = \operatorname{Re} w = \frac{1}{\pi} \operatorname{Arg} \zeta = \frac{1}{\pi} \operatorname{Arg} \left[i \left(\frac{i - z}{i + z} \right) \right].$$

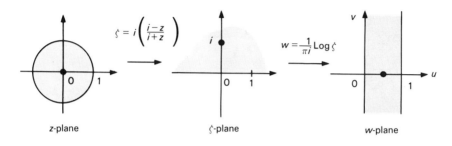

z-plane · ζ-plane · w-plane

FIGURE 6.4.

To solve Dirichlet's problem using Poisson's integral formula, observe that

$$u(z) = \frac{1}{2\pi} \int_{-\pi/2}^{\pi/2} \frac{1 - |z|^2}{|e^{i\phi} - z|^2} \, d\phi,$$

since $U(\phi) = 0$ in the left half plane. Setting $z = re^{i\theta}$,

$$u(z) = \frac{1 - r^2}{2\pi} \int_{-\pi/2}^{\pi/2} \frac{d(\phi - \theta)}{1 + r^2 - 2r \cos(\phi - \theta)}$$

$$= \frac{1}{\pi} \tan^{-1} \left(\frac{1 + r}{1 - r} \tan \frac{\phi - \theta}{2} \right) \Big|_{-\pi/2}^{\pi/2}.$$

Hence,

$$\tan \pi u(z) = \frac{\dfrac{1 + r}{1 - r} \left[\tan \left(\dfrac{\pi}{4} - \dfrac{\theta}{2} \right) + \tan \left(\dfrac{\pi}{4} + \dfrac{\theta}{2} \right) \right]}{1 - \left(\dfrac{1 + r}{1 - r} \right)^2 \tan \left(\dfrac{\pi}{4} - \dfrac{\theta}{2} \right) \tan \left(\dfrac{\pi}{4} + \dfrac{\theta}{2} \right)}$$

$$= \frac{1 - r^2}{2r} \left[\frac{\tan^2 \dfrac{\theta}{2} + 1}{\tan^2 \dfrac{\theta}{2} - 1} \right] = \frac{1 - r^2}{-2r \cos \theta}.$$

To check that these answers are equivalent, observe that

$$\text{Arg}\left[i\left(\frac{i-z}{i+z}\right)\right] = \text{Arg}\left[\frac{-(z+\bar{z})+i(1-|z|^2)}{|i+z|^2}\right]$$

and, using the identity $\text{Arg}(x + iy) = \tan^{-1}(y/x)$ in the right half plane,

$$\text{Arg}\left[i\left(\frac{i-z}{i+z}\right)\right] = \tan^{-1}\left[\frac{1-|z|^2}{-(z+\bar{z})}\right] = \tan^{-1}\left(\frac{1-r^2}{-2r\cos\theta}\right).$$

EXERCISES

1. Let $u(z)$ be harmonic in $|z - \zeta| < R$. Prove the **area mean value theorem**

$$u(\zeta) = \frac{1}{\pi R^2} \iint_{|z-\zeta|<R} u(z)r\, dr\, d\theta.$$

(*Hint:* Use the mean value theorem for harmonic functions.)

2. Use the area mean value theorem to prove the maximum principle for harmonic functions.

3. If u is harmonic in a simply connected region G, show that grad u is analytic. Then, using the development in Section 5.5, prove that

$$\int_\gamma u_n\, ds = 0,$$

where γ is any pws closed curve in G and u_n is the directional derivative of u in the outward normal direction.

4. Show that Poisson's integral formula for the upper half plane $y > 0$ is

$$u(z) = \frac{y}{\pi}\int_{-\infty}^{\infty} \frac{u(t)\, dt}{(t-x)^2 + y^2}, \qquad z = x + iy.$$

5. Show that Poisson's integral formula for the first quadrant is

$$u(z) = \frac{4xy}{\pi}\left[\int_0^{\infty} \frac{tu(t)\, dt}{t^4 - 2t^2(x^2-y^2) + (x^2+y^2)^2}\right.$$
$$\left. + \int_0^{\infty} \frac{tu(it)\, dt}{t^4 + 2t^2(x^2-y^2) + (x^2+y^2)^2}\right].$$

6. Find the temperature at any point z in the upper half plane if the temperature (in degrees) on the real axis is given by

$$u(t) = \begin{cases} 1 - |t|, & |t| \leqslant 1, \\ 0, & |t| \geqslant 1. \end{cases}$$

7. Prove that any solution of Dirichlet's problem continuous on the closure of a simply connected region G must be unique. (*Hint:* Use the maximum principle.)

8. Suppose $u(z)$ and $v(z)$ are harmonic on a region G, continuous on its closure, and satisfy $u(z) \leqslant v(z)$ on the boundary of G. Show that $u(z) \leqslant v(z)$ for every z in G.

9. Let G be the region $0 < |z| < 1$. Show that there is no function $u(z)$ harmonic in G with boundary values $u(e^{i\theta}) = 0$, $u(0) = a > 0$. (*Hint:* Apply Exercise 8 to the functions

$$u_r(z) = a \, \frac{\log |z|}{\log r},$$

harmonic in $r < |z| < 1$.)

10. Prove **Harnack's inequality**: If $u(z)$ is harmonic and nonnegative in $|z| < R$, then

$$u(0) \, \frac{R - |z|}{R + |z|} \leqslant u(z) \leqslant u(0) \, \frac{R + |z|}{R - |z|}.$$

11. If $f(z) = u(z) + iv(z)$ is analytic in a region containing $|z| \leqslant R$, prove **Schwarz's formula**

$$f(z) = \frac{1}{2\pi} \int_0^{2\pi} \frac{Re^{i\phi} + z}{Re^{i\phi} - z} \, u(Re^{i\phi}) \, d\phi + iv(0).$$

12. Show that Schwarz's formula can be rewritten in the form

$$f(z) = \frac{1}{\pi i} \int_{|\zeta| = R} \frac{u(\zeta)}{\zeta - z} \, d\zeta - \overline{f(0)}.$$

(*Hint:* Apply the mean value theorem to $u(0) = iv(0) + \overline{f(0)}$.)

13. Let $U(\phi)$ be continuous for $0 \leqslant \phi \leqslant 2\pi$ except for a finite number of jumps. Prove that

$$g(z) = \frac{z}{\pi} \int_0^{2\pi} \frac{U(\phi)}{Re^{i\phi} - z} \, d\phi$$

is analytic in $|z| < R$. (*Hint:* Rewrite Schwarz's formula.)

14. Show, using the method in Poisson's theorem, that

$$u(z) = \frac{y}{\pi} \int_{-\infty}^{\infty} \frac{u(t)}{(t-x)^2 + y^2} \, dt, \qquad z = x + iy,$$

is a harmonic function in the upper half plane even if $u(t)$ is discontinuous at finitely many values of t.

15. Use Exercise 14 to find a function $u(z)$ that is harmonic in the upper half plane and that satisfies the boundary values

$$u(t) = \begin{cases} 1 & \text{if} \quad |t| < 1, \\ 0 & \text{if} \quad |t| \geqslant 1. \end{cases}$$

16. Repeat Exercise 15 for the boundary values

$$u(t) = \begin{cases} 1 & \text{if} \quad -1 \leqslant t \leqslant 0, \\ 1-t & \text{if} \quad\ \ 0 \leqslant t \leqslant 1, \\ 0 & \text{elsewhere.} \end{cases}$$

6.3 APPLICATIONS

In Sections 5.5 and 5.7, we studied three analogous examples of steady-state vector fields occurring in nature: fluid flow, heat flow, and the electrostatic field. The vector fields, assumed to be two-dimensional and irrotational, were examined within a region G containing no sources and sinks. In this section these problems will be extended to include sources and vortices in the region G. The theory will be developed for fluid flow, and the analogies with the other two fields will be presented in Table 6.1 at the end of the section.

Recall that the velocity vector $V(z)$ of the field equals $\overline{w'(z)}$, where the analytic function $w(z)$ is the complex potential of the flow. Thus, the potential function u and the stream function v are conjugate harmonic functions, and the problem of finding the streamline reduces to Dirichlet's problem. Observe by the maximum principle that *if an equipotential line forms a closed curve* γ, *then either* γ *encloses a singularity of* $u(z)$ *or* $u(z)$ *is constant in* G. Of course, *there are no closed streamlines*, since the flow is irrotational. Furthermore, *neither streamlines nor equipotential lines begin or end at an interior point* z_0 *of* G, otherwise a sufficiently small disk centered at z_0 lies in G and its boundary meets the streamline $v(z) = k$ at only one point. By continuity, the remaining boundary points satisfy either $v(z) > k$ or $v(z) < k$, violating the mean value theorem. Thus, distinct streamlines can meet only at boundary points of G (for example, at sources) or at infinity. We illustrate with a problem in heat flow.

EXAMPLE 1.

Let the plate G be a disk of radius R with boundary temperatures $1°$ in the upper half plane and $0°$ in the lower half plane. Find the temperature at all points of G and describe the isotherms.

SOLUTION: Applying Poisson's integral formula, we have for $z = re^{i\theta}$, $r < R$,

$$u(z) = \frac{1}{2\pi} \int_0^\pi \text{Re}\left(\frac{Re^{i\phi} + z}{Re^{i\phi} - z}\right) d\phi$$

$$= \frac{1}{2\pi} \int_0^\pi \frac{R^2 - r^2}{R^2 + r^2 - 2Rr\cos(\phi - \theta)} d(\phi - \theta)$$

$$= \frac{1}{\pi} \tan^{-1}\left(\frac{R + r}{R - r} \tan\frac{\phi - \theta}{2}\right)\Big|_0^n.$$

Thus,

$$\tan \pi u(z) = \frac{\dfrac{R + r}{R - r}\left(\tan\dfrac{\pi - \theta}{2} + \tan\dfrac{\theta}{2}\right)}{1 - \left(\dfrac{R + r}{R - r}\right)^2 \tan\dfrac{\pi - \theta}{2}\tan\dfrac{\theta}{2}}$$

$$= \frac{R^2 - r^2}{-2Rr\sin\theta}.$$

But since $\tan^{-1} y/x = \text{Arg}(x + iy)$,

$$\pi u(z) = \text{Arg}\{i[R^2 - |z|^2 + R(z - \bar{z})]\}$$

$$= \text{Arg}\left[i\left(\frac{R + z}{R - z}\right)\right],$$

implying that the temperature is given by

$$u(z) = \frac{1}{\pi} \text{Arg}\, i\frac{R + z}{R - z}.$$

The isotherms satisfy

$$\text{Arg } i \, \frac{R + z}{R - z} = \text{constant},$$

and $i(R + z)/(R - z)$ maps $|z| < R$ onto the upper half plane, so the isotherms correspond to the arcs of the family of circles passing through the points $\pm R$ lying in $|z| < R$ (see Figure 6.5).

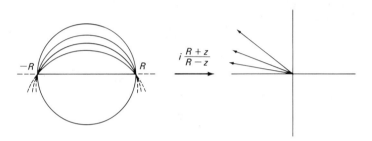

FIGURE 6.5. Family of circles through $\pm R$

Assume that a **point source** (or **sink**) is located at the origin. Then the flow Q across any Jordan curve around the origin is a nonzero constant, and if γ is the circle $|z| = r$, the normal velocity component V_n is constant in each direction, as the streamlines are radial at the origin. Thus,

$$Q = \int V_n \, ds = V_n \int_0^{2\pi} r \, d\theta = 2\pi r V_n$$

and

$$V(z) = V_n \cdot \frac{z}{|z|} = \frac{Q}{2\pi} \frac{z}{|z|^2},$$

since $z/|z|$ is the unit normal vector. Since $V(z) = \overline{w'(z)}$,

$$w(z) = \frac{Q}{2\pi} \log(z) + c, \qquad c \text{ complex},$$

hence the potential function and stream function are given by

$$u(z) = \frac{Q}{2\pi} \log |z|, \qquad v(z) = \frac{Q}{2\pi} \arg z,$$

respectively, up to an arbitrary real constant. Note that $v(z)$ is multi-valued, and both functions are harmonic in any simply connected region not containing the origin. If $Q > 0$, we have a source of *strength Q* at $z = 0$, and if $Q < 0$, we have a sink.

If the source is not at the origin, but at the point z_0, the complex potential is

$$w(z) = \frac{Q}{2\pi} \log(z - z_0) + c. \tag{1}$$

On the other hand, the vector field may not be irrotational. This may occur, for example, from the action of a cylindrical rotor, so that in any plane normal to its axis, the streamlines are concentric circles centered on the rotor. Such a field is called a **plane vortex field**.

If a **point vortex** is located at the origin, the circulation Γ along any Jordan curve γ is a nonzero constant ($\Gamma > 0$ when the flow is counterclockwise). Along a circle $|z| = r$, the tangential velocity component V_s is constant, so

$$\Gamma = \int_\gamma V_s \, ds = 2\pi r V_s$$

and

$$V(z) = V_s \cdot \frac{iz}{|z|} = \frac{i\Gamma}{2\pi} \frac{z}{|z|^2},$$

since $iz/|z|$ is the unit tangential vector. Then, except for an arbitrary constant,

$$w(z) = \frac{-i\Gamma}{2\pi} \log z = \frac{\Gamma}{2\pi} \arg z + i \frac{\Gamma}{2\pi} \log \frac{1}{|z|} \tag{2}$$

is the complex potential of this field. As a point source may also be a vortex, we combine equations (1) and (2) (at z_0), obtaining

$$w(z) = \frac{\Gamma + iQ}{2\pi i} \log(z - z_0) + c \tag{3}$$

as the complex potential of a **vortex source** located at z_0 with *intensity* Γ and *strength Q*. We obtain the complex potential of a system of vortex sources $\Gamma_1 + iQ_1, \ldots, \Gamma_k + iQ_k$ located at $z_1, \ldots, z_k$ by adding up the individual complex potentials

$$w(z) = \frac{1}{2\pi i} \sum_{j=1}^{k} (\Gamma_j + iQ_j) \log(z - z_j), \tag{4}$$

as the vector field is obtained by superposition. Furthermore, this result and the usual limiting procedure can be used to obtain the complex potential of a line L of sources, provided the flow function $Q(\zeta)$ is integrable:

$$w(z) = \frac{1}{2\pi} \int_L Q(\zeta) \log(z - \zeta) \, ds, \qquad \zeta \text{ on } L. \qquad (5)$$

EXAMPLE 2.

If the system consists of *two sources*, each of strength Q, located at z_1 and z_2, the complex potential is given by

$$w(z) = \frac{Q}{2\pi} \log(z - z_1)(z - z_2).$$

The equipotential lines, satisfying

$$|z - z_1| \, |z - z_2| = \text{constant},$$

are known as **lemniscates** and are shown in Figure 6.6. The lemniscate, shaped like ∞, is given by the equation

$$|z - z_1| \, |z - z_2| = \frac{|z_1 - z_2|^2}{4}.$$

Note that $(z_1 + z_2)/2$ is a stagnation point.

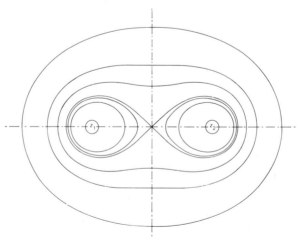

FIGURE 6.6. Lemniscates

EXAMPLE 3.

A system consisting of a *source* and a *sink* of strengths Q and $-Q$ situated at z_1 and z_2, respectively, has a complex potential given by the equation

$$w(z) = \frac{Q}{2\pi} \log \frac{z - z_1}{z - z_2}.$$

The equipotential lines satisfy

$$\left| \frac{z - z_1}{z - z_2} \right| = \text{constant}$$

and form the circles of Apollonius indicated as solid lines in Figure 6.7. The streamlines are the family of circles passing through z_1 and z_2.

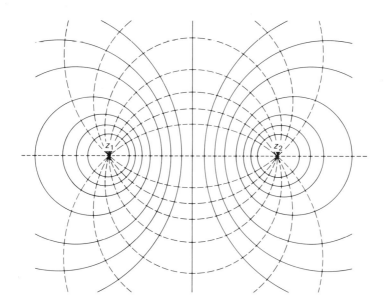

FIGURE 6.7. Circles of Apollonius

Let $z_1 = -h$, $z_2 = 0$. Then

$$w(z) = \frac{Q}{2\pi} \log \frac{z + h}{z} = \frac{p}{2\pi} \log \left(1 + \frac{h}{z} \right)^{1/h}, \qquad p = Qh.$$

If we now permit the source to approach the sink, simultaneously increasing Q so that p remains constant, we obtain in the limit a **point doublet** of moment p located at 0 whose streamlines are directed along the positive real axis. Its complex potential is given by

$$w(z) = \frac{p}{2\pi} \lim_{h \to 0} \log \left(1 + \frac{h}{z}\right)^{1/h} = \frac{p}{2\pi} \log e^{1/z} = \frac{p}{2\pi z}, \quad (6)$$

hence

$$u = \frac{p}{2\pi} \frac{x}{x^2 + y^2}, \qquad v = \frac{-p}{2\pi} \frac{y}{x^2 + y^2}.$$

Then

$$\left(x - \frac{p}{4\pi u}\right)^2 + y^2 = \left(\frac{p}{4\pi u}\right)^2, \qquad x^2 + \left(y + \frac{p}{4\pi v}\right)^2 = \left(\frac{p}{4\pi v}\right)^2,$$

and the equipotential lines and streamlines are the families of circles indicated in Figure 6.8.

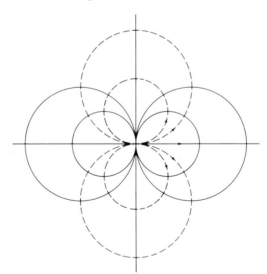

FIGURE 6.8. Point doublet (dipole) at the origin

The procedure above also holds for z_1 complex, but now the moment of the doublet is complex with argument $\pi + \arg z_1$ coinciding with the direction of the streamlines at the origin.

EXAMPLE 4.

We consider the problem of complete streamlining of the exterior of the unit disk so that the velocity vector tends to 1 at ∞.

As was shown in Example 3, Section 5.5 if the flow is symmetric with the x-axis, the complex potential is given by

$$w_1(z) = z + \frac{1}{z},$$

since $V_1(z) = 1 - (1/\bar{z}^2)$. Dropping the assumption of symmetry, observe that the flow might also be subject to a vortex flow centered at the origin of intensity Γ, with complex potential

$$w_2(z) = \frac{\Gamma}{2\pi i} \log z,$$

since its corresponding velocity vector

$$V_2(z) = \frac{i\Gamma}{2\pi\bar{z}}$$

vanishes at ∞. By superposition, the equation of the complex potential is given by

$$w(z) = z + \frac{1}{z} + \frac{\Gamma}{2\pi i} \log z.$$

The magnitude of the velocity satisfies

$$|\overline{V(z)}| = |w'(z)| = \left| 1 - \frac{1}{z^2} + \frac{\Gamma}{2\pi i z} \right|,$$

vanishing at the zeros z_s (stagnation points) of the equation

$$z^2 + \frac{\Gamma}{2\pi i} z - 1 = 0.$$

That is,

$$z_s = \frac{\Gamma i \pm \sqrt{16\pi^2 - \Gamma^2}}{4\pi}. \tag{7}$$

If $|\Gamma| < 4\pi$, then $|z_s| = \sqrt{\Gamma^2 + 16\pi^2 - \Gamma^2}/4\pi = 1$ and

$$\tan \text{Arg } z_s = \frac{\pm\Gamma}{\sqrt{16\pi^2 - \Gamma^2}},$$

and if $|\Gamma| > 4\pi$, the stagnation points are on the imaginary axis and satisfy

$$|z_s| = \frac{\Gamma \pm \sqrt{\Gamma^2 - 16\pi^2}}{4\pi}.$$

Thus, only one stagnation point is outside the unit circle. A sketch of the streamlines is shown in Figure 6.9.

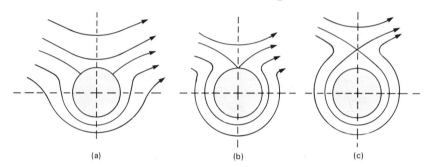

(a)   (b)   (c)

FIGURE 6.9. Complete streamlining of the exterior of a disk with a point vortex at its center: (a) $0 < \Gamma < 4\pi$; (b) $\Gamma = 4\pi$; (c) $\Gamma > 4\pi$

To completely streamline a region G, we need merely map G conformally onto the exterior of the unit disk, $f: G \to \{|z| > 1\}$; then the composite function $w \circ f$ is the complex potential of G. Of particular interest in aerodynamics is the complete streamlining of the **Joukowsky profile**, given by the mapping $z = \zeta + 1/\zeta$, mapping given circles as shown in Figure 6.10. The profiles can be made to

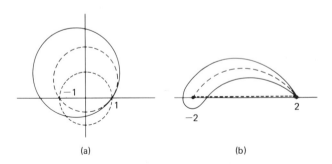

(a)   (b)

FIGURE 6.10. Joukowsky profile: (a) ζ-plane; (b) z-plane

approximate cross sections of airfoils, and the lift of the airfoil can be evaluated.

We can now incorporate the information from this section into our analogy between fluid flow, heat flow, and electrostatics (see Table 6.1).

Table 6.1 Steady-State Vector Fields

Fluid Flow	Heat Flow	Electrostatic Field
$w(z) = \dfrac{\Gamma + iQ}{2\pi i} \log(z - z_0)$	$w(z) = \dfrac{Q}{2\pi k} \log \dfrac{1}{z - z_0}$	$iw(z) = \dfrac{Q}{2\pi} i \log \dfrac{1}{z - z_0}$
Vortex source of strength Q and intensity Γ at z_0	*Source* of strength Q at z_0	*Charge* of magnitude $Q/2\pi$ at z_0
$w(z) = \dfrac{p}{2\pi} \dfrac{1}{z - z_0}$	$w(z) = \dfrac{-p}{2\pi k} \dfrac{1}{z - z_0}$	$iw(z) = \dfrac{-ip}{2\pi} \dfrac{1}{z - z_0}$
Doublet of moment p at z_0	*Doublet* of moment p at z_0	*Dipole* of moment $p/2\pi$ at z_0

EXERCISES

1. Find the potential function for the plane electrostatic field in $|z| < 1$ bordered by electrodes represented by the semicircles $e^{i\theta}$, $|\theta| < \pi/2$, and $e^{i\theta}$, $|\theta - \pi| < \pi/2$, with potentials of u_0 and u_1, respectively.

2. Find the temperature of a plate Q in the shape of the upper half plane, given boundary temperatures of $100°$ on $|x| > 1$ and $0°$ on $|x| < 1$.

3. Find the complex potential and the streamlines for the plane flow of a fluid in the upper half plane when there is a source of strength Q at i and a sink of equal strength at 0.

4. What is the complex potential for the plane flow of a fluid with a sink of strength Q at -1 and a vortex source of strength Q and intensity Γ at 0?

In Exercises 5–8, given the complex potential of the flow of a fluid, construct the equipotential lines and streamlines, and find the velocity vector V, the stagnation points, the strength and intensity of the vortex sources, the moments of the doublets, and the behavior of the flow at ∞.

5. $w(z) = \dfrac{Q}{2\pi} \log \left(z + \dfrac{1}{z}\right)$

6. $w(z) = \log \left(1 + \dfrac{4}{z^2}\right)$

7. $w(z) = \log \left(z^2 - \dfrac{1}{z^2}\right)$

8. $w(z) = az + \dfrac{Q}{2\pi} \log \dfrac{1}{z}, \qquad a, Q > 0$

9. A point *multiplet* (multipole) is a generalization of a doublet (dipole), obtained by taking a sink of strength Q at the origin together with n sources of strength Q/n symmetrically distributed on a circle of radius r, and holding Qr fixed as r tends to 0. Show that its complex potential is given by

$$w(z) = \frac{-p}{2\pi n} \frac{1}{z^n}, \qquad |p| = Qr,$$

and that the streamlines are directed along the arguments of the nth roots of p. Such a multiplet is said to have *order* $2n$.

Sketch the image of the circles described in Exercises 10 and 11 under the mapping $z = \zeta + (1/\zeta)$. (*Hint:* Show that

$$\frac{z - 2}{z + 2} = \left(\frac{\zeta - 1}{\zeta + 1}\right)^2 .)$$

10. $|\zeta - i| = \sqrt{2}$ **11.** $|\zeta + 1 - i| = \sqrt{5}$

6.4 FOURIER SERIES

Poisson's integral formula is intimately related to the notion of a Fourier series. We have seen that if $U(\phi)$ is piecewise continuous for $0 \leqslant \phi \leqslant 2\pi$, then the function

$$u(z) = \frac{1}{2\pi} \int_0^{2\pi} U(\phi) \, \mathrm{Re} \left(\frac{Re^{i\phi} + z}{Re^{i\phi} - z}\right) d\phi$$

is harmonic in $|z| < R$ and has boundary values $u(Re^{i\phi}) = U(\phi)$ at all points of continuity of U. In practice, it is often easiest to obtain

$u(z)$ by expanding the right side of the above equation into an infinite series. Then

$$u(z) = \text{Re} \left[\frac{1}{2\pi} \int_0^{2\pi} U(\phi) \left[1 + 2 \sum_{n=1}^{\infty} \left(\frac{z}{Re^{i\phi}} \right)^n \right] d\phi \right],$$

and since the series converges uniformly in $|z| \leqslant \rho < R$, term-by-term integration is permissible, yielding the **Fourier series**

$$u(z) = \frac{1}{2\pi} \int_0^{2\pi} U(\phi)\, d\phi + 2\ \text{Re} \sum_{n=1}^{\infty} \left(\frac{1}{2\pi} \int_0^{2\pi} U(\phi) e^{-in\phi}\, d\phi \right) \left(\frac{z}{R} \right)^n$$

$$= c_0 + 2\ \text{Re} \left[\sum_{n=1}^{\infty} r^n c_n e^{in\theta} \right], \qquad z = re^{i\theta}, \tag{1}$$

where

$$R^n c_n = \frac{1}{2\pi} \int_0^{2\pi} U(\phi) e^{-in\phi}\, d\phi \tag{2}$$

is the nth **Fourier coefficient** of $U(\phi)$.

EXAMPLE 1.

Find a harmonic function for $|z| < R$ that has the boundary values $U(\phi) = \cos\phi,\ 0 \leqslant \phi \leqslant 2\pi$.

SOLUTION: First calculate the Fourier coefficients of $U(\phi)$:

$$2\pi R^n c_n = \int_0^{2\pi} \cos\phi\, e^{-in\phi}\, d\phi$$

$$= \int_0^{2\pi} \frac{e^{-i(n-1)\phi} + e^{-i(n+1)\phi}}{2}\, d\phi$$

$$= \begin{cases} \dfrac{-1}{2i} \left[\dfrac{e^{-i(n-1)\phi}}{n-1} + \dfrac{e^{-i(n+1)\phi}}{n+1} \right]_0^{2\pi} = 0, & n \neq 1, \\[3ex] \dfrac{1}{2} \left[\phi - \dfrac{e^{-2i\phi}}{2i} \right]_0^{2\pi} = \pi, & n = 1. \end{cases}$$

Thus, $u(z) = 2\ \text{Re}\,(rc_1 e^{i\theta}) = \text{Re}\,(z/R)$.

EXAMPLE 2.

A plate shaped like a disk of radius R is maintained at a constant temperature of $100°$ along the upper half of its boundary and at $0°$ along the bottom half. Find the temperature at any point z in the plate.

SOLUTION: Here $U(\phi) = 100$ for $0 \leqslant \phi < \pi$ and $U(\phi) = 0$ for $\pi \leqslant \phi < 2\pi$, so the Fourier coefficients satisfy $c_0 = 50$ and

$$2\pi R^n c_n = 100 \int_0^\pi e^{-in\phi}\, d\phi = \frac{100}{in}\,[1 - e^{-in\pi}], \qquad n > 0.$$

Therefore,

$$u(z) = 50 + 2\,\mathrm{Re}\left[\sum_{n=0}^\infty (re^{i\theta})^{2n+1}\, c_{2n+1}\right]$$

$$= 50 + \frac{200}{\pi}\,\mathrm{Re}\left[\frac{1}{i}\sum_{n=0}^\infty \frac{(z/R)^{2n+1}}{2n+1}\right].$$

Using the method in Example 1 of Section 3.2, we see that the sum

$$\sum_{n=0}^\infty \frac{z^{2n+1}}{2n+1} = \int_0^z \frac{dz}{1-z^2} = \frac{1}{2}\,\mathrm{Log}\left(\frac{1+z}{1-z}\right).$$

Hence,

$$u(z) = 50 + \frac{200}{\pi}\,\mathrm{Re}\left[\frac{1}{2i}\,\mathrm{Log}\left(\frac{R+z}{R-z}\right)\right]$$

$$= 50 + \frac{100}{\pi}\,\mathrm{Arg}\left(\frac{R+z}{R-z}\right)$$

$$= \frac{100}{\pi}\,\mathrm{Arg}\left[i\left(\frac{R+z}{R-z}\right)\right].$$

A similar connection exists between the Fourier series and the Laurent series of a function $f(z)$ analytic in an annulus $r_1 < |z| < r_2$. Here

$$f(z) = \sum_{n=-\infty}^\infty c_n z^n, \tag{3}$$

where

$$R^n c_n = \frac{R^n}{2\pi i} \int_{|z|=R} \frac{f(z)}{z^{n+1}} \, dz$$

$$= \frac{1}{2\pi} \int_0^{2\pi} f(Re^{i\phi}) e^{-in\phi} \, d\phi, \qquad r_1 < R < r_2.$$

Observe that

$$\int_0^{2\pi} \left| \sum_{n=-k}^k r^n c_n e^{in\phi} \right|^2 \, d\phi = \int_0^{2\pi} \sum_{n=-k}^k \sum_{m=-k}^k r^{n+m} c_n \bar{c}_m e^{i(n-m)\phi} \, d\phi$$

$$= 2\pi \sum_{n=-k}^k r^{2n} \left| c_n \right|^2,$$

since

$$\int_0^{2\pi} e^{i(n-m)\phi} \, d\phi = \frac{e^{i(n-m)\phi}}{i(n-m)} \bigg|_0^{2\pi} = 0, \qquad m \neq n.$$

Since the Laurent series representation converges uniformly in $r_1 < \rho_1 \leq |z| \leq \rho_2 < r_2$, we can interchange limits and integrals, obtaining **Parseval's identity**

$$\int_0^{2\pi} \left| f(re^{i\phi}) \right|^2 \, d\phi = \lim_{k \to \infty} \int_0^{2\pi} \left| \sum_{n=-k}^k r^n c_n e^{in\phi} \right|^2 \, d\phi$$

$$= 2\pi \sum_{n=-\infty}^\infty r^{2n} \left| c_n \right|^2.$$

If $z = e^{i\phi}$, the series in equations (1) and (3) may both be written in the form

$$\sum_{n=-\infty}^\infty c_n e^{in\phi} \tag{4}$$

by letting $c_{-n} = \bar{c}_n$, since $2 \, \mathrm{Re}(c_n e^{in\phi}) = c_n e^{in\phi} + \bar{c}_n e^{-in\phi}$.

EXAMPLE 3.

Let $f(z) = \sum_{n=-\infty}^\infty c_n z^n$ be analytic and one-to-one in a region containing the annulus $r \leq |z| \leq R$. Show that the area of the image of the annulus is

$$\pi \sum_{n=-\infty}^\infty n \, |c_n|^2 \, (R^{2n} - r^{2n}).$$

SOLUTION: In Exercise 11 of Section 5.1, we saw that the local change of scale of areas produced by the mapping $f(z)$ is $|f'(z)|^2$. Hence, the area of the image of the annulus is

$$\iint_{r \leqslant |z| \leqslant R} |f'(z)|^2 \, dx \, dy = \int_r^R \int_0^{2\pi} |f'(re^{i\phi})|^2 \, r \, d\phi \, dr.$$

Using the method used in proving Parseval's identity, we have

$$\int_0^{2\pi} |f'(re^{i\phi})|^2 \, d\phi = \lim_{k \to \infty} \int_0^{2\pi} \left| \sum_{n=-k}^{k} nc_n z^{n-1} \right|^2 d\phi$$

$$= 2\pi \sum_{n=-\infty}^{\infty} n^2 \, |c_n|^2 r^{2(n-1)}.$$

Hence, since the convergence is uniform,

$$\iint_{r \leqslant |z| \leqslant R} |f'(z)|^2 \, dx \, dy = 2\pi \sum_{n=-\infty}^{\infty} n^2 \, |c_n|^2 \int_r^R r^{2n-1} \, dr$$

$$= \pi \sum_{n=-\infty}^{\infty} n \, |c_n|^2 \, (R^{2n} - r^{2n}).$$

The Fourier series in equation (1) need not converge to $U(\phi)$. Consider, for example, a function $U_1(\phi)$ differing from $U(\phi)$ at only one point. Both functions have the same Fourier series, but it cannot represent them both at every point. In fact, continuous functions exist whose Fourier series diverge at all rational numbers ϕ in the interval $[0, 2\pi)$. The problem of convergence is of fundamental importance in the study of Fourier series.

Before studying the question of convergence, it is useful to define one-sided limits and derivatives. For $\epsilon > 0$, the limits

$$U(\phi + 0) = \lim_{\epsilon \to 0} U(\phi + \epsilon), \qquad U(\phi - 0) = \lim_{\epsilon \to 0} U(\phi - \epsilon)$$

are the right- and left-side limits, and

$$U'(\phi + 0) = \lim_{\epsilon \to 0} \frac{U(\phi + \epsilon) - U(\phi + 0)}{\epsilon},$$

$$U'(\phi - 0) = \lim_{\epsilon \to 0} \frac{U(\phi - \epsilon) - U(\phi - 0)}{-\epsilon}$$

are the right- and left-side derivatives, respectively, of the function U at ϕ. Observe that if U is continuous at ϕ, both one-sided limits coincide with $U(\phi)$, and if U is differentiable at ϕ, both one-sided derivatives agree with $U'(\phi)$.

A real function $U(\phi)$ is said to be *piecewise smooth* (pws) on $[a, b]$ if it has a continuous derivative at all but finitely many points at which the one-sided limits and derivatives of U exist.

Our next theorem settles the convergence problem for a useful class of functions.

THEOREM

Let $U(\phi)$ be a pws function on $[0, 2\pi]$ with period 2π, and let

$$c_n = \frac{1}{2\pi} \int_0^{2\pi} U(\theta)e^{-in\theta}\, d\theta.$$

Then

$$\lim_{k \to \infty} \sum_{n=-k}^{k} c_n e^{in\phi} = \tfrac{1}{2}\left[U(\phi + 0) + U(\phi - 0)\right].$$

(Note that if the Fourier series (4) converges, it agrees with the limit above, but the latter exists even when (4) diverges.)

PROOF†: If $U(\phi)$ is differentiable on $a < \phi < b$, then the integral

$$\int_a^b U(\phi)e^{ik\phi}\, d\phi = \frac{U(\phi)e^{ik\phi}}{ik}\bigg|_a^b - \frac{1}{ik}\int_a^b U'(\phi)e^{ik\phi}\, d\phi$$

vanishes as $k \to \infty$, since the last integral is bounded. Thus, the integral over the interval $[0, 2\pi]$ will also vanish as $k \to \infty$. Now

$$s_k(\phi) = \sum_{n=-k}^{k} c_n e^{in\phi}$$

$$= \frac{1}{2\pi} \int_0^{2\pi} U(\theta)\left[\sum_{n=-k}^{k} e^{in(\phi-\theta)}\right]\, d\theta$$

$$= \frac{1}{2\pi} \int_0^{2\pi} U(\theta)\left[\frac{e^{-ik(\phi-\theta)} - e^{i(k+1)(\phi-\theta)}}{1 - e^{i(\phi-\theta)}}\right]\, d\theta$$

$$= \frac{1}{2\pi} \int_0^{2\pi} \frac{\sin(k + \tfrac{1}{2})(\phi - \theta)}{\sin\tfrac{1}{2}(\phi - \theta)} U(\theta)\, d\theta,$$

so setting $t = \theta - \phi$, integrating over the interval $[-\pi, \pi]$, and dividing the range of integration into halves, we can write

$$s_k (\phi) = \frac{1}{2\pi} \int_0^\pi \frac{\sin(k + \frac{1}{2})t}{\sin \frac{1}{2}t} \left[U(\phi + t) + U(\phi - t) \right] dt$$

(see Figure 6.11).

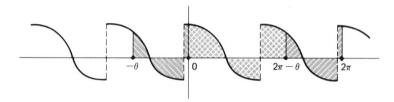

FIGURE 6.11.

In particular, if $U(\phi) = 1$ for all ϕ, then $c_0 = 1$ and $c_n = 0$ for $n \neq 0$, so that

$$1 = \frac{1}{2\pi} \int_0^\pi \frac{\sin(k + \frac{1}{2})t}{\sin \frac{1}{2}t} \cdot 2 \ dt.$$

Multiplying this identity by $[U(\phi + 0) + U(\phi - 0)]/2$, we have

$$s_k (\phi) - \frac{U(\phi + 0) + U(\phi - 0)}{2}$$

$$= \frac{1}{2\pi} \int_0^\pi \frac{\sin(k + \frac{1}{2})t}{\sin \frac{1}{2}t} \left[U(\phi + t) - U(\phi + 0) + U(\phi - t) - U(\phi - 0) \right] dt.$$

Since the one-sided derivatives of U exist, the function

$$\frac{t}{\sin \frac{1}{2}t} \left[\frac{U(\phi + t) - U(\phi + 0)}{t} + \frac{U(\phi - t) - U(\phi - 0)}{t} \right]$$

is piecewise smooth for $0 \leqslant t \leqslant \pi$, so the first observation in this proof applies and the integral

$$\int_0^\pi \frac{\sin(k + \frac{1}{2})t}{\sin \frac{1}{2}t} \left[U(\phi + t) - U(\phi + 0) + U(\phi - t) - U(\phi - 0) \right] dt$$

vanishes as $k \to \infty$. Thus,

$$\lim_{k \to \infty} s_k (\phi) = \frac{1}{2} \left[U(\phi + 0) + U(\phi - 0) \right]. \blacksquare$$

EXAMPLE 4.

Show that

$$\frac{\pi - 2}{4} = \frac{1}{1 \cdot 3} - \frac{1}{3 \cdot 5} + \frac{1}{5 \cdot 7} - \frac{1}{7 \cdot 9} + \ldots$$

by calculating the Fourier series of the function

$$U(\phi) = \begin{cases} \sin \phi, & 0 \leqslant \phi \leqslant \pi, \\ 0, & \pi \leqslant \phi \leqslant 2\pi. \end{cases}$$

SOLUTION: $U(\phi)$ is piecewise smooth and

$$2\pi c_n = \int_0^\pi \sin \phi e^{-in\phi} \, d\phi = \frac{1}{2i} \int_0^\pi e^{i(1-n)\phi} - e^{-i(1+n)\phi} \, d\phi,$$

implying $c_{2k} = [\pi(1 - 4k^2)]^{-1}$, $c_{\pm 1} = \pm[4i]^{-1}$, the remaining coefficients being zero. Since $c_{2k} = c_{-2k}$, $c_{-1} = -c_1$, we have

$$c_{2k} e^{2k\phi i} + c_{-2k} e^{-2k\phi i} = 2c_{2k} \cos 2k\phi,$$
$$c_1 e^{i\phi} + c_{-1} e^{-i\phi} = 2ic_1 \sin \phi,$$

and

$$U(\phi) = \frac{\sin \phi}{2} + \frac{1}{\pi} + \frac{2}{\pi} \sum_{k=1}^\infty \frac{\cos 2k\phi}{(1 - 2k)(1 + 2k)}.$$

In particular, for $\phi = \pi/2$, $U(\pi/2) = 1$, so

$$1 = \frac{1}{2} + \frac{1}{\pi} - \frac{2}{\pi} \sum_{k=1}^\infty \frac{(-1)^k}{(2k - 1)(2k + 1)}.$$

Hence,

$$\frac{\pi - 2}{4} = \frac{1}{1 \cdot 3} - \frac{1}{3 \cdot 5} + \frac{1}{5 \cdot 7} - \frac{1}{7 \cdot 9} + \ldots$$

EXERCISES

1. Let the plate G shaped like the unit disk have temperature $u(e^{i\phi}) = \phi$ degrees for $0 \leqslant \phi < 2\pi$. Show that the temperature in G for $z \neq 1$ is given by

$$u(z) = \pi + 2 \operatorname{Arg}(1 - z).$$

2. Find the temperature in $|z| < 1$, given the temperature on the boundary is $u(e^{i\phi}) = \cosh \phi$.
3. Find the Fourier series for $U(\phi) = \pi$ on $0 \leqslant \phi \leqslant \pi$ and vanishing on $\pi < \phi < 2\pi$.
4. Find the Fourier series for $U(\phi) = \phi^2$ on $0 \leqslant \phi < 2\pi$.
5. Use Parseval's identity to prove Liouville's theorem. (*Hint:* Show that $|c_n| \leqslant Mr^{-n}$, for all n, where $|f(z)| \leqslant M$.)
6. Applying Parseval's identity to the function $U(\phi) = \phi$, show that

$$\frac{\pi^2}{6} = \sum_{n=1}^{\infty} \frac{1}{n^2} .$$

7. Show that

$$\frac{\pi^4}{90} = \sum_{n=1}^{\infty} \frac{1}{n^4} .$$

8. Applying Parseval's identity to the function $f(z) = (1 - z)^{-1}$, prove that

$$\frac{1}{2\pi} \int_0^{2\pi} \frac{d\phi}{1 - 2r \cos \phi + r^2} = \frac{1}{1 - r^2}, \qquad 0 \leqslant r < 1.$$

9. Applying Parseval's identity to the function

$$f(z) = 1 + z + \ldots + z^{n-1},$$

prove that

$$\int_0^{2\pi} \left(\sin \frac{n\phi}{2} \bigg/ \sin \frac{\phi}{2} \right)^2 d\phi = 2\pi n.$$

(*Hint:* $f(z) = (z^n - 1)/(z - 1)$ for $z \neq 1$.)

10. For computational purposes, the coefficients of the Fourier series in equation (4) are approximated by sums of the form

$$Nc_n = \sum_{k=0}^{N-1} U\left(\frac{2\pi k}{N}\right) e^{-2\pi i k n/N} .$$

If $N = N_1 \cdot N_2$, $k = k_1 N_2 + k_2$, and $n = n_2 N_1 + n_1$, $0 \leqslant k_j, n_j < N_j$, show that

$$Nc_n = \sum_{k_2=0}^{N_2-1} W_{N_2}^{n_2 k_2} \left\{ W_N^{n_1 k_2} \sum_{k_1=0}^{N_1-1} U\left(\frac{2\pi k}{N}\right) W_{N_1}^{n_1 k_1} \right\},$$

where $W_N = e^{-2\pi i/N}$. Thus, $c_n = c_{n_1, n_2}$ is obtained as a transform of the N_2 coefficients

$$c_{n_1,k_2} = W^{n_1 k_2} \sum_{k_1=0}^{N_1-1} U\left(\frac{2\pi k}{N}\right) W_{N_1}^{n_1 k_1}, \quad 0 \leqslant k_2 < N_2.$$

11. Extend Exercise 10 to the situation where $N = N_1 \cdot N_2 \cdot \ldots \cdot N_m$. This is the procedure that is used in the **fast Fourier transform**.

6.5 FOURIER TRANSFORMS

The Fourier series of a function $U(\phi)$ of period 2π may be written in the form

$$\sum_{n=-\infty}^{\infty} c_n e^{in\phi}, \qquad c_n = \frac{1}{2\pi} \int_{-\pi}^{\pi} U(\phi) e^{-in\phi} \, d\phi.$$

Similarly, if $U(\phi)$ has period $2\pi\lambda$, setting $\psi = \phi/\lambda$ yields a function of period 2π; hence, $U(\phi)$ has the Fourier series

$$\sum_{n=-\infty}^{\infty} c_n e^{in\psi} = \sum_{n=-\infty}^{\infty} c_n e^{in\phi/\lambda},$$

where

$$c_n = \frac{1}{2\pi} \int_{-\pi}^{\pi} U(\lambda\psi) e^{-in\psi} \, d\psi = \frac{1}{2\pi\lambda} \int_{-\pi\lambda}^{\pi\lambda} U(\phi) e^{-in\phi/\lambda} \, d\phi.$$

However, many interesting functions are not periodic, for example, a single unrepeated pulse. We might hope to approximate this situation by a function consisting of identical pulses each a distance $2\pi\lambda$ apart, investigating the effect on its Fourier series as $\lambda \to \infty$ (see Figure 6.12). Letting $t_n = n/\lambda$, defining

$$u(t) = \frac{1}{\sqrt{2\pi}} \int_{-\pi\lambda}^{\pi\lambda} U(\phi) e^{-it\phi} \, d\phi,$$

and observing that $t_{n+1} - t_n = 1/\lambda$, we can write the Fourier series in the form

$$\sum_{n=-\infty}^{\infty} \frac{u(t_n)}{\lambda\sqrt{2\pi}} \, e^{it_n\phi} = \frac{1}{\sqrt{2\pi}} \sum_{n=-\infty}^{\infty} u(t_n) e^{it_n\phi} (t_{n+1} - t_n),$$

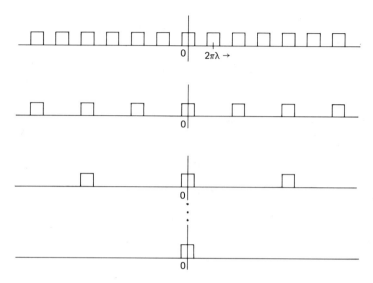

FIGURE 6.12. Pulse trains of decreasing frequency

very similar in appearance to the sums by which Riemann integrals are defined. Letting $\lambda \to \infty$ and ignoring all technical difficulties, we formally obtain the expressions

$$\hat{U}(\phi) = \frac{1}{\sqrt{2\pi}} \int_{-\infty}^{\infty} u(t)e^{it\phi}\, dt, \qquad u(t) = \frac{1}{\sqrt{2\pi}} \int_{-\infty}^{\infty} U(\phi)e^{-it\phi}\, d\phi.$$

The similarity between the formulas for $\hat{U}$ and u is unmistakable; they are said to constitute a **Fourier transform pair**, and $u(t)$ is called the **Fourier transform** of $U(\phi)$. As in Section 6.4, the main problem is to discover under what circumstances the values $\hat{U}(\phi)$ and $U(\phi)$ coincide, as then $\hat{U}$ provides an **inversion** formula for the Fourier transform u. This has the effect of doubling the size of a given table of integrals, for if a closed form solution is known for the Fourier transform $u(t)$, one is also known for its inverse. The next theorem provides useful conditions under which $\hat{U}(\phi)$ and $U(\phi)$ agree, but is by no means the best theorem of this type.

FOURIER INTEGRAL THEOREM

If $U(\phi)$ is piecewise smooth and $|U(\phi)|$ is integrable on $-\infty < \phi < \infty$, then

$$\text{PV } \hat{U}(\phi) = \text{PV } \frac{1}{\sqrt{2\pi}} \int_{-\infty}^{\infty} u(t)e^{it\phi} \, dt = \frac{1}{2}[U(\phi+0) + U(\phi-0)].$$

PROOF†: Since $|U(\phi)|$ is integrable, the integral

$$\int_{-\infty}^{\infty} U(\phi)e^{it(\theta-\phi)} \, d\phi$$

converges uniformly with respect to t over any finite range. We may therefore integrate with respect to t over the interval $(-T, T)$, and invert the order of integration. Thus,

$$\int_{-T}^{T} \int_{-\infty}^{\infty} U(\phi)e^{it(\theta-\phi)} \, d\phi \, dt = \int_{-\infty}^{\infty} U(\phi) \int_{-T}^{T} e^{it(\theta-\phi)} \, dt \, d\phi$$

$$= 2 \int_{-\infty}^{\infty} U(\phi) \frac{\sin T(\theta-\phi)}{\theta-\phi} \, d\phi,$$

and $\Phi > |\theta| + 1$ (θ fixed) can be chosen such that

$$\int_{|\phi|>\Phi} |U(\phi)| \, d\phi < \frac{\epsilon}{4}.$$

Then $\Phi - |\theta| > 1$, implying that

$$\left| \int_{|\phi|>\Phi} U(\phi) \frac{\sin T(\theta-\phi)}{\theta-\phi} \, d\phi \right| \leq \int_{|\phi|>\Phi} |U(\phi)| \, d\phi < \frac{\epsilon}{4}.$$

As in the first part of the proof of the convergence theorem for Fourier series in Section 6.4,

$$\int_{\theta+\delta}^{\Phi} \frac{U(\phi)}{\theta-\phi} \sin T(\theta-\phi) \, d\phi \to 0 \quad \text{as} \quad T \to \infty, \tag{1}$$

and similarly for the integral on $[-\Phi, \theta-\delta]$, implying that for large T,

$$\left| \int_{-T}^{T} \int_{-\infty}^{\infty} U(\phi)e^{it(\theta-\phi)} \, d\phi \, dt - 2\int_{\theta-\delta}^{\theta+\delta} U(\phi) \frac{\sin T(\theta-\phi)}{\theta-\phi} \, d\phi \right| < \epsilon.$$

But changing variables, we find

$$\int_{\theta-\delta}^{\theta+\delta} U(\phi) \frac{\sin T(\theta-\phi)}{\theta-\phi} \, d\phi = \int_{0}^{\delta} \frac{\sin T\phi}{\phi} [U(\theta+\phi) + U(\theta-\phi)] \, d\phi,$$

and it follows that

$$\text{PV } \hat{U}(\theta) = \lim_{T \to \infty} \frac{1}{\pi} \int_0^\delta \frac{\sin T\phi}{\phi} \left[U(\theta + \phi) + U(\theta - \phi) \right] d\phi. \quad (2)$$

Since the one-sided derivatives of U exist, the function

$$\frac{U(\theta + \phi) - U(\theta + 0) + U(\theta - \phi) - U(\theta - 0)}{\phi}$$

is piecewise smooth. By (1) the integral

$$\int_0^\delta \sin T\phi \left[\frac{U(\theta + \phi) + U(\theta - \phi) - U(\theta + 0) - U(\theta - 0)}{\phi} \right] d\phi \to 0$$

as $T \to \infty$, implying that

$$\text{PV } \hat{U}(\theta) = \lim_{T \to \infty} \frac{U(\theta + 0) + U(\theta - 0)}{\pi} \int_0^\delta \frac{\sin T\phi}{\phi} d\phi$$

$$= \frac{U(\theta + 0) + U(\theta - 0)}{\pi} \lim_{T \to \infty} \int_0^{T\delta} \frac{\sin \psi}{\psi} d\psi$$

$$= \frac{1}{2} \left[U(\theta + 0) + U(\theta - 0) \right]$$

by Dirichlet's integral (Exercise 16, Section 2.2 or Example 1, Section 4.4). ∎

EXAMPLE 1.

   Suppose $U(\phi) = e^{-|\phi|}$; then $|U(\phi)|$ is integrable and $U(\phi)$ is piecewise smooth. Its Fourier transform

$$u(t) = \frac{1}{\sqrt{2\pi}} \left[\int_{-\infty}^0 e^{(-it+1)\phi} \, d\phi + \int_0^\infty e^{(-it-1)\phi} \, d\phi \right]$$

$$= \frac{2}{\sqrt{2\pi}(1 + t^2)}$$

satisfies

$$e^{-|\phi|} = \frac{1}{\pi} \int_{-\infty}^\infty \frac{e^{it\phi}}{1 + t^2} \, dt = \frac{2}{\pi} \int_0^\infty \frac{\cos t\phi}{1 + t^2} \, dt.$$

Compare this result with Example 1 in Section 4.3.

EXAMPLE 2.

Separating integrals as in the computation above, we obtain

$$\int_{-\infty}^{\infty} e^{-y\,|t|-i(\phi-x)t}\,dt = \frac{2y}{(\phi-x)^2 + y^2},$$

transforming Poisson's integral formula for the upper half plane (Exercise 4, Section 6.2) into

$$U(z) = \frac{y}{\pi} \int_{-\infty}^{\infty} \frac{U(\phi)}{(\phi-x)^2 + y^2}\,d\phi$$

$$= \frac{1}{2\pi} \int_{-\infty}^{\infty} \int_{-\infty}^{\infty} U(\phi)e^{-y|t|-i(\phi-x)t}\,dt\,d\phi.$$

Reversing the order of integration and letting $u(t)$ be the Fourier transform of $U(\phi)$ yields

$$U(z) = \frac{1}{\sqrt{2\pi}} \int_{-\infty}^{\infty} u(t)e^{-y\,|t|+ixt}\,dt$$

$$= \left\{ \mathrm{Re}\,\frac{2}{\sqrt{2\pi}} \int_{0}^{\infty} u(t)e^{izt}\,dt \right\}, \qquad (3)$$

since $\overline{u(t)} = u(-t)$ for real-valued functions $U(\phi)$ and

$$2\,\mathrm{Re}\,u(t)e^{izt} = u(t)e^{izt} + \overline{u(t)}e^{-\bar{i}zt}$$

$$= u(t)e^{(-y+ix)t} + u(-t)e^{(y+ix)(-t)}.$$

Formula (3) is the analog for the half plane of the Fourier series expansion of Poisson's integral formula in the disk.

EXERCISES

Find the Fourier transforms of the functions given in Exercises 1–4. (*Hint:* Use the contour integrals in Section 4.3.)

1. $\dfrac{b}{x^2 + b^2}$

2. $\dfrac{x}{x^2 + b^2}$

3. $\dfrac{x^2}{(x^2 + b^2)^2}$

4. $\dfrac{1}{x^4 + b^4}$

Find the Fourier transforms of the functions given in Exercises 5–8. (*Hint:* Use the integrals in Sections 2.2 and 4.5.)

5. e^{-kx^2}

6. xe^{-kx^2}

7. $\dfrac{1}{\sinh x}$

8. $\dfrac{x}{\sinh x}$

9. Suppose $U(\phi) = 1/\sqrt{\phi}$ on $0 < \phi < \infty$ and vanishes on $-\infty < \phi \leqslant 0$. Find a function harmonic in the upper half plane having $U(\phi)$ as its boundary values.

10. Let the plate G shaped like the upper half plane have a temperature of $1°$ on the interval $[-1, 1]$ and $0°$ on the remainder of the real axis. Find the temperature at every point of G.

11. Suppose $U = \hat{U}$ at almost every point in $(-\infty, \infty)$. Without worrying about convergence, show that

$$\int_{-\infty}^{\infty} U(\phi)\overline{V(\phi)}\, d\phi = \int_{-\infty}^{\infty} u(t)\overline{v(t)}\, dt,$$

where u, v are the Fourier transforms of U, V, respectively. Then obtain *Parseval's identity* for integrals

$$\int_{-\infty}^{\infty} |U(\phi)|^2 \, d\phi = \int_{-\infty}^{\infty} |u(t)|^2 \, dt.$$

12. Show that

$$\int_{-\infty}^{\infty} e^{-2|\phi|} \, d\phi = \frac{2}{\pi} \int_{-\infty}^{\infty} \frac{dt}{(1 + t^2)^2}.$$

(*Hint:* Use Exercise 11.)

6.6 LAPLACE TRANSFORMS

Fourier transform methods often cannot be used in analyzing functions that are not absolutely integrable on $(-\infty, \infty)$. For example, the Heaviside function

$$H(\phi - a) = \begin{cases} 1, & \phi > a, \\ 0, & \phi < a, \end{cases}$$

does not have a Fourier transform, as the integral

$$\int_a^\infty e^{-it\phi}\, d\phi$$

diverges. This is because the multiplier $e^{-it\phi}$ does not tend to zero as $\phi \to \infty$, leading us to try multipliers of the form $e^{-s\phi} = e^{-(q+it)\phi}$, which vanish when $q > 0$ as $\phi \to \infty$. The function

$$\mathcal{L}_2\{U(\phi)\}(s) = \int_{-\infty}^\infty U(\phi)e^{-s\phi}\, d\phi \tag{1}$$

is called the **two-sided Laplace transform** of the function $U(\phi)$. Writing $s = q + it$, equation (1) becomes

$$\int_{-\infty}^\infty U(\phi)e^{-q\phi}\, e^{-it\phi}\, d\phi,$$

which is the Fourier transform of the function $\sqrt{2\pi}U(\phi)e^{-q\phi}$.

Instead of developing an analysis of the two-sided Laplace transform, it is more convenient to write the integral in (1) in two parts as

$$\int_{-\infty}^\infty U(\phi)e^{-s\phi}\, d\phi = \int_{-\infty}^0 U(\phi)e^{-s\phi}\, d\phi + \int_0^\infty U(\phi)e^{-s\phi}\, d\phi$$

$$= \int_0^\infty U(-\phi)e^{s\phi}\, d\phi + \int_0^\infty U(\phi)e^{-s\phi}\, d\phi.$$

Then a study of the properties of the integral

$$\mathcal{L}\{U(\phi)\}(s) = \int_0^\infty U(\phi)e^{-s\phi}\, d\phi, \tag{2}$$

called the (one-sided) **Laplace transform** of $U(\phi)$, enables us to investigate the behavior of the two-sided Laplace transform, since

$$\mathcal{L}_2\{U(\phi)\}(s) = \mathcal{L}\{U(-\phi)\}(-s) + \mathcal{L}\{U(\phi)\}(s).$$

The one-sided Laplace transform defined in equation (2) has many properties similar to those of power series. We shall prove the existence of a half plane of convergence analogous to the notion of the radius of convergence in Abel's theorem. The two-sided Laplace transform will converge in a strip $a < \text{Re } s < b$ $(a \leqslant b)$ in analogy with the development of the Laurent series.

THEOREM

Suppose $U(\phi)$ is piecewise smooth and of **exponential order** (that is, there are real constants a and Φ such that $e^{-a\phi}\,|\,U(\phi)\,|$ is bounded for all $\phi > \Phi$). Then the Laplace transform $\mathcal{L}\{U\}(s)$ is analytic on Re $s > a$ (a *half plane of convergence*).

PROOF: Letting M be a finite bound for $e^{-a\phi}|U(\phi)|$ on $\phi > \Phi$, we have

$$\int_0^\infty |U(\phi)e^{-s\phi}|\,d\phi \leqslant \int_0^\Phi |U(\phi)e^{-s\phi}|\,d\phi + M \int_\Phi^\infty |e^{-(s-a)\phi}|\,d\phi.$$

But $U(\phi)$ is bounded on $[0, \Phi]$, since it is piecewise continuous, implying that the first integral is finite, and

$$\int_\Phi^\infty |e^{-(s-a)\phi}|\,d\phi = \int_\Phi^\infty e^{-(\mathrm{Re}\,s-a)\phi}\,d\phi = \frac{e^{-(\mathrm{Re}\,s-a)\Phi}}{\mathrm{Re}\,s - a} < \infty,$$

so the Laplace transform of $U(\phi)$ converges absolutely on Re $s > a$ (see Figure 6.13).

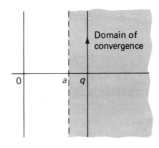

FIGURE 6.13. Half plane of convergence

This last equation implies that the Laplace transform converges uniformly in any closed set D lying entirely in the half plane of convergence, because

$$\left|\int_\phi^\infty U(\phi)e^{-s\phi}\,d\phi\right| \leqslant M \frac{e^{-(\mathrm{Re}\,s-a)\phi}}{\mathrm{Re}\,s - a}$$

when $\phi > \Phi$, and ϕ can be chosen so as to make the right side of the equation above smaller than a preselected $\epsilon > 0$, for all s in D.

For integers $n \geqslant 0$, the functions

$$u_n(s) = \int_0^n U(\phi)e^{-s\phi} \, d\phi$$

are analytic in the region Re $s > a$ because

$$u_n'(s) = -\int_0^n \phi \, U(\phi)e^{-s\phi} \, d\phi.$$

Hence, by Weierstrass's theorem, the series of analytic functions

$$\sum_{n=0}^{\infty} [u_{n+1}(s) - u_n(s)] = \int_0^{\infty} U(\phi)e^{-s\phi} \, d\phi = \mathcal{L}\{U(\phi)\}(s)$$

is analytic in Re $s > a$. In particular,

$$\frac{d}{ds} \mathcal{L}\{U(\phi)\} = -\int_0^{\infty} \phi \, U(\phi)e^{-s\phi} \, d\phi = -\mathcal{L}\{\phi \, U(\phi)\}. \quad \blacksquare \qquad (3)$$

EXAMPLE 1.

Show that

$$\mathcal{L}\{e^{-z\phi}\} = \frac{1}{s+z}$$

for Re $s > -$Re z.

SOLUTION: In the region Re $s > -$Re z, we have

$$\mathcal{L}\{e^{-z\phi}\} = \int_0^{\infty} e^{-z\phi} e^{-s\phi} \, d\phi = \frac{e^{-(s+z)\phi}}{-(s+z)} \bigg|_0^{\infty} = \frac{1}{s+z}.$$

EXAMPLE 2.

Verify that

$$\mathcal{L}\{\phi^n\} = \frac{n!}{s^{n+1}}$$

for Re $s > 0$, $n = 0, 1, 2, \ldots$.

SOLUTION: By integrating by parts repeatedly, we obtain

$$\mathcal{L}\{\phi^n\} = \int_0^\infty \phi^n e^{-s\phi}\, d\phi$$

$$= \frac{\phi^n e^{-s\phi}}{-s}\bigg|_0^\infty + \frac{n}{s}\int_0^\infty \phi^{n-1} e^{-s\phi}\, d\phi$$

$$= \frac{n}{s}\left[\frac{\phi^{n-1}e^{-s\phi}}{-s}\bigg|_0^\infty + \frac{n-1}{s}\int_0^\infty \phi^{n-2} e^{-s\phi}\, d\phi\right]$$

$$\vdots$$

$$= \frac{n!}{s^n}\left[\int_0^\infty e^{-s\phi}\, d\phi\right] = \frac{n!}{s^{n+1}}\,.$$

We could also set $z = 0$ in Example 1 and use equation (3) repeatedly to get the result.

Two other results that are useful in calculating Laplace transforms are the shifting theorems. Observe that

$$\int_0^\infty (U(\phi)e^{-z\phi})e^{-s\phi}\, d\phi = \int_0^\infty U(\phi)e^{-(s+z)\phi}\, d\phi,$$

yielding the **first shifting theorem**

$$\mathcal{L}\{U(\phi)e^{-z\phi}\}(s) = \mathcal{L}\{U(\phi)\}(s+z),$$

where the last expression means that each s in $\mathcal{L}\{U\}$ is replaced by $s + z$. If $a \geq 0$, we have

$$\int_0^\infty U(\phi)H(\phi - a)e^{-s\phi}\, d\phi = \int_a^\infty U(\phi)e^{-s\phi}\, d\phi.$$

Substituting $\phi = \theta + a$ in the right side of this equation, we get

$$\int_0^\infty U(\phi)H(\phi - a)e^{-s\phi}\, d\phi = e^{-as}\int_0^\infty U(\theta + a)e^{-s\theta}\, d\theta.$$

This equation can be rewritten as the **second shifting theorem**

$$\mathcal{L}\{U(\phi)H(\phi - a)\} = e^{-as}\,\mathcal{L}\{U(\phi + a)\}.$$

The next two examples illustrate the use of the shifting theorems.

EXAMPLE 3.

Verify for any integer $n \geqslant 0$,

$$\mathcal{L}\{\phi^n \, e^{-z\phi}\} = \frac{n!}{(s+z)^{n+1}}, \qquad \text{Re } s > -\text{Re } z.$$

SOLUTION: We could use equation (3) repeatedly to get this result, but it is much easier to use the first shifting theorem when we already have the result of Example 2. Thus,

$$\mathcal{L}\{\phi^n \, e^{-z\phi}\} = \mathcal{L}\{\phi^n\} \, (s+z) = \frac{n!}{(s+z)^{n+1}}.$$

EXAMPLE 4.

Show that if $a > 0$, then

$$\mathcal{L}\{e^{-z\phi} \, H(\phi - a)\} = \frac{e^{-a(s+z)}}{s+z}, \qquad \text{Re } s > -\text{Re } z.$$

SOLUTION: By the second shifting theorem and Example 1,

$$\mathcal{L}\{e^{-z\phi} \, H(\phi - a)\} = e^{-as} \, \mathcal{L}\{e^{-z(\phi+a)}\}$$
$$= e^{-a(s+z)} \, \mathcal{L}\{e^{-z\phi}\}.$$

In particular, observe that

$$\mathcal{L}\{H(\phi - a)\} = \frac{e^{-as}}{s}, \qquad \text{Re } s > 0,$$

by setting $z = 0$.

Note that the Laplace transform is linear:

$$\mathcal{L}\{aU(\phi) + bV(\phi)\} = \int_0^\infty [aU(\phi) + bV(\phi)] \, e^{-s\phi} \, d\phi$$

$$= a \int_0^\infty U(\phi)e^{-s\phi} \, d\phi + b \int_0^\infty V(\phi)e^{-s\phi} \, d\phi$$

$$= a\mathcal{L}\{U(\phi)\} + b\mathcal{L}\{V(\phi)\}.$$

EXAMPLE 5.

Show that when Re $s > |\text{Im } z|$,

$$\mathcal{L}\{\cos z\phi\} = \frac{s}{s^2 + z^2} \qquad \text{and} \qquad \mathcal{L}\{\sin z\phi\} = \frac{z}{s^2 + z^2}.$$

SOLUTION: Observe that

$$\mathcal{L}\{\cos z\phi\} = \mathcal{L}\{\tfrac{1}{2}(e^{iz\phi} + e^{-iz\phi})\}$$

$$= \tfrac{1}{2}\left(\frac{1}{s - iz} + \frac{1}{s + iz}\right) = \frac{s}{s^2 + z^2},$$

$$\mathcal{L}\{\sin z\phi\} = \mathcal{L}\left\{\frac{1}{2i}(e^{iz\phi} - e^{-iz\phi})\right\}$$

$$= \frac{1}{2i}\left(\frac{1}{s - iz} - \frac{1}{s + iz}\right) = \frac{z}{s^2 + z^2}.$$

DIFFERENTIATION PROPERTY

If $U(\phi)$ and $U'(\phi)$ are pws functions of exponential order, then

$$\mathcal{L}\{U'(\phi)\} = s\,\mathcal{L}\{U(\phi)\} - U(0^+), \tag{4}$$

where $U(0^+)$ is the right-side limit of U.

PROOF: Since U and U' are of exponential order, all terms in equation (4) exist on Re $s > a$. To prove (4), integrate by parts:

$$\int_0^\infty U'(\phi)e^{-s\phi}\,d\phi = U(\phi)e^{-s\phi}\,\Big|_0^\infty + s\int_0^\infty U(\phi)e^{-s\phi}\,d\phi. \quad \blacksquare$$

EXAMPLE 6.

Verify that

$$\mathcal{L}\{\sin^2 z\phi\} = \frac{2z^2}{s(s^2 + 4z^2)}, \qquad \text{Re } s > 2|\text{Im } z|.$$

SOLUTION: Since

$$\frac{d}{d\phi}\left(\frac{\phi}{2} - \frac{\sin 2z\phi}{4z}\right) = \tfrac{1}{2}(1 - \cos 2z\phi) = \sin^2 z\phi,$$

by the differentiation property and linearity,

$$\mathcal{L}\{\sin^2 z\phi\} = s\mathcal{L}\left\{\frac{\phi}{2} - \frac{\sin 2z\phi}{4z}\right\}$$

$$= s\left(\frac{1}{2s^2} - \frac{1}{2(s^2 + 4z^2)}\right) = \frac{2z^2}{s(s^2 + 4z^2)}$$

when $\operatorname{Re} s > |\operatorname{Im} 2z|$.

The **convolution** of two functions $U(\phi)$ and $V(\phi)$ is the function

$$U * V(\phi) = \int_0^\phi U(t)V(\phi - t)\, dt.$$

Note that the roles of U and V can be interchanged without changing the value of the convolution. If the functions U, V are absolutely integrable on $(0, \infty)$, the convolution satisfies the identity

$$\mathcal{L}\{U * V\} = \mathcal{L}\{U\}\mathcal{L}\{V\}, \tag{5}$$

that is, *the Laplace transform of the convolution is the product of the transforms of the functions.* The hypotheses above are sufficient to permit reversing the order of integration

$$\mathcal{L}\{U * V\} = \int_0^\infty \left[\int_0^\phi U(t)V(\phi - t)\, dt\right] e^{-s\phi}\, d\phi$$

$$= \int_0^\infty \left[\int_0^\infty U(t)V(\phi - t)H(\phi - t)\, dt\right] e^{-s\phi}\, d\phi$$

$$= \int_0^\infty U(t)\left[\int_0^\infty V(\phi - t)H(\phi - t)e^{-s\phi}\, d\phi\right] dt$$

$$= \int_0^\infty U(t)\mathcal{L}\{V(\phi - t)H(\phi - t)\}\, dt,$$

which by the second shifting theorem yields

$$\mathcal{L}\{U * V\} = \mathcal{L}\{V\}\int_0^\infty U(t)e^{-ts}\, dt = \mathcal{L}\{U\}\mathcal{L}\{V\}.$$

The next example gives an indication of the importance of the notion of convolution.

EXAMPLE 7.

Obtain a particular solution of the differential equation
$$U''(\phi) + 2wU'(\phi) + (w^2 + z^2)U(\phi) = V(\phi)$$
with $U(\phi) = U'(\phi) = 0$.

SOLUTION: Using the differentiation property, we obtain
$$[s^2 + 2ws + (w^2 + z^2)] \, \mathcal{L}\{U(\phi)\} = \mathcal{L}\{V(\phi)\}.$$
But by the first shifting theorem, we have
$$\mathcal{L}\{e^{-w\phi} \sin z\phi\} = \frac{z}{(s + w)^2 + z^2}.$$
Thus,
$$\mathcal{L}\{U(\phi)\} = \mathcal{L}\{z^{-1} e^{-w\phi} \sin z\phi\}\mathcal{L}\{V(\phi)\},$$
and we find the solution
$$U(\phi) = \frac{1}{z} \int_0^\phi e^{-wt} \sin(zt) V(\phi - t) \, dt.$$

This last example motivates the important concept of a transfer function. Many physical systems may be thought of as devices that transform a given input function V into an output function U. Assuming all initial conditions are zero when $\phi = 0$ and taking the Laplace transform of the equations describing the system, we obtain the expression

$$\mathcal{L}\{U(\phi)\} = \frac{\mathcal{L}\{V(\phi)\}}{Z(s)},$$

where $Z(s)^{-1}$, the **transfer function**, is independent of V. Let U_H be the output function when $V(\phi) = H(\phi)$. Then, using Example 4, we find

$$Z(s) \, \mathcal{L}\{U_H\} = \mathcal{L}\{H(\phi)\} = \frac{1}{s}$$

or

$$\mathcal{L}\{U\} = \frac{s\mathcal{L}\{V\}}{sZ(s)} = s\mathcal{L}\{U_H\}\mathcal{L}\{V\} = s\mathcal{L}\{U_H * V\}.$$

So, by the differentiation property,

$$U(\phi) = (U_H * V)'(\phi) = \int_0^\phi U_H(t)V'(\phi - t)\, dt + U_H(\phi)V(0). \quad (6)$$

By commuting $\mathcal{L}\{U_H\}$ and $\mathcal{L}\{V\}$ above, it also follows that

$$U(\phi) = (V * U_H)'(\phi) = \int_0^\phi V(t)U_H'(\phi - t)\, dt, \quad (7)$$

since the initial conditions imply $U_H(0) = 0$. Equations (6) and (7) are called **Duhamel's formulas,** and they express the response of a system to an input function $V(\phi)$ in terms of the experimentally accessible response to the Heaviside function.

EXERCISES

Verify the Laplace transforms and regions of convergence in Exercises 1–13.

1. $\mathcal{L}\{\cosh z\phi\} = \dfrac{s}{s^2 - z^2}$, $\quad$ Re $s > |$ Re $z|$

2. $\mathcal{L}\{\sinh z\phi\} = \dfrac{z}{s^2 - z^2}$, $\quad$ Re $s > |$ Re $z|$

3. $\mathcal{L}\{\phi \sin z\phi\} = \dfrac{2sz}{(s^2 + z^2)^2}$, $\quad$ Re $s > |$ Im $z|$

4. $\mathcal{L}\{\phi \cos z\phi\} = \dfrac{s^2 - z^2}{(s^2 + z^2)^2}$, $\quad$ Re $s > |$ Im $z|$

5. $\mathcal{L}\{\phi \sinh z\phi\} = \dfrac{2sz}{(s^2 - z^2)^2}$, $\quad$ Re $s > |$ Re $z|$

6. $\mathcal{L}\{\phi \cosh z\phi\} = \dfrac{s^2 + z^2}{(s^2 - z^2)^2}$, $\quad$ Re $s > |$ Re $z|$

7. $\mathcal{L}\{e^{-w\phi} \sin z\phi\} = \dfrac{z}{(s + w)^2 + z^2}$, $\quad$ Re$(s + w) > |$ Im $z|$

8. $\mathcal{L}\{e^{-w\phi} \cos z\phi\} = \dfrac{s+w}{(s+w)^2 + z^2}$, $\text{Re}(s+w) > |\text{Im } z|$

9. $\mathcal{L}\{\phi^2 \sin z\phi\} = \dfrac{2z(3s^2 - z^2)}{(s^2 + z^2)^3}$, $\text{Re } s > |\text{Im } z|$

10. $\mathcal{L}\{\phi^2 \cos z\phi\} = \dfrac{2s(s^2 - 3z^2)}{(s^2 + z^2)^3}$, $\text{Re } s > |\text{Im } z|$

11. $\mathcal{L}\{\cos^2 z\phi\} = \dfrac{2z^2 + s^2}{s(s^2 + 4z^2)}$, $\text{Re } s > 2 |\text{Im } z|$

12. $\mathcal{L}\{H(\phi - a) \sin z\phi\} = \dfrac{e^{-as}}{s^2 + z^2} (z \cos za + s \sin za)$,

$\text{Re } s > |\text{Im } z|$

13. $\mathcal{L}\{H(\phi - a) \cos z\phi\} = \dfrac{e^{-as}}{s^2 + z^2} (s \cos za - z \sin za)$,

$\text{Re } s > |\text{Im } z|$

14. Solve the differential equation
$$U''(\phi) + 3U'(\phi) + 2U(\phi) = \sin \phi, \qquad U(0) = U'(0) = 0,$$
using Laplace transforms.

15. Solve the differential equation
$$U''(\phi) + U(\phi) = \phi \sin \phi, \qquad U(0) = 0, \quad U'(0) = 1,$$
using Laplace transforms.

16. Solve the system of differential equations
$$U'(\phi) = U(\phi) - V(\phi) + \sin \phi, \qquad U(0) = 0,$$
$$V'(\phi) = U(\phi) + V(\phi) + e^\phi, \qquad V(0) = 1,$$
using Laplace transforms.

17. Find the Laplace transform of the general solution of the differential equation
$$\phi U''(\phi) + U'(\phi) + \phi U(\phi) = 0.$$

18. Give an example of a pws function that is not of exponential order.

19. Give an example of a function of exponential order that is not piecewise smooth.

20. If $U(\phi)$ is piecewise smooth and of exponential order, show that

$$\mathcal{L}\left\{\int_c^\phi U(\phi)\, d\phi\right\} = \frac{1}{s}\,\mathcal{L}\{U(\phi)\} + \frac{1}{s}\int_c^0 U(\phi)\, d\phi,$$

and use this to find the Laplace transform of the **sine integral**

$$\mathrm{Si}(\phi) = \int_0^\phi \frac{\sin\phi}{\phi}\, d\phi.$$

If $U(\phi)$ and $U'(\phi)$ are piecewise smooth and of exponential order, prove the conditions given in Exercises 21–23, provided the domain of convergence of $U'(\phi)$ includes the closed right half plane.

21. $\lim\limits_{s\to\infty} \mathcal{L}\{U(\phi)\} = 0$

22. $\lim\limits_{s\to\infty} s\mathcal{L}\{U(\phi)\} = U(0^+)$

23. $\lim\limits_{s\to 0^+} s\mathcal{L}\{U(\phi)\} = \lim\limits_{\phi\to\infty} U(\phi)$

24. Can the functions

$$\frac{s}{s-1}, \quad \frac{1}{\sqrt{s}}, \quad e^{s^{1/2}}$$

be the Laplace transforms of functions $U(\phi)$ that together with $U'(\phi)$ are piecewise smooth and of exponential order?

In Exercises 25–27, prove convolution is distributive, commutative, and associative.

25. $U * (V_1 + V_2) = U * V_1 + U * V_2$
26. $U * V = V * U$
27. $(U * V) * W = U * (V * W)$

6.7 THE INVERSE LAPLACE TRANSFORM

In this section we shall discuss a very powerful technique for determining the function $U(\phi)$ when we know only the Laplace transform

$$u(s) = \mathcal{L}\{U(\phi)\}(s) = \int_0^\infty U(\phi)e^{-s\phi}\, d\phi \tag{1}$$

of that function. We shall assume that $U(\phi)$ is of exponential order with $e^{-a\phi}\,|U(\phi)|$ bounded for $\phi > \Phi$.

If we write $s = q + it$, the Laplace transform becomes

$$u(s) = \int_0^\infty (U(\phi)e^{-q\phi})e^{-it\phi} \, d\phi, \qquad q > a,$$

which is the Fourier transform of the function

$$P(\phi) = \begin{cases} \sqrt{2\pi}U(\phi)e^{-q\phi}, & \phi \geqslant 0, \\ 0, & \phi < 0. \end{cases} \tag{2}$$

Since $q > a$, $|P(\phi)|$ is integrable, so the Fourier integral theorem applies, implying that at all points $\phi > 0$ of continuity of P, we have

$$P(\phi) = \text{PV } \hat{P}(\phi) = \frac{1}{\sqrt{2\pi}} \text{ PV} \int_{-\infty}^\infty u(q + it)e^{it\phi} \, dt$$

or

$$U(\phi) = \frac{1}{2\pi} \text{ PV} \int_{-\infty}^\infty u(q + it)e^{(q+it)\phi} \, dt$$

$$= \frac{1}{2\pi i} \text{PV} \int_{q-i\infty}^{q+i\infty} u(s)e^{s\phi} \, ds, \qquad s = q + it, \quad \phi > 0. \tag{3}$$

This last equation is the **inversion formula for Laplace transforms.** We write $\mathcal{L}^{-1}\{u(s)\} = U(\phi)$ at all points $\phi > 0$ of continuity of U, and call U the **inverse** transform of u. In particular, note that different continuous pws functions of exponential order have different Laplace transforms.

Suppose we wish to find the inverse Laplace transform of a single-valued function $u(s)$, Re $s > a$. Suppose, in addition, that $u(s)$ vanishes as $s \to \infty$ in $\mathfrak{M}$. Observe that for $s = q + Re^{i\theta}$, $q > a$,

$$|e^{s\phi}| = e^{q\phi + R\phi \cos\theta} \to 0$$

as $R \to \infty$, provided $\phi \cos\theta < 0$. Thus, the contour integrals over the curves indicated in Figure 6.14,

$$\frac{-1}{2\pi i} \int_{\gamma_1} u(s)e^{s\phi} \, ds, \qquad \phi < 0, \tag{4}$$

$$\frac{1}{2\pi i} \int_{\gamma_2} u(s)e^{s\phi} \, ds, \qquad \phi > 0, \tag{5}$$

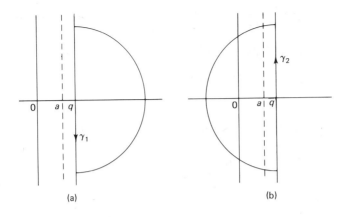

FIGURE 6.14.   (a) $\phi < 0$; (b) $\phi > 0$

converge to $\mathcal{L}^{-1}\{u(s)\}$ as $R \to \infty$. Since $u(s)$ is analytic in Re $s > a$, equation (4) vanishes by Cauchy's theorem, implying that $U(\phi) = 0$ if $\phi < 0$. Finally, the residue theorem implies that

$$U(\phi) = \sum_{\text{Re } s < q} \text{Res } u(s)e^{s\phi}, \quad \text{if} \quad \phi > 0. \tag{6}$$

Since the exponential function $e^{s\phi}$ is entire, one need only use the singularities of the Laplace transform $u(s) = \mathcal{L}\{U(\phi)\}$ in the half plane Re $s \leqslant a$. Equation (6) provides a useful evaluation of the inversion formula (3) for single-valued functions $u(s)$ that vanish as $s \to \infty$.

EXAMPLE 1.

Find the inverse of the Laplace transform

$$u(s) = \frac{1}{(s + 1)(s + 2)}, \quad \text{Re } s > 0.$$

SOLUTION: The transform $u(s)$ has poles at $s = -1$ and $s = -2$. Using the residue formula in equation (6), we get

$$U(\phi) = \text{Res}_{-1} \frac{e^{s\phi}}{(s + 1)(s + 2)} + \text{Res}_{-2} \frac{e^{s\phi}}{(s + 1)(s + 2)}$$

$$= e^{-\phi} - e^{-2\phi}.$$

EXAMPLE 2.

Obtain the inverse Laplace transform of the function

$$u(s) = \frac{2s + 3}{s^2 + 4}, \qquad \text{Re } s > 0,$$

SOLUTION: Here the poles are at $\pm 2i$, so that

$$U(\phi) = \text{Res}_{2i} \frac{(2s + 3)e^{s\phi}}{s^2 + 4} + \text{Res}_{-2i} \frac{(2s + 3)e^{s\phi}}{s^2 + 4}$$

$$= \left(\frac{4i + 3}{4i}\right) e^{2i\phi} + \left(\frac{-4i + 3}{-4i}\right) e^{-2i\phi}$$

$$= 2 \cos 2\phi + \tfrac{3}{2} \sin 2\phi.$$

EXAMPLE 3.

Invert the Laplace transform

$$u(s) = \frac{e^{-2s}}{(s + 3)^2}, \qquad \text{Re } s > 0.$$

SOLUTION: Equation (6) cannot be used because $u(s)$ does not vanish as $s \to \infty$ along the negative real axis. However, the method we used in obtaining equation (6) can still be employed. Integrating

$$u(s)e^{s\phi} = \frac{e^{(\phi - 2)s}}{(s + 3)^2}$$

over the two contours shown in Figure 6.15, note that for $s = Re^{i\theta}$,

$$|e^{(\phi - 2)s}| = e^{R (\phi - 2)\cos \theta} \to 0$$

as $R \to \infty$ if $(\phi - 2) \cos \theta < 0$. By Cauchy's theorem,

$$U(\phi) = \frac{-1}{2\pi i} \lim_{R \to \infty} \int_{\gamma_1} \frac{e^{(\phi - 2)s}}{(s + 3)^2} \, ds = 0 \qquad \text{if } \phi < 2.$$

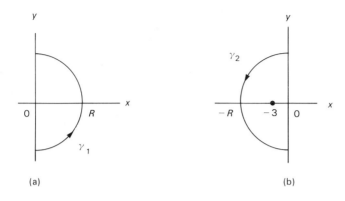

FIGURE 6.15. (a) $\phi < 2$; (b) $\phi > 2$

For $\phi > 2$, the residue theorem yields

$$U(\phi) = \frac{1}{2\pi i} \lim_{R \to \infty} \int_{\gamma_2} \frac{e^{(\phi - 2)s}}{(s+3)^2} \, ds$$

$$= \operatorname{Res}_{-3} \frac{e^{(\phi - 2)s}}{(s+3)^2} = (\phi - 2)e^{-3(\phi - 2)}.$$

Hence, $u(\phi) = (\phi - 2)e^{-3(\phi - 2)} H(\phi - 2)$.

To invert a Laplace transform $u(s)$ that is multivalued, special care must be taken to avoid crossing branch cuts.

EXAMPLE 4.

Find the inverse of the Laplace transform

$$u(s) = \frac{1}{\sqrt{s}}, \qquad \operatorname{Re} s > 0.$$

SOLUTION: Cauchy's theorem indicates that both integrals

$$\frac{1}{2\pi i} \int_{\gamma_i} \frac{e^{s\phi}}{\sqrt{s}} \, ds = 0,$$

where γ_1 and γ_2 are the contours shown in Figure 6.16. Since

$$|e^{(s-1)\phi}| = e^{R\phi \cos \theta} \to 0,$$

as $R \to \infty$ when $\phi \cos \theta < 0$, the function $U(\phi) = 0$ for $\phi < 0$.

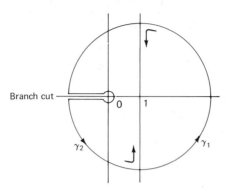

FIGURE 6.16.

For $\phi > 0$, note that the integral vanishes as $R \to \infty$ on the large semicircle in γ_2 while

$$\left| \frac{1}{2\pi i} \int_{|s|=r} \frac{e^{s\phi}}{\sqrt{s}} \, ds \right| \leq \frac{1}{2\pi} \int_0^{2\pi} \sqrt{r} \, e^{r\phi \cos\theta} \, d\theta \to 0,$$

as $r \to 0$ on the small circle on γ_2. Thus,

$$U(\phi) = \lim_{R \to \infty} \frac{1}{2\pi i} \int_{1-iR}^{1+iR} \frac{e^{s\phi}}{\sqrt{s}} \, ds$$

$$= \lim_{\substack{R \to \infty \\ r \to 0}} \left\{ \frac{1}{2\pi i} \int_{-R}^{-r} \frac{e^{t\phi} \, dt}{\sqrt{|t|e^{-i\pi}}} - \frac{1}{2\pi i} \int_{-R}^{-r} \frac{e^{t\phi} \, dt}{\sqrt{|t|e^{i\pi}}} \right\}.$$

Setting $x = -t$ yields

$$U(\phi) = \lim_{\substack{R \to \infty \\ r \to 0}} \frac{1}{\pi} \int_r^R \frac{e^{-x\phi} \, dx}{\sqrt{x}} = \frac{1}{\pi} \int_0^\infty \frac{e^{-x\phi} \, dx}{\sqrt{x}},$$

and using the definition of the gamma function in Exercise 14 of Section 4.5, we have

$$U(\phi) = \frac{\Gamma(\tfrac{1}{2})}{\pi\sqrt{\phi}} = \frac{1}{\sqrt{\pi\phi}}.$$

EXAMPLE 5.

Solve the initial value problem

$$U''(\phi) + 2wU'(\phi) + w^2 U(\phi) = -\sin w\phi,$$

$$U(0) = 0, \qquad U'(0) = \frac{1}{2w}.$$

SOLUTION: Using the differentiation property of Section 6.6,

$$\mathcal{L}\{U''(\phi)\} = s\mathcal{L}\{U'(\phi)\} - \frac{1}{2w} = s^2 \mathcal{L}\{U(\phi)\} - \frac{1}{2w},$$

$$\mathcal{L}\{U'(\phi)\} = s\mathcal{L}\{U(\phi)\},$$

so the Laplace transform of the differential equation becomes

$$(s^2 + 2ws + w^2)\mathcal{L}\{U(\phi)\} - \frac{1}{2w} = \frac{-w}{s^2 + w^2}$$

or

$$\mathcal{L}\{U(\phi)\} = \frac{s - w}{2w(s + w)(s^2 + w^2)}.$$

By the inversion theorem for Laplace transforms,

$$U(\phi) = \sum_{s \leqslant 0} \text{Res } \mathcal{L}\{U(\phi)\}(s)\, e^{s\phi}$$

so that

$$U(\phi) = \frac{-e^{-w\phi}}{2w^2} + \frac{e^{iw\phi}}{4w^2} + \frac{e^{-iw\phi}}{4w^2} = \frac{\cos w\phi - e^{-w\phi}}{2w^2}.$$

EXAMPLE 6.

The thermal diffusion equation in a semiinfinite conducting rod is of the form

$$\frac{\partial U}{\partial t} = \delta \frac{\partial^2 U}{\partial x^2}, \tag{7}$$

where δ is the coefficient of diffusivity of heat in the rod, x is the location on the rod, t is time, and U is the temperature. Assume that we are given the following initial and boundary conditions:

$$U(x, 0) = 0, \qquad 0 \leqslant x < \infty,$$
$$U(0, t) = c \neq 0, \qquad t > 0, \qquad\qquad (8)$$
$$\lim_{x \to \infty} U(x, t) = 0, \qquad t > 0.$$

Find the temperature U at any point x for any time t.

SOLUTION: Treat x as a parameter and define

$$\mathcal{L}\{U(x, t)\}(s) = \int_0^\infty e^{-st}\, U(x, t)\, dt.$$

By the differentiation property, equation (7) becomes

$$s\mathcal{L}\{U\} - U(x, 0) = \mathcal{L}\left\{\frac{\partial U}{\partial t}\right\} = \delta\,\frac{\partial^2}{\partial x^2}\,\mathcal{L}\{U\},$$

by interchanging the operations of taking Laplace transforms and differentiating with respect to x. Because this interchange may not be valid, we must check the answer to verify that it is a solution to the problem. Setting $u = \mathcal{L}\{U\}$, we have the differential equation

$$\frac{\partial^2 u}{\partial x^2} = \frac{s}{\delta}\, u, \qquad\qquad (9)$$

where we treat x as the independent variable and s as a parameter. The boundary conditions in (8) can be rewritten in the form

$$u(0, s) = \mathcal{L}\{U(0, t)\} = \int_0^\infty ce^{-st}\, dt = \frac{c}{s} \qquad\qquad (10)$$

and

$$\lim_{x \to \infty} u(x, s) = \int_0^\infty \lim_{x \to \infty} U(x, t)e^{-st}\, dt = 0, \qquad\qquad (11)$$

provided the interchange of integral and limit is valid. The general solution for equation (9) has the form

$$u = c_1 e^{\sqrt{s/\delta}\, x} + c_2 e^{-\sqrt{s/\delta}\, x},$$

so that $c_1 = 0$ by equation (11) and $c_2 = c/s$ by equation (10). Hence, $\mathcal{L}\{U\} = ce^{-\sqrt{s/\delta}x}/s$, and we may use the inversion theorem for Laplace transforms or Appendix 2 to obtain

$$U(x, t) = c\left(1 - \frac{2}{\sqrt{\pi}} \int_0^{x/2\sqrt{\delta t}} e^{-v^2} \, dv\right). \tag{12}$$

To check that equation (12) is indeed the solution, note that

$$\int_0^\infty e^{-v^2} \, dv = \frac{\sqrt{\pi}}{2}$$

by Example 3 of Section 2.2, so the initial and boundary conditions (8) hold. Moreover,

$$\frac{\partial U}{\partial x} = \frac{-c}{\sqrt{\pi \delta t}} e^{-x^2/4\delta t}$$

and

$$\delta \frac{\partial^2 U}{\partial x^2} = \frac{cx}{2\sqrt{\pi\delta}} \frac{e^{-x^2/4\delta t}}{t^{3/2}} = \frac{\partial U}{\partial t}.$$

EXERCISES

Find the inverses of the Laplace transforms given in Exercises 1–10. Assume that each transform is defined in the half plane Re $s > a$ and b is real.

1. $\dfrac{1}{(s + a)^3}$

2. $\dfrac{1}{(s^2 + a^2)^2}$

3. $\dfrac{s}{(s^2 + a^2)^2}$

4. $\dfrac{1}{(s + a)^4}$

5. $\dfrac{1}{s(s^2 + a^2)}$

6. $\dfrac{s}{s^3 + a^3}$

7. $\dfrac{1}{(s^3 + a^3)}$

8. $\dfrac{e^{-bs}}{(s + a)^3}$

9. $\dfrac{e^{-bs}}{s^2 + a^2}$

10. $\dfrac{e^{-bs}}{s(s^2 + a^2)}$

Use the methods in Examples 3 and 4 to invert the Laplace transforms in Exercises 11–18. Assume $a, b > 0$.

11. $\dfrac{1}{\sqrt[3]{s}}$, $\operatorname{Re} s > 0$

12. $\dfrac{1}{\sqrt[4]{s}}$, $\operatorname{Re} s > 0$

13. $\operatorname{Log}\left(\dfrac{s-a}{s-b}\right)$, $\operatorname{Re} s > \max(a, b)$

14. $\tan^{-1}\dfrac{a}{s}$, $\operatorname{Re} s > 0$

15. $\tfrac{1}{4}\operatorname{Log}\left(1 + \dfrac{4a^2}{s^2}\right)$, $\operatorname{Re} s > 0$

16. $e^{-a\sqrt{s}}$, $\operatorname{Re} s > 0$

17. $\dfrac{e^{-a\sqrt{s}}}{\sqrt{s}}$, $\operatorname{Re} s > 0$

18. $\dfrac{1}{2s}\operatorname{Log}(1 + s^2)$, $\operatorname{Re} s > 0$

19. Solve the equation

$$U(\phi) = \phi^2 + \int_0^\phi \sin(\phi - t)\, U(t)\, dt.$$

20. Solve the equation

$$U(\phi) = e^{-\phi} - 2\int_0^\phi \cos(\phi - t)\, U(t)\, dt.$$

21. Solve the integrodifferential equation

$$U'(\phi) + \int_0^\phi U(t)\, dt = e^{-a\phi},   \phi > 0,$$

given that $U(0) = c\,(\neq 0)$, using Laplace transforms.

22. Find the solution of the delay differential equation

$$U''(\phi) = U(\phi - 1) - U'(\phi - 1)$$

given $U(\phi) = 1$ for $-1 \leqslant \phi \leqslant 0$.

23. Solve the convolution equation

$$U(\phi) = 1 + \int_0^\phi (\phi - t)\, U(t)\, dt.$$

24. Find a solution of the wave equation
$$U_{tt} = a^2 U_{xx},$$
having initial and boundary conditions
$$U(x, 0) = 0, \quad U_t(x, 0) = 0, \quad U(0, t) = \sin\frac{a\pi t}{b}, \quad U(b, t) = 0,$$
where a and b are fixed constants.

25. Find an expression for the solution of the wave equation
$$U_{tt} = a^2 U_{xx}, \qquad x, t > 0,$$
on a semiinfinite string, given the initial and boundary conditions
$$U(x, 0) = f(x), \qquad x \geqslant 0,$$
$$U_t(x, 0) = 0, \qquad x \geqslant 0,$$
$$U(0, t) = 0, \qquad t \geqslant 0,$$
$$\lim_{x \to \infty} U(x, t) = 0, \qquad t \geqslant 0.$$

26. A finite string, subject to a forcing function $f(x, t)$, satisfies the equation of motion
$$U_{xx} - \frac{1}{a^2} U_{tt} = f(x, t).$$
The initial conditions
$$U(x, 0) = g(x), \qquad 0 \leqslant x \leqslant L,$$
$$U_t(x, 0) = h(x), \qquad 0 \leqslant x \leqslant L,$$
are given functions and the string is fixed at its ends:
$$U(0, t) = U(L, t) = 0.$$
Use Laplace transforms to obtain an expression for the solution of this problem.

27. Solve the partial differential equation
$$U_t = \delta U_{xx} + \mu U_x, \qquad t > 0, \quad x > 0,$$
given the initial and boundary conditions
$$U(x, 0) = 0, \qquad x \geqslant 0,$$
$$U(0, t) = c(\neq 0), \qquad t > 0,$$
$$\lim_{x \to \infty} U(x, t) = \lim_{x \to \infty} U_x(x, t) = 0, \qquad t > 0.$$

NOTES

SECTION 6.2

A more thorough discussion of Dirichlet's problem in the complex plane may be found in [A, pp. 237-253]. Dirichlet's problem in three-dimensional space is studied in potential theory; a classic in this area is [Ke]. The hypothesis on $U(\phi)$ in Poisson's theorem may be relaxed substantially [Hf, Chapter 3, and H, Chapter 19].

SECTION 6.3

A complete treatment of the Joukowski (or more properly Zhukovsky) profile is found in [R, pp. 115-121].

SECTION 6.4

An example of a continuous function whose Fourier series diverges at the rational numbers in $[0, 2\pi]$ is found in [J, p. 546]. A theorem by Y. Katznelson [Studia Math., 26 (1966), 301-304] shows that for any set S of measure zero, there is a continuous function whose Fourier series diverges on that set. Conversely, by a result of L. Carleson [Acta Math., 116 (1966), 135-157], the Fourier series of a continuous function converges except on a set of measure zero. A comprehensive discussion of the convergence problem can be found in [Hf]. Term-by-term integrations and differentiations may be performed on Fourier series, obtaining the Fourier series of the indefinite integral or derivative, if the functions concerned are piecewise smooth.

SECTION 6.5

Alternative definitions for Fourier transforms are frequently encountered. All such definitions are equivalent up to a rotation and magnification by $\sqrt{2\pi}$.

SECTION 6.6

Tables of Laplace transforms may be found in many mathematical handbooks. One such table is in the appendix.

SECTION 6.7

For a proof of the uniqueness of the Laplace transform for continuous functions, see [M, p. 412].

The development of an inversion formula for two-sided Laplace transforms is hampered by their nonuniqueness. Any inversion formula for the two-sided Laplace transform must take the region of convergence into consideration.

A.1 TABLE OF CONFORMAL MAPPINGS

z-plane	Mapping Function	w-plane
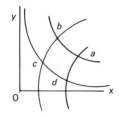	$w = z^2$ $\{xy = c\} \rightarrow \{v = 2c\}$ $\{x^2 - y^2 = c\} \rightarrow \{u = c\}$	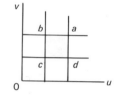
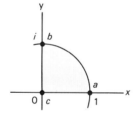	$w = z^2$	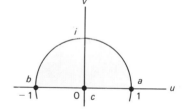
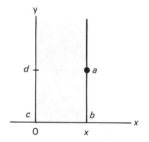	$w = z^2$	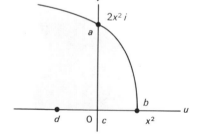

z-plane	Mapping Function	w-plane

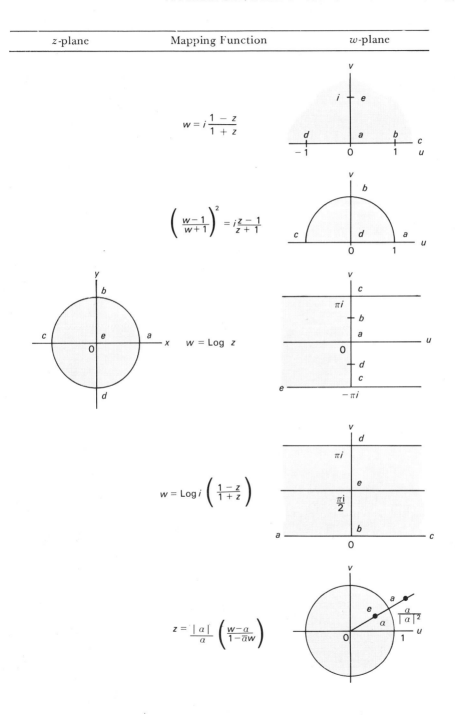

z-plane	Mapping Function	w-plane

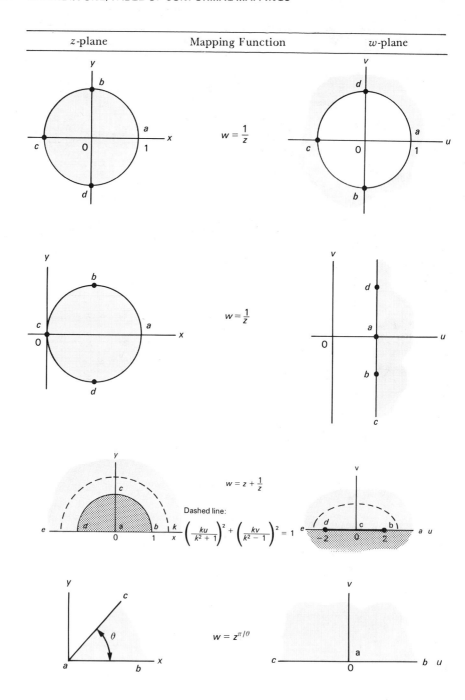

$$w = \frac{1}{z}$$

$$w = \frac{1}{z}$$

$$w = z + \frac{1}{z}$$

Dashed line:

$$\left(\frac{ku}{k^2 + 1}\right)^2 + \left(\frac{kv}{k^2 - 1}\right)^2 = 1$$

$$w = z^{\pi/\theta}$$

z-plane	Mapping Function	w-plane

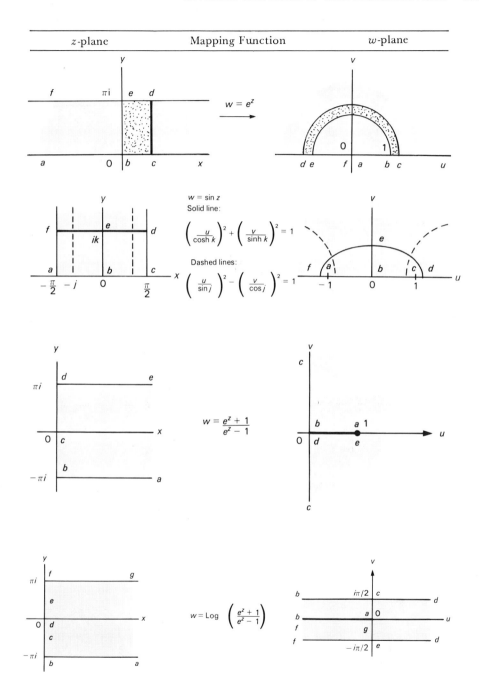

$w = e^z$

$w = \sin z$

Solid line:

$$\left(\frac{u}{\cosh k}\right)^2 + \left(\frac{v}{\sinh k}\right)^2 = 1$$

Dashed lines:

$$\left(\frac{u}{\sin j}\right)^2 - \left(\frac{v}{\cos j}\right)^2 = 1$$

$$w = \frac{e^z + 1}{e^z - 1}$$

$$w = \text{Log}\left(\frac{e^z + 1}{e^z - 1}\right)$$

z-plane	Mapping Function	w-plane

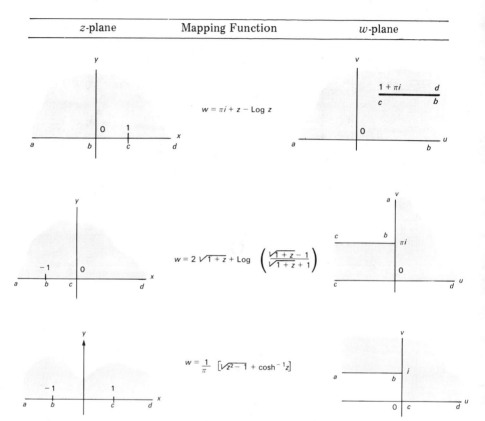

$$w = \pi i + z - \text{Log } z$$

$$w = 2\sqrt{1+z} + \text{Log } \left(\frac{\sqrt{1+z}-1}{\sqrt{1+z}+1} \right)$$

$$w = \frac{1}{\pi} \left[\sqrt{z^2-1} + \cosh^{-1} z \right]$$

$A.2$ TABLE OF LAPLACE TRANSFORMS

$u(\phi)$	$\mathcal{L}\{u(\phi)\}(s)$	Domain of Convergence
ϕ^z $(\mathrm{Re}\,z > -1)$	$\Gamma(z+1)/s^{z+1}$	$\mathrm{Re}\,s > 0$
$e^{z\phi}$	$1/(s-z)$	$\mathrm{Re}\,s > \mathrm{Re}\,z$
$\sin z\phi$	$z/(s^2+z^2)$	$\mathrm{Re}\,s > |\,\mathrm{Im}\,z\,|$
$\cos z\phi$	$s/(s^2+z^2)$	$\mathrm{Re}\,s > |\,\mathrm{Im}\,z\,|$
$e^{w\phi}\sin z\phi$	$z/[(s-w)^2+z^2]$	$\mathrm{Re}\,s > \mathrm{Re}\,w + |\,\mathrm{Im}\,z\,|$
$e^{w\phi}\cos z\phi$	$(s-w)/[(s-w)^2+z^2]$	$\mathrm{Re}\,s > \mathrm{Re}\,w + |\,\mathrm{Im}\,z\,|$
$\phi \sin z\phi$	$2sz/(s^2+z^2)^2$	$\mathrm{Re}\,s > |\,\mathrm{Im}\,z\,|$
$\phi \cos z\phi$	$(s^2-z^2)/(s^2+z^2)^2$	$\mathrm{Re}\,s > |\,\mathrm{Im}\,z\,|$
$\sin^2 z\phi$	$2z^2/s(s^2+4z^2)$	$\mathrm{Re}\,s > 2\,|\,\mathrm{Im}\,z\,|$
$\cos^2 z\phi$	$(s^2+2z^2)/s(s^2+4z^2)$	$\mathrm{Re}\,s > 2\,|\,\mathrm{Im}\,z\,|$
$(e^{z\phi}-e^{w\phi})/\phi$	$\log[(s-w)/(s-z)]$	$\mathrm{Re}\,s > \max\{|\,\mathrm{Re}\,z\,|, |\,\mathrm{Re}\,w\,|\}$
$\sin z\phi/\phi$	$\tan^{-1} z/s$	$\mathrm{Re}\,s > |\,\mathrm{Im}\,z\,|$
$\sin^2 z\phi/\phi$	$\frac{1}{4}\log(1+4z^2/s^2)$	$\mathrm{Re}\,s > 2\,|\,\mathrm{Im}\,z\,|$
$e^{-z\phi^2}$	$\frac{1}{2}\sqrt{\pi/z}\,e^{s^2/4z}[1-\mathrm{erf}\,(s/2\sqrt{z})]$	e
$\log \phi$	$(\log s - \gamma)/s,\ \gamma = 0.5772\ldots$	$\mathrm{Re}\,s > 0$
$H(\phi - a)$	e^{-as}/s	$\mathrm{Re}\,s > 0$
$\dfrac{e^{a\phi}(1-2a\phi)}{\sqrt{\pi\phi}}$	$\dfrac{s}{(s-a)^{3/2}}$	$\mathrm{Re}\,s > 0$
$\dfrac{1}{\sqrt{\pi\phi}}\cos(2\sqrt{a\phi})$	$\dfrac{e^{-a/s}}{\sqrt{s}}$	$\mathrm{Re}\,s > 0$

$u(\phi)$	$\mathcal{L}\{u(\phi)\}$ (s)	Domain of Convergence
$\dfrac{1}{\sqrt{\pi\phi}}\ \cosh\ (2\sqrt{a\phi})$	$\dfrac{e^{a/s}}{\sqrt{s}}$	$\mathrm{Re}\ s > 0$
$\dfrac{a}{2\sqrt{\pi\phi^3}}\ e^{-a^2/4\phi}$	$e^{-a\sqrt{s}}$	$\mathrm{Re}\ s > 0$
$1 - \dfrac{2}{\sqrt{\pi}}\ \displaystyle\int_0^{a/2\sqrt{\phi}} e^{-v^2}\,dv$	$\dfrac{e^{-a\sqrt{s}}}{s}$	$\mathrm{Re}\ s > 0$

For a more extensive list of Laplace transforms, see [M, pp. 428–434].

A.3 LINE INTEGRALS AND GREEN'S THEOREM

Line integrals are a natural generalization of the concept of a definite integral

$$\int_a^b f(x)\, dx. \tag{1}$$

In this appendix we will give a brief development of line integrals of real-valued functions $f(x, y)$ along pws curves in the Euclidean plane.

The term line integral is a misnomer as we usually evaluate the integrals along curves. To understand what a line integral measures, recall that the definite integral in (1) is obtained by dividing the interval $a \leqslant x \leqslant b$ into n subintervals of lengths $\Delta x_1, \Delta x_2, \ldots, \Delta x_n$, selecting a point x_k in each subinterval, and evaluating the limit of the Riemann sum

$$\sum_{k=1}^{n} f(x_k)\, \Delta x_k$$

as all the lengths Δx_k tend to 0.

A similar process can be carried out for a real-valued function $f(x, y)$ defined on a smooth curve γ in the Euclidean plane: Divide γ into n subarcs of arc length $\Delta s_1, \Delta s_2, \ldots, \Delta s_n$, select a point (x_k, y_k) in each subarc, and evaluate the limit of the sum

$$\sum_{k=1}^{n} f(x_k, y_k)\, \Delta s_k \tag{2}$$

as all the arc lengths Δs_k approach 0 (see Figure A.1).

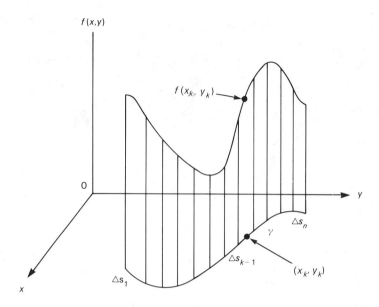

FIGURE A.1. Riemann sum for a line integral

In the same way that the definite integral (1) can be interpreted as the area under the graph f from a to b, the line integral,

$$\int_\gamma f(x, y) \, ds = \lim_{\max \Delta s_k \to 0} \sum_k f(x_k, y_k) \, \Delta s_k, \qquad (3)$$

can be interpreted as the area under the graph of f along γ.

It can be shown that if f is continuous on the smooth curve γ, then the limit in (3) exists [B, p. 301]. Indeed, the limit (3) will exist under much weaker hypotheses (see [S]).

Line integrals are seldom evaluated using the Riemann sum in equation (3). The following two examples illustrate the usual procedure in determining the value of a line integral.

EXAMPLE 1.

Evaluate

$$\int_\gamma xy \, ds$$

where γ is the arc along $y = x^2$ going from $(0, 0)$ to $(1, 1)$.

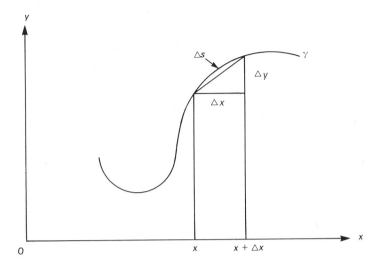

FIGURE A.2. Arc length

SOLUTION: From the Pythagorean theorem (see Figure A.2), the change in arc length Δs corresponding to a change in the independent variable from x to $x + \Delta x$ is approximately

$$(\Delta s)^2 \approx (\Delta x)^2 + (\Delta y)^2.$$

Dividing both sides by $(\Delta x)^2$ and taking the limit as Δx tends to 0, we have

$$\left(\frac{ds}{dx}\right)^2 = 1 + \left(\frac{dy}{dx}\right)^2, \tag{4}$$

where $s(x)$ is the arc length function that measures the length of the arc along the graph of γ going from $(0, 0)$ to (x, x^2). Changing variables, we have

$$\int_\gamma xy\,ds = \int_0^1 xy\sqrt{1 + (y'(x))^2}\,dx$$

$$= \int_0^1 x^3\sqrt{1 + 4x^2}\,dx.$$

Substituting $u = 1 + 4x^2$ and changing variables, we get

$$\int_0^1 x^3 \sqrt{1 + 4x^2} \, dx = \frac{1}{32} \int_1^5 (u - 1)\sqrt{u} \, du$$

$$= \frac{1}{16} \left[\frac{u^{5/2}}{5} - \frac{u^{3/2}}{3} \right] \Big|_1^5$$

$$= \frac{5^{5/2} + 1}{120}.$$

If the curve γ is described in parametric form we can proceed as in the next example.

EXAMPLE 2.

Calculate the value of the line integral

$$\int_\gamma \sqrt{y} \, ds$$

where γ is the parametric curve

$$x = t - \sin t, \, y = 1 - \cos t, \, 0 \leqslant t \leqslant 2\pi.$$

SOLUTION: Rewrite equation (4) in the form

$$\left(\frac{ds}{dt} \right)^2 = \left(\frac{dx}{dt} \right)^2 + \left(\frac{dy}{dt} \right)^2$$

and change all variables in the line integral into functions of t to obtain

$$\int_\gamma \sqrt{y} \, ds = \int_0^{2\pi} \sqrt{1 - \cos t} \sqrt{(1 - \cos t)^2 + \sin^2 t} \, dt.$$

Simplifying the second radical, we have

$$\sqrt{1 - 2 \cos t + \cos^2 t + \sin^2 t} = \sqrt{2 - 2 \cos t}.$$

Hence, multiplying the two radicals, we get

$$\int_\gamma \sqrt{y}\, ds = \sqrt{2} \int_0^{2\pi} (1 - \cos t)\, dt$$

$$= \sqrt{2}\, (t - \sin t)\Big|_0^{2\pi} = 2\sqrt{2}\pi.$$

The value of the line integral $\int_\gamma f(x,\, y)\, ds$ for a pws curve γ is *independent* of the parametrization of the curve γ. Any change in parameters for a smooth curve is determined by a continuously differentiable increasing function $t = t\,(T)$ mapping the interval $A \leqslant T \leqslant B$ onto $a \leqslant t \leqslant b$. Changing variables and using the chain rule, we have

$$\int_\gamma f(x,\, y)\, ds = \int_a^b f(x(t),\, y(t))\, s'(t)\, dt$$

$$= \int_A^B f(x(T),\, y(T))\, \frac{ds}{dt} \frac{dt}{dT}\, dT$$

$$= \int_A^B f(x(T),\, y(T))\, s'(T)\, dT.$$

EXAMPLE 3.

Find the value of the line integral

$$\int_\gamma x^3\, ds$$

where γ is the right half of the unit circle.

SOLUTION: To illustrate the fact that line integrals are independent of parametrization, we will use two different parametrizations of the right half of the unit circle. It is important that both parametrizations have the same orientation along γ.

(i) Set $y = t$ and $x = \sqrt{1 - t^2}$ with $-1 \leqslant t \leqslant 1$. Then

$$\frac{dx}{dt} = \frac{-t}{\sqrt{1 - t^2}}, \quad \frac{dy}{dt} = 1, \quad \text{and} \quad \frac{ds}{dt} = \frac{1}{\sqrt{1 - t^2}},$$

so that

$$\int_\gamma x^3 \, ds = \int_{-1}^{1} (1 - t^2) \, dt$$

$$= \left. t - \frac{t^3}{3} \right|_{-1}^{1} = \frac{4}{3}.$$

(ii) Let $x = \cos t$, $y = \sin t$ with $-\pi/2 \leqslant t \leqslant \pi/2$. Then

$$\frac{ds}{dt} = \sqrt{\sin^2 t + \cos^2 t} = 1,$$

so that

$$\int_\gamma x^3 \, ds = \int_{-\pi/2}^{\pi/2} \cos^3 t \, dt = \left. \left(\sin t - \frac{\sin^3 t}{3} \right) \right|_{-\pi/2}^{\pi/2} = \frac{4}{3}.$$

There are two other types of line integrals of the function $f(x, y)$ along γ that can be defined. If we replace Δs_k by Δx_k or Δy_k in equation (3), we obtain the **line integrals of f along γ with respect to x or y:**

$$\int_\gamma f(x, y) \, dx = \lim_{\max \Delta x_k \to 0} \sum_k f(x_k, y_k) \, \Delta x_k,$$

$$\int_\gamma f(x, y) \, dy = \lim_{\max \Delta y_k \to 0} \sum_k f(x_k, y_k) \, \Delta y_k.$$

These line integrals may be viewed as projections of the line integral (3) onto the xz-plane or yz-plane, respectively (see Figure A.3).

The evaluation of these line integrals is similar to that done above.

EXAMPLE 4.

Evaluate $\int_\gamma y(1 - x) \, dy$ along the portion γ of the unit circle lying in the first quadrant.

SOLUTION: Parametrizing γ by $x = \cos t$, $y = \sin t$, $0 \leqslant t \leqslant \pi/2$, we have

$$\int_\gamma y(1-x)\,dy = \int_0^{\pi/2} \sin t\,(1 - \cos t)\cos t\,dt$$

$$= \frac{1}{3}\cos^3 t - \frac{1}{2}\cos^2 t\,\Big|_0^{\pi/2} = \frac{1}{6}.$$

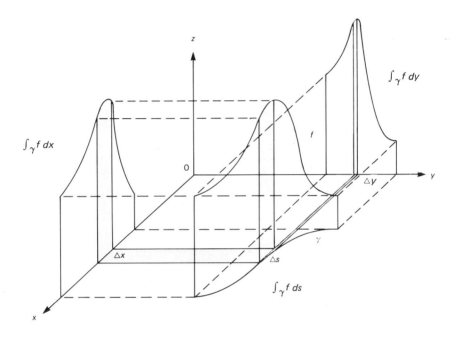

FIGURE A.3. Projections

In applications, line integrals often occur in the combination

$$\int_\gamma p(x, y)\,dy + \int_\gamma q(x, y)\,dx = \int_\gamma p(x, y)\,dy + q(x, y)\,dx, \quad (5)$$

where p and q are continuous functions in a region D containing the curve γ. When γ is a simple closed curve and the functions p and q have continuous partial derivatives in D, there is an important relationship between the line integral around γ and the double integral over the region G inside γ.

GREEN'S THEOREM

Let G be the region inside a pws simple closed curve γ. If $p(x, y)$, $q(x, y)$, $\partial p/\partial x$, and $\partial q/\partial y$ are continuous at all points of $G \cup \gamma$, then

$$\int_\gamma p\, dy + q\, dx = \iint_G \left(\frac{\partial p}{\partial x} - \frac{\partial q}{\partial y} \right) dx dy,$$

provided the line integral is taken in the positive (counterclockwise) direction around γ.

PROOF: Assume initially that γ has the property that any line parallel to either coordinate axis intersects γ in at most two points. Draw the horizontal and vertical lines that circumscribe γ (see Figure A.4). Then the arcs ABC and CDA along γ can be defined by single

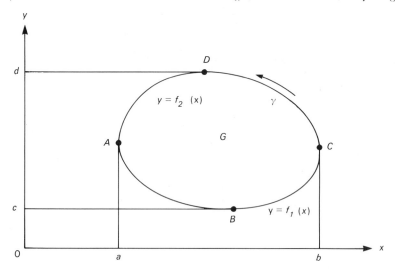

FIGURE A.4.

valued functions of x in the interval $a \leq x \leq b$. Denote these functions by $y = f_1(x)$ and $y = f_2(x)$, respectively. Observe that the line integral

$$\int_\gamma q\, dx = \int_{ABC} q\, dx + \int_{CDA} q\, dx$$

$$= \int_a^b q(x, f_1(x))\, dx - \int_a^b q(x, f_2(x))\, dx,$$

since CDA is traversed from right to left.

Because the region G lies between the curves ABC and CDA, we have

$$\iint_G -\frac{\partial q}{\partial y}\,dxdy = -\int_a^b \int_{f_1(x)}^{f_2(x)} \frac{\partial q}{\partial y}\,dy\,dx$$

$$= -\int_a^b \left[q(x,y) \Big|_{y=f_1(x)}^{y=f_2(x)} \right]\,dx$$

$$= \int_a^b \left[q(x, f_1(x)) - q(x, f_2(x)) \right]\,dx.$$

Hence,

$$\int_\gamma q\,dx = \iint_G -\frac{\partial q}{\partial y}\,dxdy.$$

Similarly, the arcs DAB and BCD can be defined by single-valued functions $x = g_1(y)$ and $x = g_2(y)$ in $c \leqslant y \leqslant d$, with

$$\int_\gamma p\,dy = \int_c^d p(g_2(y), y)\,dy - \int_c^d p(g_1(y), y)\,dy$$

and

$$\iint_G \frac{\partial p}{\partial x}\,dxdy = \int_c^d \int_{g_1(y)}^{g_2(y)} \frac{\partial p}{\partial x}\,dxdy$$

$$= \int_c^d \left[p(x, y) \Big|_{x=g_1(y)}^{x=g_2(y)} \right]\,dy = \int_\gamma p\,dy.$$

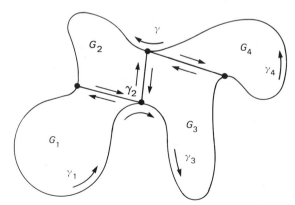

FIGURE A.5.

This establishes Green's theorem for the special curves we have been considering. Green's theorem can be extended to curves not satisfying this special property by dividing the region G into subregions G_i whose boundaries γ_i do have this property (see Figure A.5). Although this assertion is intuitively clear from examples such as that in Figure A.5, it requires a difficult proof that is beyond the scope of this book. Applying Green's theorem to each of the subregions G_i, and adding them together, we note that the line integral along the subarcs of γ_i that are *not* subarcs of γ will cancel in pairs, since each subarc is traversed twice in opposite directions. Hence, we have

$$\iint_G \left(\frac{\partial p}{\partial x} - \frac{\partial q}{\partial y} \right) dxdy = \sum_i \iint_{G_i} \left(\frac{\partial p}{\partial x} - \frac{\partial q}{\partial y} \right) dxdy$$

$$= \sum_i \int_{\gamma_i} pdy + qdx$$

$$= \int_{\gamma} pdy + qdx. \quad \blacksquare$$

EXAMPLE 5.

Evaluate the line integral

$$\int_{\gamma} (x - 2y) \, dx + x \, dy,$$

where γ is the unit circle, using Green's theorem.

SOLUTION: Here $p(x, y) = x$ and $q(x, y) = x - 2y$, so that $\partial p/\partial x = 1$, $\partial q/\partial y = -2$ and Green's theorem yields

$$\int_{\gamma} (x - 2y) \, dx + x \, dy = \iint_{x^2+y^2<1} [1 - (-2)] \, dxdy = 3\pi,$$

since the unit circle has area equal to π.

EXERCISES

In Exercises 1–8, evaluate the given line integral along the indicated curve.

1. $\int_{\gamma} x \, ds$, where γ is the line $y = x$, $0 \leqslant x \leqslant 2$.
2. $\int_{\gamma} x \, ds$, where γ is the curve $y = x^2$, $0 \leqslant x \leqslant 1$.

3. $\int_\gamma y\, ds$, where γ is the curve $x^2 = y^3$, $0 \leqslant y \leqslant 1$.

4. $\int_\gamma x^2 y^2\, ds$, where γ is the unit circle.

5. $\int_\gamma (x^2 - y^2)\, dx - 2xy\, dy$, where γ is the curve $y = x^2$, $-1 \leqslant x \leqslant 1$.

6. The line integral in 5 where γ is the unit circle.

7. $\int_\gamma xy\, dx + x^2\, dy$, for γ defined by $x = 3t - 1$, $y = 3t^2 - 2t$, $1 \leqslant t \leqslant 2$.

8. $\int_\gamma \dfrac{y^2}{1 + x^2}\, dx + 2y\, \tan^{-1} x\, dy$, where γ is the closed curve

$$x^{2/3} + y^{2/3} = 1.$$

9. Use Green's theorem to evaluate exercise 6.

10. Use Green's theorem to evaluate exercise 8.

11. Show that the area bounded by the pws simple closed curve γ is given by

$$\frac{1}{2} \int_\gamma x\, dy - y\, dx.$$

12. Use Green's theorem to evaluate the line integral

$$\int_\gamma 2xy\, dx + (x^2 + y^2)\, dy$$

where γ is any pws simple closed curve.

13. Show that

$$\int_\gamma p\, dy + q\, dx = 0$$

for any pws simple closed curve γ if p, q, $\partial p / \partial x = \partial q / \partial y$ are continuous on and inside γ.

14. Let $(\bar{x}, \bar{y})$ be the coordinates of the centroid of a region G bounded by a pws simple closed curve γ. Prove that

$$\bar{x} = \frac{1}{2A} \int_\gamma x^2\, dy, \quad \bar{y} = \frac{-1}{2A} \int_\gamma y^2\, dx$$

where A is the area of G.

15. Show, using the notation of exercise 14, that $A\bar{y} = \int_\gamma xy\, dy$. Can $\int_\gamma xy\, dx$ be expressed in terms of $\bar{x}$?

16. Using the notation in exercise 14, prove that

$$\iint_G \left(\frac{\partial^2 F}{\partial x^2} + \frac{\partial^2 F}{\partial y^2} \right) dx\, dy = \int_\gamma \frac{dF}{dn}\, ds$$

where dF/dn is the directional derivative of F in the direction of the outer normal to γ.

17. Prove that

$$\iint_G \left(\frac{\partial f}{\partial x} \frac{\partial g}{\partial y} - \frac{\partial f}{\partial y} \frac{\partial g}{\partial x} \right) dx\, dy = \int_\gamma f\, dg.$$

REFERENCES

[A]   Ahlfors, L. V. *Complex Analysis*, 2d ed. McGraw-Hill, New York, 1966.

[B]   Buck, R. C. *Advanced Calculus*. McGraw-Hill, New York, 1965.

[CKP] Carrier, G. F., Krook, M., and Pearson, C. E. *Functions of a Complex Variable*. McGraw-Hill, New York, 1966.

[H]   Hille, E. *Analytic Function Theory*, Vols. I and II. Ginn (Blaisdell), Boston, Mass., 1959.

[Hf]   Hoffman, K. *Banach Spaces of Analytic Functions*. Prentice-Hall, Englewood Cliffs, N.J., 1962.

[Ho]   Hormander, L. *An Introduction to Complex Analysis in Several Variables*. Van Nostrand-Reinhold, Princeton, N.J., 1966.

[J]   James, R. C. *Advanced Calculus*. Wadsworth, Belmont, Calif., 1966.

[Ke]   Kellogg, O. D. *Foundations of Potential Theory*. Dover, New York, 1954.

[Kn]   Knopp, K. *Theory of Functions*, Parts I and II. Dover, New York, 1947.

[Ko]   Kober, H. *Dictionary of Conformal Representations*, 2d ed. Dover, New York, 1957.

[L]   Lang, S. *Complex Analysis*. Addison Wesley, Reading, Mass., 1977.

[M]   Moretti, G. *Functions of a Complex Variable*. Prentice-Hall, Englewood Cliffs, N.J., 1964.

[MT]   Milne-Thomson, L. M. *Theoretical Hydrodynamics*. Macmillan, London, 1938.

[R]   Rothe, R., Ollendorff, F., and Pohlhausen, K. *Theory of Functions*. Dover, New York, 1961.

[S]   Saks, S. *Theory of the Integral*, 2d rev. ed. Dover, New York, 1964.

[Sp]   Springer, G. *Introduction to Riemann Surfaces*. Addison-Wesley, Reading, Mass., 1957.

[T]   Titchmarsh, E. C. *The Theory of Functions*, 2d ed. Oxford Univ. Press, London and New York, 1939.

[V]   Veech, W. A. *A Second Course in Complex Analysis.* Benjamin, New York, 1967.

[W]   Whyburn, G. T. *Topological Analysis*, rev. ed. Princeton Univ. Press, Princeton, N.J., 1964.

SOLUTIONS TO ODD NUMBERED PROBLEMS

CHAPTER 1

SECTION 1.1

1. $2 + i; -2 + i; 2i; \frac{1}{2}i$
3. $1 + 2i; 1; -1 + i; 1 - i$
5. $2; 2i; 2; i$
7. $7 + i; 3 - i; 10 + 5i; 2 - i$
9. $7 - i; -1 - 3i; 14 - 5i; \frac{10}{17} - \frac{11}{17}i$
11. $5 + 4i; 3 + 6i; 9 + i; -\frac{1}{2} + \frac{9}{2}i$
13. $-2i$
15. $-3 - 4i$
17. $-\frac{7}{10} + \frac{19}{10}i$
19. -10
21. $iz = ix - y$
23. If $z_2 \neq 0$, then $z_2 \bar{z}_2 = (\text{length } z_2)^2 \neq 0$, But $z_1 z_2 \bar{z}_2 = 0$, so divide both sides by $(\text{length } z_2)^2$ to get $z_1 = 0$.
25. $\text{Im } z_1 = -\text{Im } z_2$ so $\text{Im } z_1 z_2 = \text{Im } z_2 \cdot (\text{Re } z_1 - \text{Re } z_2) = 0$. If $\text{Re } z_1 = \text{Re } z_2$, then $\text{Re } z_1 z_2 > 0$. Hence $\text{Im } z_2 = 0$.
27. $(x_1 - x_2) - i(y_1 - y_2) = (x_1 - iy_1) - (x_2 - iy_2)$

29. $\dfrac{x_1 \pm iy_1}{x_2 \pm iy_2} \cdot \dfrac{x_2 \mp iy_2}{x_2 \mp iy_2} =$

$\dfrac{x_1 x_2 + y_1 y_2}{x_2{}^2 + y_2{}^2} \pm i \dfrac{x_2 y_1 - x_1 y_2}{x_2{}^2 + y_2{}^2}$

31. $(A + B)^3 = A^3 + B^3 + 3AB(A + B) = -b + 3ABw$

$AB = \sqrt[3]{\dfrac{b^2}{4} - D^2} = \dfrac{-a}{3}$, so

$w^3 + aw + b = (w - A - B) \cdot$
$(w^2 + (A + B)w + a + (A + B)^2) = 0$
with $(A + B)^2 - 4[a + (A + B)^2] = -3(A - B)^2$

33. Use commutativity, associativity, and the distributive laws of real numbers to verify that complex addition and multiplication also satisfy these properties. Identity and inverse laws were shown in text.

35. Suppose z_1 and z_2 are both multiplicative inverses of z. Then $z_1 z = 1 = z z_2$ and $z_1 = z_1(z z_2) = (z_1 z)z_2 = z_2$.

SECTION 1.2

1. 1; $\dfrac{\pi}{2} + 2\pi k$;

$\cos\left(\dfrac{\pi}{2} + 2\pi k\right) + i \sin\left(\dfrac{\pi}{2} + 2\pi k\right)$

3. $\sqrt{2}$; $\dfrac{\pi}{4} + 2\pi k$; $\sqrt{2}\left[\cos\left(\dfrac{\pi}{4} + 2\pi k\right)\right.$

$\left. + i \sin\left(\dfrac{\pi}{4} + 2\pi k\right)\right]$

5. 5, $\tan^{-1}\left(\dfrac{3}{4}\right) + 2\pi k \approx 0.6435 + 2\pi k$; $5\left[\cos(\tan^{-1}\left(\dfrac{3}{4}\right) + 2\pi k) + i \sin(\tan^{-1}\left(\dfrac{3}{4}\right) + 2\pi k)\right]$

7. $\sqrt{53}$; $\tan^{-1}\left(\dfrac{7}{2}\right) + 2\pi k \approx 1.2925 + 2\pi k$; $\sqrt{53}\left[\cos(\tan^{-1}\left(\dfrac{7}{2}\right) + 2\pi k) + i \sin(\tan^{-1}\left(\dfrac{7}{2}\right) + 2\pi k)\right]$

9. $\sqrt{29}$; $\tan^{-1}\left(\dfrac{2}{5}\right) + 2\pi k \approx 0.3805 + 2\pi k$; $\sqrt{29}\left[\cos(\tan^{-1}\left(\dfrac{2}{5}\right) + 2\pi k) + i \sin(\tan^{-1}\left(\dfrac{2}{5}\right) + 2\pi k)\right]$

11. $256(-1 + i)$

13. -2^{18}

15. $2^{13}(-\sqrt{3} + i)$

17. $\sqrt[4]{2}\left(\cos\dfrac{\pi}{8} + i \sin\dfrac{\pi}{8}\right)$;

$\sqrt[4]{2}\left(\cos\dfrac{9\pi}{8} + i \sin\dfrac{9\pi}{8}\right)$

19. $\sqrt{2}\left(\cos\dfrac{7\pi}{12} + i \sin\dfrac{7\pi}{12}\right)$;

$\sqrt{2}\left(\cos\dfrac{19\pi}{12} + i \sin\dfrac{19\pi}{12}\right)$

21. $\sqrt[3]{2}\left(\cos\dfrac{\pi}{9} + i \sin\dfrac{\pi}{9}\right)$;

$\sqrt[3]{2}\left(\cos\dfrac{7\pi}{9} + i \sin\dfrac{7\pi}{9}\right)$;

$\sqrt[3]{2}\left(\cos\dfrac{13\pi}{9} + i \sin\dfrac{13\pi}{9}\right)$

23. $\dfrac{1 + i}{\sqrt{2}}$; $\dfrac{-1 + i}{\sqrt{2}}$; $\dfrac{-1 - i}{\sqrt{2}}$; $\dfrac{1 - i}{\sqrt{2}}$

25. $|z - i| + |z - 1| = 2$; $3x^2 + 2xy + 3y^2 - 4x + 4y = 0$

27. $||z - a| - |z - b|| = c$

29. $|z_1 - z_2|^2 - |z_2 - z_3|^2 =$
$[|z_1|^2 + |z_2|^2 - z_1\bar{z}_2 - z_2\bar{z}_1]$
$- [|z_2|^2 + |z_3|^2 - z_2\bar{z}_3 - z_3\bar{z}_2]$
$= -\bar{z}_2(z_1 - z_3) - z_2(\bar{z}_1 - \bar{z}_3) =$
$(\overline{z_1 + z_3})(z_1 - z_3) + (z_1 + z_3)$
$(\overline{z_1 - z_3}) = |z_1|^2 - |z_3|^2 -$
$\bar{z}_1 z_3 + \bar{z}_3 z_1 + |z_1|^2 - |z_3|^2 +$
$z_3\bar{z}_1 - z_1\bar{z}_3 = 0$

31. $|z_1| - |z_2| = |(z_1 - z_2) + z_2| -$
$|z_2| \leqslant |z_1 - z_2| + |z_2| - |z_2| =$
$|z_1 - z_2|$. Similarly $|z_2| - |z_1|$
$\leqslant |z_1 - z_2|$

33. $|z_1 \pm z_2|^2 = (z_1 \pm z_2)(\bar{z}_1 \pm \bar{z}_2) =$
$|z_1|^2 + |z_2|^2 \pm (z_1\bar{z}_2 + z_2\bar{z}_1) =$
$|z_1|^2 + |z_2|^2 \pm 2\,\mathrm{Re}\,z_1\bar{z}_2$

35. $|z - a|^2 = |z|^2 + |a|^2 - 2\,\mathrm{Re}\,z\bar{a}$;
$|1 - az|^2 = 1 + |a|^2 |z|^2 - 2\,\mathrm{Re}\,z\bar{a}$,
and $0 < (1 - |z|^2)(1 - |a|^2)$
$= 1 - |z|^2 - |a|^2 + |a|^2 |z|^2$

37. $P(\bar{z}_0) = \overline{P(z_0)} = \bar{0} = 0$

39. $|z|^n (\cos n\theta + i \sin n\theta) = z^n = 1 =$
$\cos 2\pi k + i \sin 2\pi k$ so $|z| = 1$ and
$\theta = 2\pi k/n$. But

$\dfrac{2\pi(k + n)}{n} = \dfrac{2\pi k}{n} + 2\pi$ so $z_{k+n} = z_k$.

41. $1 + z + \ldots + z^{n-1} = \dfrac{1 - z^n}{1 - z} = 0$

for any nth root of unity z_k since $z_k^n = 1$. Since the first n of the nth roots of unity are all different (see Exercise 39) and the polynomial $1 - z^n$ has exactly n roots, one of which is 1, $1 + z + \ldots + z^{n-1}$ must have each of the other nth roots of unity as a root.

43. $0 \leqslant \Sigma_k (|a_k|^2 - 2|a_k z_k| \lambda +$
$|z_k|^2 \lambda^2) = \Sigma_k |a_k|^2 - \dfrac{\Sigma_k |a_k z_k|}{\Sigma_k |z_k|^2}$

$$+ \left(\Sigma_k \, |z_k|^2 \right) \left(\lambda - \frac{\Sigma_k \, |a_k z_k|}{\Sigma_k |z_k|^2} \right)^2$$

and set $\lambda = \Sigma_k \, |a_k z_k| \, / \, \Sigma_k \, |z_k|^2$

45. $|(1 - z) P(z) | \geqslant a_0 - [(a_0 - a_1) | z| + (a_1 - a_2) | z|^2 + \ldots + (a_{n-1} - a_n) | z|^n + a_n | z|^{n-1}]$ with equality iff $z \geqslant 0$ (see Exercise 36). $P(1) \neq 0$ and $|(1 - z) P(z)| > 0$ for remaining points in $|z| \leqslant 1$.

SECTION 1.3

1. open; bounded; simply connected
3. open; unbounded; not connected
5. closed; unbounded; simply connected
7. closed; unbounded; connected but not simply connected
9. open; bounded; connected but not simply connected
11. circle $|z + 3| = 2$; lines Im $z = \pm 1$; parabola $|z| = $ Re $z + 2$; ellipse $| z + 1| + | z + i| = 2$; ellipses $|z - 1| + |z + 1| = 3$ and $|z - 1| + |z + 1| = 2\sqrt{2}$
13. Let $S_1, \ldots, S_n$ be closed sets, then $\mathcal{C} - S_k$ is open. But $\cap_k (\mathcal{C} - S_k) = \mathcal{C} - \cup_k S_k$ is open
15. If $z \in \cup_\alpha S_\alpha$ with each S_α open, then z is an interior point of some S_α. Hence an ϵ-neighborhood of z is contained in S_α and therefore in $\cup_\alpha S_\alpha$.
17. Assume the closure $\overline{S}$ is disconnected. Then there are open sets G and H such that $\overline{S} \subset G \cup H$, $G \cap H$ is empty, and $G \cap \overline{S}$ and $H \cap \overline{S}$ are nonempty. Since S is connected it lies in only one of these sets, say G. Then S is contained in the closed set $\mathcal{C} - H$, so $\overline{S} \cap H$ is empty, a contradiction.

19. $z = 0$
21. $z = \infty$. No.

SECTION 1.4

1. $| 2z - 2| = 2| z - 1| < 2\delta$ so set $\epsilon = 2\delta$
3. $|z + i| < \delta$ so set $\epsilon = \delta$
5. $|(2z - 3) - (-1 + 2i)| = | 2z - 2 - 2i| = 2| z - (1 + i)| < 2\delta = \epsilon$
7. $\left| \dfrac{z^2 - 4}{z - 2} - 4 \right| = \dfrac{| z^2 - 4z + 4|}{|z - 2|} = |z - 2| < \delta = \epsilon$
9. $\left| \dfrac{z^3 - 1}{z - 1} - 3 \right| = \dfrac{| z^3 - 3z + 2|}{|z - 1|} = |z^2 + z - 2| = |(z - 1)^2 + 3(z - 1)| < \delta \, (\delta + 3)$. Let $\delta < 1$ and select $\epsilon = 4\delta$.
11. $| $ Re $z - $ Re $a| = |$ Re $(z - a)| \leqslant |z - a| < \delta = \epsilon$ so $\lim_{z \to a}$ Re $z = $ Re a
13. $|\bar{z} - \bar{a}| = |z - a| < \delta = \epsilon$ so $\lim_{z \to a} \bar{z} = \bar{a}$.
15. $\lim_{z \to a}$ Re $f(z) = $ Re$(\lim_{z \to a} f(z)) = $ Re $f(a)$
17. $||f(z)| - |f(a)|| < |f(z) - f(a)| < \epsilon$
19. $\lim_{z \to \pm 1} \dfrac{z^3 - 1}{z^2 - 1} = \lim_{z \to \pm 1} \dfrac{z^2 + z + 1}{z + 1} = \frac{3}{2}$, if $z = 1$; undefined, if $z = -1$
21. $f(z) = \bar{z}$ for $z \neq 0$ so is continuous in $\mathcal{C} - \{0\}$ by exercise 13. Since $\lim_{z \to 0} \bar{z} = 0$ define $f(0) = 0$ to make it continuous in $\mathcal{C}$.
23. $f(z) = \dfrac{x^2 - y^2}{x^2 + y^2}$ is continuous in $\mathcal{C} - \{0\}$ since quotient of continuous functions with nonzero denominator. $\lim_{z \to 0} f(z)$ doesn't exist

because we get 1 if we approach along real axis and -1 along imaginary axis. Hence function cannot be made continuous at $z = 0$.

25. $czw + dw = az + b$ so $z = \dfrac{b - dw}{cw - a}$

for $z \neq -d/c$ and $w \neq a/c$. The point $-d/c$ maps to ∞ and ∞ maps to a/c.

27. $|P(z)| > |a_0| - [|a_1| + \ldots + |a_n|] \geqslant 0$ in $|z| < 1$

$$\lim_{z \to 0} \frac{(\mathrm{Re}\, z)^2}{z} = 0, \text{ since } \frac{|\mathrm{Re}\, z|^2}{|z|} <$$

$|z| < \delta = \epsilon$.

19. $f_x = y$ but $-if_y = 2y - ix$, so can have derivative only at $z = 0$, where $f'(0) = \lim\limits_{z \to 0} \dfrac{z\, \mathrm{Im}\, z}{z} = 0$.

21. Use the chain rule in exercise 20.

23. $rv_r = r(v_x x_r + v_y y_r) = xv_x + yv_y = -xu_y + yu_x = -u_\theta$, since $x_\theta = -y$ and $y_\theta = x$. Similarly for the other identity.

SECTION 1.5

1. $-if_y = e^x (\cos y + i \sin y) = f_x$

3. $-if_y = \cos x \cosh y - i \sin x \sinh y$
$= f_x$

5. $54z^2 - \dfrac{z}{2} + 4$

7. $\dfrac{-2}{(z - 1)^2}$

9. $\dfrac{[f(z + h) \pm g(z + h)] - [f(z) \pm g(z)]}{h} =$

$\dfrac{f(z + h) - f(z)}{h} \pm \dfrac{g(z + h) - g(z)}{h}$

11. $(P / Q)' = (QP' - PQ') / Q^2$

13. $f_x = 1$ but $-if_y = 0$

15. $f_x = \dfrac{x}{|z|}$ and $-if_y = \dfrac{-iy}{|z|}$, for

$z \neq 0$, and $\lim\limits_{z \to 0} \dfrac{|z|}{z}$ doesn't exist,

as it approaches 1 as $z \to 0$ on the positive real axis and -1 as $z \to 0$ along the negative real axis.

17. $f_x = 2x$ but $-if_y = 0$, so has derivatives only along imaginary axis because

$f'(iy) = \lim\limits_{z \to 0} \dfrac{f(z + iy) - f(iy)}{z} =$

SECTION 1.6

1. See solution to exercise 1, Section 1.4, and apply the theorem on sufficient conditions for analyticity.

3. See solution to exercise 3, Section 1.4, and apply the theorem on sufficient conditions for analyticity.

5. $-if_y = 2[x \cos(x^2 - y^2) \cosh 2xy + y \sin(x^2 - y^2) \sinh 2xy] + 2i[y \cos(x^2 - y^2) \cosh 2xy - x \sin(x^2 - y^2) \sinh 2xy] = f_x$ and apply theorem on sufficient conditions for analyticity.

7. $-if_y = \dfrac{(y^2 - x^2) + 2ixy}{(x^2 + y^2)^2} \cdot$

$\cos \dfrac{x}{x^2 + y^2} \cosh \dfrac{y}{x^2 + y^2} -$

$\dfrac{2xy + i(y^2 - x^2)}{(x^2 + y^2)^2} \sin \dfrac{x}{x^2 + y^2} \cdot$

$\sinh \dfrac{y}{x^2 + y^2} = f_x$ holds for

$z \neq 0$, implying sufficient conditions for analyticity in $z \neq 0$.

9. $\dfrac{f(z)}{z} = \left(\dfrac{\bar{z}}{z}\right)^2$ has value 1 on the

real axis and -1 on the imaginary
axis, thus has no derivative at

$z = 0$. But $u = \dfrac{x^3 - 3xy^2}{x^2 + y^2}$,

$v = \dfrac{y^3 - 3x^2 y}{x^2 + y^2}$ satisfy $u_x = 1$

$= v_y$, $u_y = 0 = -v_x$ at $z = 0$.

11. Im $(f + \bar{f}) = 0$ and $f + \bar{f}$ is analytic
so zero derivative theorem implies
$f + \bar{f} = 2 \operatorname{Re} f$ is constant, and thus
f is constant.

13. The Cauchy-Riemann equations
yield the system
$$2uu_x + u_y = 0$$
$$u_x - 2uu_y = 0$$
If $1 + 4u^2 \neq 0$, then $u_x = u_y = 0$.

15. Im $f = 0$, so apply zero derivative
theorem.

17. If $f' = f_x = x$, then $f = \dfrac{x^2}{2} + g(y)$,

but $f' = -if_y = -ig'$ is not a function of x.

SECTION 1.7

1. -1

3. $\dfrac{e^{-1}}{\sqrt{2}}(1 + i)$

5. $-i$

7. $-i$

9. $2\pi k - i \log 2$, $k = 0, \pm 1, \pm 2, \ldots$

11. $\pm 1, \pm i$

13. $-2^{14}(1 + i)$

15. -2^{18}

17. $2^{15} i$

19. $-2^{13}(1 + \sqrt{3}i)$

21. Use the identity in the solution to
exercise 41 of section 1.2 to get

$$\sin \dfrac{(n + 1)x}{2} \cos \dfrac{nx}{2} \Big/ \sin \dfrac{x}{2}.$$

23. $\sin \dfrac{(n + 1)x}{2} \sin \dfrac{nx}{2} \Big/ \sin \dfrac{x}{2}$

25. Use exercise 21 of section 1.5.

27. $\left\{e^{-\pi} < |w| < e^{\pi}\right\} - \left\{\operatorname{Re} w < 0\right\}$

29. $f(z) = e^{2\pi z}$

SECTION 1.8

1. $i(e - e^{-1})/2$

3. $\dfrac{(e + e^{-1})}{2} \cos 1 + i \dfrac{(e - e^{-1})}{2} \sin 1$

5. $\dfrac{(e + e^{-1})}{2} \cos 1 - i \dfrac{(e - e^{-1})}{2} \sin 1$

7. $(e^{-1} - e) / 2$

9. $e^{2iz} = i$, so $z = \dfrac{\pi}{4} + \pi k$, $k = 0$,

$\pm 1, \pm 2, \ldots$.

11. $e^z = 2 \pm \sqrt{3}$, so $z = \log |2 \pm \sqrt{3}| + 2\pi k i$, $k = 0, \pm 1, \pm 2, \ldots$

13. No, because $e^z \neq 0$.

15. $e^{i\bar{z}} + \overline{e^{-iz}} = e^{-i\bar{z}} + e^{i\bar{z}}$

17. $\dfrac{e^{iz_1} + e^{-iz_1}}{2} \dfrac{e^{iz_2} + e^{-iz_2}}{2} +$

$\dfrac{e^{iz_1} - e^{-iz_1}}{2i} \dfrac{e^{iz_2} - e^{-iz_2}}{2i} =$

$\dfrac{e^{iz_1} e^{-iz_2} + e^{-iz_1} e^{iz_2}}{2}$

19. $\dfrac{e^{2iz} - e^{-2iz}}{2i} = 2 \dfrac{(e^{iz} - e^{-iz})}{2i}$.

$\dfrac{(e^{iz} + e^{-iz})}{2}$ and $\dfrac{e^{2iz} + e^{-2iz}}{2} =$

$\left(\dfrac{e^{iz} + e^{-iz}}{2}\right)^2 + \left(\dfrac{e^{iz} - e^{-iz}}{2i}\right)^2$

(continues on p. 326)

The last identity follows by definition of tan 2z.

21. Use the identities $\cos z = \cos x \cosh y - i \sin x \sinh y$, $\cosh^2 y - \sinh^2 y = 1$.

23. $(e^z + e^{-z})^2 - (e^z - e^{-z})^2 = 4$, last two follows from definition

25. $(e^{z_1} + e^{-z_1})(e^{z_2} + e^{-z_2}) + (e^{z_1} - e^{-z_1})(e^{z_2} - e^{-z_2}) = 2(e^{z_1} e^{z_2} + e^{-z_1} e^{-z_2})$

27. Use exercises 20, 21, and 26.

29. Use quotient formula for derivatives and exercise 28.

31. Use exercise 26 and location of zeros of sin z and cos z.

33. Add definitions for cos z and i sin z.

35. Line segment $t + iy$, $|t| < \pi/2$ maps onto upper half of ellipse
$$\frac{u^2}{\cosh^2 y} + \frac{v^2}{\sinh^2 y} = 1, \text{ for } y > 0.$$

Line segment $x + it$, $t > 0$ maps onto upper half of hyperbola
$$\frac{u^2}{\sin^2 x} - \frac{v^2}{\cos^2 x} = 1$$

lying in same quadrant as line segment.

37. Use same procedure as in exercises 34–36 to show that the strip $0 < x < \pi$, $y > 0$ maps onto the lower half plane. Then consider the action on the other half of this strip.

SECTION 1.9

1. $i \arg i = i\left(\dfrac{\pi}{2} + 2\pi k\right)$, $k = 0, \pm 1, \pm 2, \ldots$

3. $i \arg(-1) = i(\pi + 2\pi k)$, $k = 0, \pm 1, \pm 2, \ldots$

5. $e^{-\arg i} = e^{-\pi/2 + 2\pi k}$, $k = 0, \pm 1, \pm 2, \ldots$

7. $\dfrac{i\pi}{2}$

9. $e^{-\pi/2}$

11. For a real, nonnegative, with $f(0) = 1$ if $a = 0$, and $f(0) = 0$ otherwise. Entire for $a = 0, a \geqslant 1$.

13. $\log |z_1| + i \arg z_1 + \log |z_2| + i \arg z_2 = \log |z_1 z_2| + i \arg(z_1 z_2)$

15. $a \log z + b \log z = (a + b)\log z$

17. $\text{Log}(-1 - i) = \log \sqrt{2} - \dfrac{3\pi i}{4}$,

$\text{Log } i = \dfrac{\pi i}{2}$ but $\text{Log } \dfrac{-1 - i}{i} =$

$\text{Log}(-1 + i) = \log \sqrt{2} + \dfrac{3\pi i}{4}$

19. $\log z^a = \log(e^{a \log z}) = a \log z$ since the exponential and logarithm are inverse functions.

21. Let $z = \cos w = (e^{iw} + e^{-iw})/2$, then $e^{2iw} - 2ze^{iw} + 1 = 0$ has roots $e^{iw} = z + (z^2 - 1)^{1/2}$ where the square root maps $[\mathbb{C} - \{0\}]^2$ onto $\mathbb{C} - \{0\}$.

23. Let $z = \cot w = i(e^{iw} + e^{-iw})/(e^{iw} - e^{-iw})$, then
$$e^{2iw} = \frac{z + i}{z - i}.$$

25. If $z = \cosh w = \frac{1}{2}(e^w + e^{-w})$, then $e^w = z + (z^2 - 1)^{1/2}$ where square root is two-valued.

27. If $z = \sin w$, then $1 = \cos w \dfrac{dw}{dz} = (1 - \sin^2 w)^{1/2} \dfrac{dw}{dz}$.

29. If $z = \tan w$, then $1 = \sec^2 w \dfrac{dw}{dz} = (1 + \tan^2 w) \dfrac{dw}{dz}$.

31. Let $z = \cosh w$, then $1 =$

$$\sinh w \,\frac{dw}{dz} = (\cosh^2 w - 1)^{1/2}\,\frac{dw}{dz}$$

33. $z^{k/2}$ maps $[\,\mathbb{C} - \{0\}]^2$ onto $[\,\mathbb{C} - \{0\}]^k$, for $k = 1, 3$ so the range spaces are different for $z^{1/2}$ and $z^{3/2}$. Hence $(-1)^{1/2} \neq (-1)^{3/2}$.

SECTION 1.10

1. $E_{R_0} = r A \operatorname{Re} e^{iwt}$
$E_{R_1} = r s^2 A \operatorname{Re} e^{i(wt - \alpha)}$,
$E_{R_2} = r^3 s^2 A \operatorname{Re} e^{i(wt - 2\alpha)}, \ldots,$
$E_{R_n} = r^{2n-1} s^2 A \operatorname{Re} e^{i(wt - n\alpha)}$,

$$\ldots \text{so } E_{\text{reflected}} = \left(1 - \frac{s^2}{r^2}\right) E_{R_0} +$$

$$\frac{s^2 A}{r} \operatorname{Re}\left\{ e^{iwt} \sum_{n=0}^{\infty} (r^2 e^{-i\alpha})^n \right\} =$$

$$\left(1 - \frac{s^2}{r^2}\right) E_{R_0} + \frac{1}{r} E_{\text{transmitted}} =$$

$$\left(2 - \frac{1}{r}\right) A \cos wt +$$

$$\frac{(1 - r^2) A \cos (wt - \beta)}{r\sqrt{1 + r^4 - 2r^2 \cos \alpha}}$$

Then write $\cos(wt - \beta) = \cos wt \cos \beta + \sin wt \sin \beta$ and put in the form $A^* \cos(wt - \gamma)$ by selecting $\cot \gamma =$

$$\frac{[(2r - 1)\sqrt{1 + r^4 - 2r^2} \cos \alpha + (1 - r^2) \cos \beta]}{(1 - r^2) \sin \beta}$$

3. $\dfrac{\partial^2 E}{\partial x^2} = f''(ct + x) = \dfrac{1}{c^2}\,\dfrac{\partial^2 E}{\partial t^2}$.

CHAPTER 2

SECTION 2.1

1. $x = a \cos t,\ y = b \sin t$

3. For example,
$$z(t) = \begin{cases} t, 0 \leqslant t \leqslant 1 \\ 2 - t + i(t - 1), 1 \leqslant t \leqslant 2 \\ i(3 - t), 2 \leqslant t \leqslant 3 \end{cases}$$

5. $z(t) = \begin{cases} 2 + e^{i\pi(t+2)} & -3 \leqslant t \leqslant -1 \\ -t, -1 \leqslant t \leqslant 1 \\ -2 + e^{i\pi(t+1)}, & 1 \leqslant t \leqslant 3 \end{cases}$

7. $(1 - i)/2; (i - 1)/2; 1$

9. $iR^2\pi; -R^2\pi; 2\pi i R^2$

11. Set $z = e^{it}$, $0 \leqslant t \leqslant \dfrac{\pi}{2}$, then

$$\int_\gamma y\, dz = i \int_0^{\pi/2} \sin t\, e^{it}\, dt = \frac{-\pi}{4} + \frac{i}{2}$$

13. $f(z) = y$ is not the derivative of an analytic function.

15. $e^i - e$

17. $e^i - e$

19. $(\cos a - \cosh a)/a$

21. $\displaystyle\int_0^{2\pi} \frac{z'(t)\, dt}{\sqrt{(z'(t))^2}} = \pm \int_0^{2\pi} dt$

23. $2 - \pi i; 2 + \pi i$

SECTION 2.2

1. Set $z = e^{it}$ to get $-\displaystyle\int_{-\pi}^{\pi} t\, dt = 0$.

For $0 \leqslant \arg z \leqslant 2\pi$ we get $-2\pi^2$

3. $\displaystyle\int_{\partial G} y\, dx + iy\, dy = \iint_G - dx\, dy$
$$= -A$$

5. $0 = \displaystyle\int_\gamma e^z\, dz = \int_0^a e^x\, dx +$

$$ai \int_0^{\pi/2} e^{ae^{it}} e^{it}\, dt - i \int_0^a e^{iy}\, dy,$$

where $e^{it} = \cos t + i \sin t$, and consider the imaginary parts of the last two integrals

7. $\text{Sin } z = \sin x \cosh y + i \cos x \sinh y$, where $x = at$ and $y = bt$, $0 \leqslant t \leqslant T$

9. $0 = \int_\gamma \dfrac{dz}{1 + z^2} = \int_{-a}^a \dfrac{dx}{1 + x^2} +$

$$i \int_0^b \frac{dy}{(1 + a^2 - y^2) + 2iay} -$$

$$i \int_0^b \frac{dy}{(1 + a^2 - y^2) - 2iay} -$$

$$\int_{-a}^a \frac{dx}{(1 + x^2 - b^2) + 2ibx}$$

Multiply numerator and denominator of last integral by complex conjugate and take the limit as $a \to \infty$.

11. Use the rectangle in Example 3 and the function $f(z) = e^{ikz}/(1 + z^2)$.
13. Integrate $f(z) = e^{-z^2}$ on the boundary of the rectangle $0 \leqslant x \leqslant a$, $0 \leqslant y \leqslant b$, and let $a \to \infty$.
15. Show that

$$\left| Ri \int_0^{\pi/8} e^{-R^2 e^{2it}} e^{it}\, dt \right| \leqslant$$

$$R \int_0^{\pi/8} e^{-R^2 \cos 2t}\, dt \to 0,$$

as $R \to \infty$.

SECTION 2.3

1. 0
3. $2\pi i/(b - a)$
5. $2\pi i \cos 1$
7. $2\pi i \sin 1$
9. $-2\pi i \sin 1$
11. $-\pi i \sin 1$

13. Break parametrization of $\gamma_1 + \gamma_2$ into two parts.
15. Use triangle inequality and property (iv).
17. 8. Use $|1 + e^{it}| = \sqrt{2(1 + \cos t)}$
19. Apply Cauchy's estimate to f with $M = (1 - r)^{-1}$ and minimize the result for all $0 \leqslant r \leqslant 1$.
21. Let $z = e^{i\theta}$, $0 \leqslant \theta \leqslant 2\pi$ and equate imaginary terms.
23. The function is analytic in each disk D and analyticity is a local property.

SECTION 2.4

1. Show f^n is constant.
3. Apply the maximum principle to $F(z)$ as defined by exercise 2. Then $|F| \leqslant 1$ on $|z| = 1$. Equality makes F constant.
5. Let $f(z) = z$ and let G be the open unit disk.
7. $|f| \neq 0$ on the boundary of G and if f has no zero in G, the maximum and minimum principles imply f is constant in G.
9. By Cauchy's theorem

$$\int_{|z|=R} \frac{a_0}{zP(z)}\, dz = 2\pi i.$$

As $R \to \infty$ we have

$$\lim_{R \to \infty} \frac{P(z)}{a_n z^n} = 1 \text{ so for large } R,$$

$$|P(z)| \geqslant |a_n z^n|/2. \text{ Then } 2\pi =$$

$$\left| \int_{|z|=R} \frac{a_0\, dz}{zP(z)} \right| \leqslant \frac{4\pi |a_0|}{|a_n| R^n} \text{ which}$$

leads to a contradiction as $R \to \infty$

SECTION 2.5

1. Use Cauchy's theorem for derivatives.

3. $\left| \int_{\Gamma_R} e^{zt} f(z)\, dz \right| \le$

   $2 \sup_{\Gamma_R} |f(z)|$

   $\int_0^{\pi/2} e^{tR \cos \theta} R\, d\theta$ if $\sigma > 0$. Let

   $g(\theta) = \cos \theta - (1 - 2\theta/\pi)$. Then $g(0) = g(\pi/2) = 0$ and $g'' < 0$. Use $\cos \theta > 1 - 2\theta/\pi$ to bound the integral and show bound tends to zero as $R \to \infty$.

5. Assume the polygonal path $\gamma - \gamma'$ in the proof of the antiderivative theorem avoids the points $z_1, \ldots, z_n$. If finitely many of the exceptional points lie inside any subrectangle formed by the curves $\gamma - \gamma'$, apply exercise 4. Then apply the Fundamental theorem.

CHAPTER 3

SECTION 3.1

1. $1 + \frac{1}{2} + \left(\frac{1}{3} + \frac{1}{4}\right) + \left(\frac{1}{5} + \frac{1}{6} + \frac{1}{7} + \frac{1}{8}\right)$ $+ \ldots$ and each group of terms in parentheses exceeds $1/2$.

3. Let $f(z) = \sin z$. Then $f^{(4n)}(0) = 0$, $f^{(4n+1)}(0) = 1$, $f^{(4n+2)}(0) = 0$, $f^{(4n+3)}(0) = -1$.

5. If $f(z) = \sinh z$, then $f^{(4n)}(0) = 0$, $f^{(4n+1)}(0) = 1$, $f^{(4n+2)}(0) = 0$, $f^{(4n+3)}(0) = 1$.

7. Use the geometric series in Example 2.

9. $\dfrac{1}{1 - i - (z - i)} = \dfrac{1}{1 - i}\left[1 + \right.$

$\dfrac{z - i}{1 - i} + \left(\dfrac{z - i}{1 - i}\right)^2 + \ldots \left.\right], |z - i|$

$< \sqrt{2}$

11. $\displaystyle\sum_{n=0}^{\infty} (-1)^n \frac{(z - \pi/2)^{2n}}{(2n)!}, |z| < \infty$

13. $\dfrac{i\pi}{2} - \displaystyle\sum_{n=1}^{\infty} \frac{(i)^n}{n} (z - i)^n, |z - i|$

< 1

15. $\log |2| + 3\pi i - \displaystyle\sum_{n=1}^{\infty} \frac{(z - 2e^{3\pi i})^n}{2^n n}$,

$|z - 2e^{3\pi i}| < 2$

17. 10

19. 6

21. No

23. $f(z) = (2 - z)^{-1}$

25. $f(z) = 0$ and $g(z) = \sin z$ are both entire.

27. Expand $e^{\alpha \mathrm{Log}(1+z)}$ in a Maclaurin series.

SECTION 3.2

1. 0

3. 1

5. 1/3

7. R

9. R^k

11. R

13. $[(1 - z)^{-1}]^n =$

$\displaystyle\sum_{n=0}^{\infty} (n + 1)(n + 2)z^n, R = 1$

15. Integrate the Maclaurin series for $\sin z/z$ to get

$\displaystyle\sum_{n=0}^{\infty} \frac{(-1)^n z^{2n+1}}{(2n + 1)!(2n + 1)}, R = \infty$

17. $f(z) = a_0 \cos z + a_1 \sin z$

19. $a_0 \displaystyle\sum_{n=0}^{\infty} (-1)^n z^{2n}/2^{2n}(n!)^2$

21. Use the Taylor series of $f(z)$ and $g(z)$ centered at z_0 to get $f(z)/g(z) =$

$$\frac{f'(z_0)(z - z_0) + f''(z_0)(z - z_0)^2/2 + \dots}{g'(z_0)(z - z_0) + g''(z_0)(z - z_0)^2/2 + \dots},$$

Then divide numerator and denominator by $z - z_0$ and take the limit as $z \to z_0$.

23. $\sqrt{3}\,(1 - z)/2\,(1 - z)^3$

25. The power series for e^{zt} is entire, so apply Weierstrass's theorem.

27. Differentiate under the integral sign since series converges uniformly.

29. $\dfrac{g(t)}{1 - zt} = \displaystyle\sum_{n=0}^{\infty} z^n\, t^n\, g(t)$ converges

uniformly on $0 \leqslant t \leqslant 1$ for $|z| < 1$. Apply Weierstrass's theorem and differentiate term-by-term to get

$$f'(z) = \int_0^1 \frac{t\, g(t)}{(1 - zt)^2}\, dt.$$

SECTION 3.3

1. $\dfrac{1}{z} - 1 + z - z^2 + z^3 - \dots =$

$$\sum_{n=0}^{\infty} (-1)^n\, z^{n-1}$$

3. $-\displaystyle\sum_{n=0}^{\infty} (1 - z)^{-(n+1)} -$

$$\frac{1}{2} \sum_{n=0}^{\infty} \left(\frac{1 - z}{2}\right)^n$$

5. $\displaystyle\sum_{n=1}^{\infty} z^{-2n-1}$

7. $\dfrac{1}{4} \displaystyle\sum_{n=-1}^{\infty} \left(\frac{z - 1}{2}\right)^n -$

$$\sum_{n=-\infty}^{-1} (z - 1)^n$$

9. $\dfrac{1/3}{z - 1} + \dfrac{2}{9} \displaystyle\sum_{n=0}^{\infty} \left(\frac{1 - z}{3}\right)^n$

11. $\dfrac{1}{3} \displaystyle\sum_{n=0}^{\infty} \left[\left(\frac{-z}{2}\right)^n - z^n\right]$

13. $\dfrac{1}{z + 2} + \displaystyle\sum_{n=1}^{\infty} \frac{3^{n-1}}{(z + 2)^{n+1}}$

15. $\displaystyle\sum_{n=-\infty}^{\infty} c_n\, z^n$, where $c_n = c_{-n}$ and

$$c_n = \sum_{k=0}^{\infty} [k!\,(n + k)!]^{-1}$$

17. $\displaystyle\sum_{n=0}^{\infty} c_{2n+1} (z^{2n+1} + z^{-2n-1})$, where

$$c_{2n+1} = \sum_{k=0}^{\infty} \frac{(-1)^{n+1}}{k!\,(n + 1 + k)!}$$

19. $\left(\displaystyle\sum_{n=0}^{\infty} (1 - z)^n\right) \cdot$

$$\left(\sum_{n=0}^{\infty} \frac{(-1)^n (z - 1)^{-(2n+1)}}{(2n + 1)!}\right) =$$

$$\sum_{n=-\infty}^{\infty} (-1)^{n+1}\, c_n (z - 1)^n \text{ where}$$

$$c_n = \sum_{k=q}^{\infty} \frac{(-1)^k}{(2k + 1)!} \text{, with}$$

$q = 0$, for $n \geqslant -1$,
$q = [\,|n/2|\,]$, for $n \leqslant -2$, where
$[m]$ is the largest integer $\leqslant m$.

21. $\displaystyle\sum_{n=0}^{\infty} c_n (z - 1)^{-n}$, where

$$c_n = \frac{(-1)^{[n/2]}}{(2\,[n/2] + 1)!}$$

23. $a_n = \dfrac{1}{2\pi i} \displaystyle\int_{|\zeta|=1} \frac{e^{z/2(\zeta - 1/\zeta)}\, d\zeta}{\zeta^{n+1}} =$

$$\frac{1}{2\pi} \int_{-\pi}^{\pi} e^{i(z \sin\theta - n\theta)} \, d\theta$$

and the imaginary part of this integral is zero.

25. $\left(z + \dfrac{1}{z}\right)^m = \sum_{n=0}^{m} \binom{m}{n} z^{m-2n}$ and

$$a_n = \frac{1}{2\pi i} \int_{|z|=1} \frac{(z + z^{-1})^m \, dz}{z^{n+1}} =$$

$$\frac{1}{\pi} \int_{-\pi}^{\pi} e^{-in\theta} \cos^m(\theta) \, d\theta. \text{ Hence}$$

$$\frac{1}{\pi} \int_{-\pi}^{\pi} \cos n\theta \, \cos^m \theta \, d\theta = \binom{m}{n}.$$

SECTION 3.4

1. $z = 0, \infty$ are removable, $z = \pm i$ are simple poles.

3. $z = 0$ is an essential singularity, $z = \infty$ is a simple pole.

5. $z = 0$ is an essential singularity, $z = \infty$ is removable.

7. $(z + 1)e^{1/(z-1)}/(z^4 + z^3)$ is an example.

9. $C(z) = \displaystyle\sum_{n=1}^{\infty} \frac{(-1)^n z^{2n}}{2n \, (2n)!}$

11. $L(z) = \displaystyle\sum_{n=1}^{\infty} \frac{(-1)^{n+1} z^n}{n^2}$

13. Let k be the order of the pole at ∞ ($k = 0$ if ∞ is a removable singularity), and apply Liouville's theorem to $z^{-k} f(z)$. Essential singularities.

15. $z = \infty$ is not an essential singularity, nor can it be a pole of order $k \geqslant 1$, as then $f(z) = z^k f_k(z)$, with f_k analytic on $\mathfrak{M}$, and zero is not omitted.

SECTION 3.5

1. Consider $-\text{Log}(1 - z)$. $R = 1/2$

3. Use $-\log(1-z)$.

5. $z^{1/2} = 1 + \frac{1}{2}(z - 1) -$

$$\sum_{n=2}^{\infty} \frac{(-1)^n \, (2n - 3)! \, (z - 1)^n}{2^{2(n-1)} \, (n - 2)! \, n!},$$

$|z - 1| < 1$. No.

7. $\left(\sin \dfrac{\pi z}{2}\right)^{1/2} = 1 -$

$$\frac{\pi^2}{8} \frac{(z - 1)^2}{2!} - \frac{\pi^4}{8^2} \frac{(z - 1)^4}{4!} + \dots.$$

No.

9. Let $\zeta = e^{2\pi i p/q}$ for any positive integers p and q, and observe that $f(t\zeta) \to \infty$ as $t \to 1^-$.

11. $\displaystyle\sum_{n=0}^{\infty} \frac{(-1)^n \, (n + 2)! \, (z - 1)^n}{n!}$,

$$f(z) = \frac{2}{z^3}$$

13. Define $F(z) = \begin{cases} f(z) \text{ on } G + \cup \gamma. \\ \overline{f(\bar{z})} \text{ on } G-. \end{cases}$

Then $F(z)$ is continuous on $G = G + \cup\gamma\cup G-$. On $G-$ we have $F(z) = u(\bar{z}) - iv(\bar{z})$ where $f = u + iv$. Then $-iF_y = v_y(\bar{z}) + iu_y(\bar{z}) = u_x(\bar{z}) - iv_x(\bar{z}) = F_x$ so F is analytic in $G-$. For any z_0 in γ consider $|z - z_0| < \rho$ interior to G. Divide circle into two semicircles and show by continuity that

$$\int_{|z-z_0|=\rho} \frac{F(z)}{z - z_0} \, dz = 0.$$

15. If $|z| = 1$ then $\left| \displaystyle\sum_{n=1}^{\infty} \frac{z^n}{n^2} \right| \leqslant$

$$\sum_{n=1}^{\infty} \frac{1}{n^2} \leqslant 1 + \sum_{n=2}^{\infty} \frac{1}{n(n - 1)} =$$

$$1 + (\tfrac{1}{1} - \tfrac{1}{2}) + (\tfrac{1}{2} - \tfrac{1}{3}) + \dots = 2 \text{ so}$$

always converges. However, term-
wise differentiation gives

$$\sum_{n=1}^{\infty} \frac{z^{n-1}}{n} = \frac{-1}{z} \text{ Log } (1 - z)$$

which diverges at $z = 1$.

and by induction

$$= \lim_{z \to r} \frac{d^{a-k}}{dz^{a-k}} \left\{ (z - r) \left(\frac{P}{Q_1} \right)^{(k+1)} + \right.$$

$$(k - a) \left(\frac{P}{Q_1} \right)^{(k)} \left. \right\} = 0.$$

CHAPTER 4

SECTION 4.1

1. $\text{Res}_i f(z) = \dfrac{-1}{2}$, $\text{Res}_{-i} f(z) = \dfrac{-1}{2}$

3. $\text{Res}_0 f(z) = 1$

5. $\text{Res}_0 f(z) = \dfrac{1}{2}$

7. $\text{Res}_0 f(z) = \dfrac{-1}{2}$

9. $\text{Res}_{k\pi i} f(z) = (-1)^k \, k\pi i$

11. $\text{Res}_{k\pi} f(z) = 1$

13. $-2\pi i$

15. 0. (Use Inside-outside theorem.)

17. $-2\pi i \, [\text{Res}_{-i} f(z)] = \pi i^n$

19. 0, since $\text{Res}_0 f(z) = 0$.

21. πi

23. $2\pi i \, [\text{Res}_{\pi/2} \tan z + \text{Res}_{-\pi/2} \tan z]$
$= -4\pi i$

25. Remove all common factors of P
and Q. Then $\left(\dfrac{P}{(z - r)^a \, Q_1} \right)' =$

$$\frac{1}{(z - r)^a} \left(\frac{P}{Q_1} \right)' - \frac{1}{(z - r)^{a+1}} \cdot$$

$$\left(\frac{aP}{Q_1} \right) \text{ so } \text{Res}_r \left(\frac{P}{Q} \right)' =$$

$$\lim_{z \to r} \frac{d^a}{dz^a} \left\{ (z - r) \left(\frac{P}{Q_1} \right)' - \frac{aP}{Q_1} \right\}$$

SECTION 4.2

1. $\dfrac{1}{4} \displaystyle\int_0^{2\pi} \dfrac{d\theta}{a + \sin^2 \theta} =$

$$i \int_{|z|=1} \frac{z \, dz}{(z^2 - 1)^2 - 4az^2}$$

and $|\sqrt{a} - \sqrt{a + 1}| < 1$

3. $-4i \displaystyle\int_{|z|=1}$

$$\frac{z \, dz}{(a^2 - b^2) z^4 + 2(a^2 + b^2) z^2 + (a^2 - b^2)}$$

$$= \frac{-4i}{a^2 - b^2} \cdot \int_{|z|=1}$$

$$\frac{z \, dz}{\left(z^2 + \dfrac{a + b}{a - b} \right) \left(z^2 + \dfrac{a - b}{a - b} \right)}$$

and $|a - b| < a + b$

5. $i \displaystyle\int_{|z|=1} \dfrac{dz}{(az - 1)(z - a)}$

7. $\dfrac{-i}{2^n} \cdot$

$$\int_{|z|=1} \frac{[(a - ib)z^2 + (a + ib)]^n \, dz}{z^{n+1}}$$

and calculate residue at $z = 0$.

9. Use $\cos ib = \cosh b$ and $\sin ib = i \sinh b$.

SECTION 4.3

1. Set $a = 0$ in the theorem of this section and evaluate the residue of $z/(z^2 + 2z + 2)^2$ at $-1 + i$.

3. $\frac{1}{2} \displaystyle\int_{-\infty}^{\infty} \frac{x^2\,dx}{(x^2 + a^2)^2} =$

   $\frac{1}{2} \operatorname{Re} \left\{ 2\pi i \operatorname{Res}_{ai} \dfrac{z^2}{(z^2 + a^2)^2} \right\}$

5. $\operatorname{Re} \left\{ 2\pi i \operatorname{Res}_i (z^2 + 1)^{-n-1} \right\}$

7. $\operatorname{Im} \left\{ 2\pi i \operatorname{Res}_{ib} \dfrac{z^3\,e^{iaz}}{(z^2 + b^2)^2} \right\}$

9. $\frac{1}{2} \operatorname{Im} \left\{ 2\pi i \left[\operatorname{Res}_{(1+i)b/\sqrt{2}} \dfrac{z\,e^{iaz}}{z^4 + b^4} \right.\right.$

   $\left.\left. + \operatorname{Res}_{(-1+i)b/\sqrt{2}} \dfrac{z\,e^{iaz}}{z^4 + b^4} \right] \right\}$

SECTION 4.4

1. Use residues at $x = \pm 1/2$

3. Residues at $x = 0, 1/2$, and write numerator as $\frac{1}{2} \sin 2\pi x$

5. Residues at $x = 0$ and $x = bi$ of $f(z) = e^{iz}/z(z^2 + b^2)$

7. Use $f(z) = (e^{iaz} - e^{ibz})/z^2$ on contour of Figure 4.5.

9. Use the identity

   $\cos(A - B) - \cos(A + B) = 2 \sin A \sin B$

   and integrate the function

   $f(z) = (e^{i(A-B)} - e^{i(A+B)})/ 2(z - a)(z - b)$

   where $A = m(z - a)$, $B = n(z - b)$ around a semicircle of radius R with half circles added at a and b of radius r.

SECTION 4.5

1. Use method of example 1. For $a = 0$ solve directly and interpret the answer as a limit as $a \to 0$.

3. See the answer of exercise 1.

5. Solve directly for $a = 0, 1, 2$. Interpret the answer for $a = 1$ as a limit.

7. Use the method of example 2.

9. Use $f(z) = z^a \log z/(z^2 + b^2)$ on Figure 4.6.

11. Use $f(z) = e^{iaz}/\sinh z$ over a rectangular contour $\left\{ z: |x| \leqslant R, 0 \leqslant y \leqslant 2\pi i \right\}$ with semicircles at 0 and $2\pi i$ removed. Pole at πi with residue $-i \sinh \pi a$.

13. Use $f(z) = e^{az}/\cosh \pi z$ over the rectangular contour $\left\{ z : |x| \leqslant a, 0 \leqslant y \leqslant 1 \right\}$. Pole at $i/2$ with residue $\cos(a/2)/\pi i$.

15. Substitute $u = x^a$ and apply exercise 14.

17. Substitute $x = b \tan\theta$ and let $b = 1$. This substitution changes the second integral into that of example 2.

SECTION 4.6

1. 0

3. 0

5. (i) 5, (ii) 8, (iii) 5, (iv) 6

7. 3. Let $f(z) = (z^2 - 1) \cdot (z^4 - 5z^2 + 5)$ and note that $|z^2 - 1| > |z|$ on the semicircle $|z| = R > 2$, $x > 0$, and on the line segment $z = iy$, $|y| \leqslant R$.

9. 0. Multiply the equation by $z^2 + 2$.

11. Apply the argument principle to $g(z) = f(z) - a$.

13. Select any interior point z_0 in G. By exercise 12, $f(z_0)$ is interior to $f(G)$ since every point in a suf-

ficiently small disk centered at $f(z_0)$ has a preimage in G. No interior point of G can have maximal absolute value as all are interior to $f(G)$.

CHAPTER 5

SECTION 5.1

1. $\mathcal{C}$
3. $z \neq 0$
5. Quadruples the angle to 2π radians.
7. Doubles the angle.
9. Set $z = r (1 + e^{i\theta})$ and square
11. $\dfrac{\partial (u, v)}{\partial (x, y)} = \begin{vmatrix} u_x & u_y \\ v_x & v_y \end{vmatrix}$ and apply Cauchy-Riemann equations.
13. Look at Liouville's theorem.
15. There exists a value θ and an $\epsilon > 0$ such that $\text{Re}\,(e^{i\theta} f'(z)) > 0$ for every z in $|z - z_0| < \epsilon$.

SECTION 5.2

1. $v < 0$
3. $|w - (1 + i)/2| > \sqrt{2}/2$, $|w - (1 - i)/2| > \sqrt{2}/2$
5. 3. Substitution in Example 5 yields $16(w^4 + 8w^3 + 3w^2 - 2w + 1) = 0$.
7. Let $w = \exp(2\pi i z/(z - 2))$.
9. Let $w = \sin^{-1}(iz^2)$.

SECTION 5.3

1. $w = \dfrac{i - 1}{2} \cdot \dfrac{z + 1}{z - 1 - i}$
3. $\dfrac{2 - w}{w - 4} = \dfrac{1 + i}{2} \cdot \dfrac{1 + z}{1 + i - z}$

5. No. It is not analytic.
7. $(3 + 4i)/25$
9. $(2 + 5i)/3$
11. $w = (-2 + 4i)(z + 1)/(5iz + 2 + i)$
13. $2 - \sqrt{3}$

15. The mapping $w = i\,\dfrac{z - (a - R)}{z - (a + R)}$ maps the circle onto the real axis. Thus

$$\left[i\,\frac{z - (a - R)}{z - (a + R)} \right] =$$

$$i\,\frac{\left[\dfrac{R^2}{\bar z - \bar a} + a \right] - (a - R)}{\left[\dfrac{R^2}{\bar z - \bar a} + a \right] - (a + R)}$$

implying that $z^* = (R^2/(\bar z - \bar a)) + a$.

SECTION 5.4

1. $w = \dfrac{2i}{\pi} \log \dfrac{1 + w}{1 - w}$
3. $w = \dfrac{1 - z^2}{1 + z^2}$
5. See Exercise 9 of Section 5.1. Inside of cardioid.
7. Maps onto itself with orientation reversed.
9. First shift left 1 unit and consider the square root with branch cut on the positive real axis. Finally

$$w = \frac{i\sqrt{z - 1} + \sqrt{2}}{\sqrt{2} - 2 - i\sqrt{z - 1}}$$

11. Suppose μ and ν are branch points of the Riemann surface of the transformation. Letting $w = f(z)$, then $\zeta(z) = (2f(z) - \mu - \nu)/(\mu - \nu)$

maps the z-plane onto a two-sheeted Riemann surface with branch points at ± 1. Writing $\zeta(z)$ as a Laurent series, we see that $Z(z) = \zeta(z) + \sqrt{\zeta^2(z)} - 1$ is analytic at ± 1 and is either analytic or has simple poles at the poles of $\zeta(z)$. Furthermore $Z(z)$ must have at least one simple pole as otherwise it is constant, so $Z(z)$ is a rational function. Finally show that $Z(z)$ is of first degree, from which the answer follows. The second alternative occurs when $\nu = \infty$.

SECTION 5.5

1. $\mathrm{Im}\,(Az^{4/3}) - A\,\mathrm{Im}\,(z^{4/3}) -$ constant. The origin is a stagnation point.
3. $\mathrm{Im}\,(Az^4) = $ constant or $x^3 y - xy^3 = $ constant. The origin is a stagnation point.
5. $V = \overline{w}' = 1 - 3\overline{z}^2$, so $|V| = 1, 2, 4$ at $0, 1, i$, respectively. $w' = 0$ when $z = \pm(3)^{-1/2}$.
7. $V = 3 + 2i\overline{z}$, so $|V| = 3, \sqrt{13}, 1$ at $0, 1, i$, respectively. $w' = 0$ when $z = -3i/2$.
9. $y + 3x^2 y - y^3 = $ constant; $y + 2(x^2 - y^2) = $ constant; $3y - (x^2 - y^2) = $ constant; $\cos x \sinh y = $ constant.
11. $e^w = (z \pm \sqrt{z^2 - a^2})/a$ so $\arg [(z \pm \sqrt{z^2 - a^2})/a] = $ constant
13. Streamlines are the confocal hyperbolas

$$\frac{x^2}{a^2 \cos^2 v} - \frac{y^2}{a^2 \sin^2 v} = 1,$$

$v = $ constant. For $v = 0, \pi$ get the flow on the edges of the aperture,

but this flow is not physically realizable because $\dfrac{1}{V} = \dfrac{dz}{dw} =$

$a \sinh w = a \sqrt{\cosh^2 w - 1} = \sqrt{z^2 - a^2}$, so that $|V|$ is infinite at $z = \pm a$.

15. $V = w' = A(1 - a^2/z^2)$ so for $z = ae^{i\theta}$ we get $|V(ae^{i\theta})| = |A(1 - e^{-2i\theta})| = 2A|\sin\theta|$. But

$$\frac{p(\infty)}{\rho} + \tfrac{1}{2}A^2 = \frac{p(z)}{\rho} + 2A^2\sin^2\theta$$

implying that

$$\frac{p(z)}{\rho} = \frac{p(\infty)}{\rho} + \tfrac{1}{2}A^2[1 - 4\sin^2\theta].$$

If $A^2 > \dfrac{2p(\infty)}{3\rho}$ and $\theta = \pm\pi/2$,

cavitation occurs.

SECTION 5.6

1. $\mathrm{Im}\,\sqrt{w^2 + 1} = $ constant are the streamlines in the w-plane.
3. Clearly maps upper half plane into square since

$$\int_0^1 \frac{dx}{\sqrt{x(x^2 - 1)}} =$$

$$i\int_0^1 \frac{dx}{\sqrt{x(1 - x^2)}} =$$

$$-\int_0^{-1} \frac{dx}{\sqrt{x(1 - x^2)}} =$$

$$\int_0^{-1} \frac{d(-x)}{\sqrt{(-x)(x^2 - 1)}}.$$

Set $x^2 = t$, use the integral in Example 4 and $\Gamma(x)\,\Gamma(1 - x) = \pi/\sin\pi x$ to get

$$\int_0^1 \frac{dx}{\sqrt{x(1-x^2)}} =$$

$$\frac{1}{2} \int_0^1 \frac{dt}{t^{3/4}\sqrt{1-t}} = \frac{\Gamma(\frac{1}{4})\,\Gamma(\frac{1}{2})}{2\Gamma\,(3/4)}.$$

5. $w = (1 - 2z)\sqrt{z-z^2} - (\sin^{-1}(2z-1))/2 - \pi/4$

7. Use the identities $\Gamma(\frac{1}{4})\,\Gamma(3/4) = \pi\sqrt{2}$ and $\Gamma(\frac{1}{2})^2 = \pi$. Then, $w =$

$$\int_0^z \frac{(z-1)^{3/4}\,dz}{\sqrt{z}} \quad \text{and}$$

$$\left| \int_0^1 \frac{(x-1)^{3/4}\,dx}{\sqrt{x}} \right| =$$

$$\left| (-1)^{-3/4}\,\frac{\Gamma(\frac{1}{2})\,\Gamma(7/4)}{\Gamma(9/4)} \right| =$$

$$\frac{12\pi\sqrt{2\pi}}{5\Gamma(\frac{1}{4})^2}$$

9. Set $s^2 = (z - a)/z$ and consider the mapping

$$w = \int_0^z \frac{\sqrt{z-a}}{\sqrt{z(z-1)}}\,dz.$$

SECTION 5.7

1. $\zeta = (z - \sqrt{3})/(z + \sqrt{3})$ maps region into the annulus $(\sqrt{3} - 1)/(\sqrt{3} + 1) < |\zeta| < 1$. The equipotential lines in the ζ plane are circles $|\zeta|$ = constant, so equipotential lines in z plane satisfy $|z - \sqrt{3}| / |z + \sqrt{3}|$ = constant.

3. $u = \frac{1}{2} - \frac{1}{\pi}\,\mathrm{Re}\,(\sin^{-1} e^z)$

5. $u = \frac{1}{\pi}\,\mathrm{Arg}\,\frac{1 + \sin(\pi z/2)}{1 - \sin(\pi z/2)}$

7. Let $z = \cosh \zeta$ and $w =$

$A\,(\cosh \zeta - i\alpha)$. The first mapping produces a family of confocal ellipses and a family of confocal hyperbolas.

SECTION 5.8

1. V is real on edges of vessel, speed is equal at ± 1 but in opposite directions and constant on free stream lines. As $V(0)$ is pure imaginary we obtain the hodograph

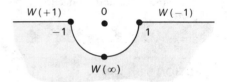

with $W = A/\overline{V} = A/w'(z)$. Then consider the succession of mappings $z_1 = 2\,\mathrm{Log}\,W + i\pi$

$$z_2 = \sin(iz_1/2)$$

$$\zeta = \frac{2}{\pi}\,\mathrm{Log}\,z_2 - i.$$

CHAPTER 6

SECTION 6.1

1. $\phi = \mathrm{Re}(e^z)$. Check that Laplace's equation holds.

3. $\phi = \mathrm{Re}(z^3)$. Check that Laplace's equation holds.

5. $f(z) = \sin z$; entire.

7. No.

9. $v = \tan^{-1}(y/x)$ + constant.

11. $v = -y/[(x-1)^2 + y^2]$ + constant

13. Consider the real part of $\log f(z)$.

15. $\frac{1}{2\pi} \int_0^{2\pi} \log |1 + re^{i\theta}|\,d\theta =$

$$\frac{1}{4\pi} \int_0^{2\pi} \log |1 + r^2 + 2r \cos \theta| \, d\theta$$

$$\to \frac{1}{2\pi} \int_0^{2\pi} \log \left(2 \cos \frac{\theta}{2}\right) d\theta \text{ and}$$

$$\int_0^{2\pi} \log \cos \frac{\theta}{2} \, d\theta =$$

$$\int_0^{2\pi} \log \sin \frac{\theta}{2} \, d\theta = 2 \int_0^{\pi} \log \sin \phi \, d\phi$$

SECTION 6.2

1. Integrate both sides of the Mean-Value theorem in Section 6.1

   getting $u(\zeta) = \dfrac{2u(\zeta)}{R^2} \displaystyle\int_0^R r \, dr =$

   $$\frac{1}{\pi R^2} \int_0^R \int_0^{2\pi} u(\zeta + re^{i\theta}) \, r \, dr \, d\theta$$

3. $\overline{\text{grad } u} = f'$, where $f = u + iv$ with v any harmonic conjugate of u.

5. Use $g(z) = (z^2 - z_0^2)/(z^2 - \bar{z}_0^2)$.

   Then $u(z) = \dfrac{1}{2\pi i} \displaystyle\int_{\partial G} u(\zeta) \cdot$

   $$\frac{2\zeta(z^2 - \bar{z}^2)}{(\zeta^2 - \bar{z}^2)(\zeta^2 - z^2)} d\zeta = \frac{4xy}{\pi} \cdot$$

   $$\left[\int_0^{\infty} t\left(\frac{u(it)}{|t^2 + z^2|^2} + \frac{u(t)}{|t^2 - z^2|^2}\right) dt\right]$$

7. If $u(z)$, $U(z)$ are both solutions of Dirichlet's problem that are continuous on $\bar{G}$, then $U(z) - u(z)$ is harmonic on the simply connected region G and continuous on $\bar{G}$. The maximum principle implies that $U - u$ attains both its maximum and minimum on ∂G. But $U - u = 0$ on ∂G, so u is unique.

9. $u_r(e^{i\theta}) = 0 = u(e^{i\theta})$ and $u_r(re^{i\theta}) = a \geqslant u(re^{i\theta})$ by the maximum

principle. Hence, by exercise 8, for $r < k < 1$

$$u(ke^{i\theta}) \leqslant u_r(ke^{i\theta}) = a \frac{\log k}{\log r} \to 0$$

as $r \to 0+$. Since k can be made arbitrarily small, u is not continuous at 0.

11. Use Cauchy's integral formula for $f(z)$ and the identity

$$\frac{1}{2\pi} \int_0^{2\pi} \frac{f(\zeta) \bar{z}}{\bar{z} - \bar{\zeta}} \, d\phi = 0.$$

13. Add and subtract

$$u(0) = \frac{1}{2\pi i} \int_0^{2\pi} u(\zeta) \frac{d\zeta}{\zeta}$$

   and apply the techniques in Poisson's theorem.

15. $u(z) = \dfrac{y}{\pi} \displaystyle\int_{-1}^1 \dfrac{d(t - x)}{(t - x)^2 + y^2} =$

   $$\frac{1}{-\pi} \arg\left(\frac{z - 1}{z + 1}\right) = \frac{1}{\pi} \arg\left(\frac{z - 1}{z + 1}\right),$$

   for $0 \leqslant \arg z \leqslant \pi$.

SECTION 6.3

1. $u(z) = \dfrac{1}{2} \left[(u_0 + u_1) + (u_0 - u_1) \cdot \dfrac{2}{\pi} \text{Arg}\left(\dfrac{i - z}{i + z}\right)\right]$

3. $w(z) = \dfrac{Q}{2\pi} \log \dfrac{z - i}{z} \cdot$ Family of

   circles through i and 0.

5. Sources at $\pm i$, sinks at 0, ∞, all of strength Q, and stagnation points at ± 1 and ∞. Equipotential lines are given by $|z + 1/z| = $ constant, streamlines by $\arg(z + 1/z) = $ constant.

7. Sinks of strength 2π at ± 1, $\pm i$, and sources of strength 4π at 0 and ∞. $\pm(1 \pm i)$ and ∞ are stagnation points. Equipotential lines satisfy

$$\left| z^2 - \frac{1}{z^2} \right| = \text{constant while}$$

streamlines are given by
$\arg(z^2 - 1/z^2) = \text{constant}.$

9. Let $w = \dfrac{-Q}{2\pi n} \log \dfrac{z^n + r}{z^n} =$

$$\frac{-P}{2\pi n} \log\left(1 + \frac{r}{z^n}\right)^{1/r} \rightarrow \frac{-p}{2\pi n} \frac{1}{z^n}.$$

SECTION 6.4

1. $c_0 = \pi$, $c_n = i/n$ so that $u(z) =$

$$\pi + 2\text{Re}\left(i \sum_{n=1}^{\infty} \frac{z^n}{n} \right) =$$

$$\pi + 2\text{Re}\left[-i\,\text{Log}(1 - z) \right] =$$

$$\pi + 2\,\text{Arg}\,(1 - z).$$

3. $u(re^{i\theta}) = \dfrac{\pi}{2} +$

$$2 \sum_{n=0}^{\infty} \frac{r^{2n+1}}{2n+1} \sin(2n+1)\theta$$

11.

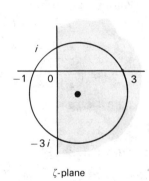

ζ-plane

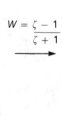

$W = \dfrac{\zeta - 1}{\zeta + 1}$

W-plane

$w = W^2$

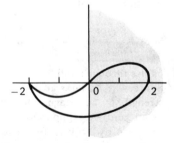

z-plane

$w = \dfrac{z - 2}{z + 2}$

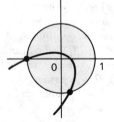

w-plane

5. Let $f(z)$ be entire and bounded by M. By Parseval's identity

$$2\pi \sum_{n=0}^{\infty} r^{2n} |c_n|^2 =$$

$$\int_0^{2\pi} |f(re^{i\phi})|^2 \, d\phi \leq 2\pi M^2,$$

hence $|c_n| \leq M/r^n \to 0$ as $r \to \infty$. Thus $c_n = 0$ for $n > 0$ and $f(z) = \sum_n c_n z^n = c_0$, a constant.

7. Let $f(re^{i\phi}) = \phi^2 - 2\pi\phi$ in Parseval's identity obtaining $c_0 = -2\pi^2/3$, $c_n = 2/n^2$ for $n > 0$

9. $c_0 = c_1 = \ldots = c_{n-1} = 1$, other $c_n = 0$ so Parseval's identity yields

$$2\pi \sum_0^{n-1} r^{2k} = \int_0^{2\pi} \left| \frac{re^{ni\phi} - 1}{re^{i\phi} - 1} \right|^2 d\phi.$$

Set $r = 1$ to get

$$2\pi n = \int_0^{2\pi} \left| \frac{e^{ni\phi/2} - e^{-ni\phi/2}}{e^{i\phi/2} - e^{-i\phi/2}} \cdot \right.$$

$$\left. \frac{e^{ni\phi/2}}{e^{i\phi/2}} \right|^2 d\phi = \int_0^{2\pi} \left(\frac{\sin n\phi/2}{\sin \phi/2} \right)^2 d\phi.$$

11. Let $N = N_1 \cdot N_2 \cdot \ldots \cdot N_j$, $k = k_1 N_2 \cdot \ldots \cdot N_j + k_2 N_3 \cdot \ldots \cdot N_j + \ldots + k_j$, and $n = n_j N_{j-1} \cdot \ldots \cdot N_1 + \ldots + n_1$, then recurrently define

$$c^{(l)} (n_1, n_2, \ldots, n_l, k_{l+1}, \ldots, k_j) =$$

$$\prod_{m=1}^{l} e^{k_{l+1} N_{l+1} \cdots N_j n_m N_{m-1} \cdots N_1}.$$

$$\sum_{k_l=0}^{N_{l-1}} c^{(l-1)} (n_1, \ldots, n_{l-1}, k_l, \ldots,$$

$$k_j) e^{n_l k_l}$$

SECTION 6.5

1. $(\pi/2)^{1/2} e^{-b|t|}$

3. $(\pi/2)^{1/2} (1 - b \, |t|) \, e^{-b|t|}/2b$

5. $e^{-t^2/4k}/\sqrt{2k}$

7. $-i(\pi/2)^{1/2} \tanh (\pi t/2)$

9. $u(t) = (1 - i \operatorname{sign} t)/2 \, |t|^{1/2}$. Set $t = s^2$ and compare with exercise 15 of section 2.2.

11. $$\frac{1}{\sqrt{2\pi}} \int_{-\infty}^{\infty} e^{-ix\phi} \, U(\phi) \overline{V(\phi)} \, d\phi =$$

$$\frac{1}{\sqrt{2\pi}} \int_{-\infty}^{\infty} e^{-ix\phi} \, \overline{V(\phi)} \cdot$$

$$\left\{ \frac{1}{\sqrt{2\pi}} \int_{-\infty}^{\infty} u(t) \, e^{it\phi} \, dt \right\} d\phi =$$

$$\frac{1}{\sqrt{2\pi}} \int_{-\infty}^{\infty} u(t) \left\{ \frac{1}{\sqrt{2\pi}} \int_{-\infty}^{\infty} \overline{V(\phi)} \right.$$

$$\left. e^{i(t-x)\phi} \, d\phi \right\} dt$$

$$= \frac{1}{\sqrt{2\pi}} \int_{-\infty}^{\infty} u(t) \, \overline{v(t - x)} \, dt.$$

Then set $x = 0$. For second part let $V = U$.

SECTION 6.6

1. $$\int_0^{\infty} \cosh z\phi \, e^{-s\phi} \, d\phi =$$

$$\frac{1}{2} \int_0^{\infty} [e^{-(s-z)\phi} + e^{-(s+z)\phi}] \, d\phi =$$

$$\frac{1}{2} \left[\frac{1}{s - z} + \frac{1}{s + z} \right] \text{ with } \operatorname{Re} s >$$

$\operatorname{Re} z, -\operatorname{Re} z$

3. $\dfrac{-d}{ds} \mathcal{L} \{\sin z\phi\} = \dfrac{-d}{ds} (z(s^2 + z^2)^{-1})$

by equation (3).

5. $\dfrac{-d}{ds} \mathcal{L} \{\sinh z\phi\} =$

$\dfrac{-d}{ds} (z(s^2 - z^2)^{-1})$

7. By first shifting theorem
$$\mathcal{L}\{e^{-w\phi}\sin z\phi\}(s) = \mathcal{L}\{\sin z\phi\}(s+w)$$
and apply example 5.

9. $\dfrac{d^2}{ds^2}\,\mathcal{L}\{\sin z\phi\} = \mathcal{L}\{\phi^2\sin z\phi\}$ by

equation (3).

11. $\cos 2z\phi = 2\cos^2 z\phi - 1$ so find
$\mathcal{L}\{(1+\cos 2z\phi)/2\}$.

13. By second shifting theorem
$\mathcal{L}\{H(\phi-a)\cos z\phi\} =$
$e^{-as}\mathcal{L}\{\cos z(\phi+a)\}$ and $\cos z(\phi+a)$
$= \cos z\phi\cos za - \sin z\phi\sin za$.

15. $U(\phi) = \sin\phi + (\phi\sin\phi - \phi^2\cos\phi)/4$

17. $\mathcal{L}\{U\} = U(0+)(s^2+1)^{-1/2}$, which
is the transform for $U =$
$U(0+)\,J_0(\phi)$, with J_0 the Bessel
function of order zero.

19. $U(\phi) = |\sin\phi^{-1}|$

21. Suppose $|U|$ is bounded by M on
$(0,\ \Phi)$ and by $Ne^{a\phi}$ for $\phi > \Phi$.
Then

$$|\mathcal{L}\{U\}| \leqslant \int_0^{\Phi} M\,|e^{-s\phi}|\,d\phi +$$

$$\int_\Phi^\infty N\,|e^{-(s-a)\phi}|\,d\phi =$$

$$M\left[\frac{1-e^{-(\mathrm{Re}\,s)\Phi}}{\mathrm{Re}\,s}\right] + N\,\frac{e^{-(\mathrm{Re}\,s-a)\Phi}}{\mathrm{Re}\,s-a}$$

$\to 0$, as $s\to\infty$

23.

$$\lim_{s\to 0+} s\mathcal{L}\{U\} = \lim_{s\to 0+}\left\{-U(\phi)e^{-s\phi}\Big|_0^\infty + \right.$$

$$\int_0^\infty U'(\phi)e^{-s\phi}\,d\phi = U(\phi) +$$

$$\int_0^\infty U'(\phi)\left\{\lim_{s\to 0+}e^{-s\phi}\right\}d\phi = \lim_{\phi\to\infty}U(\phi)$$

25. Since integrals are linear operators
we get the result.

27. $(U*V)*W =$

$$\int_0^x\left(\int_0^\phi U(t)V(\phi-t)\,dt\right)W(x-\phi)\,d\phi =$$

$$\int_0^x\left(\int_t^x V(\phi-t)W(x-\phi)\,d\phi\right)U(t)\,dt$$

$$= \int_0^x U(t)\left(\int_0^{x-t}V(s)W(x-t-s)\,ds\right)dt$$

$$= U*(V*W) \text{ by setting } s = \phi - t.$$

SECTION 6.7

1. $\phi^2\,e^{-a\phi}/2$

3. $\phi\sin a\phi/(2a)$

5. $(1-\cos a\phi)/a^2$

7. $\left[e^{-a\phi} - e^{a\phi/2}\left(\cos\dfrac{\sqrt{3}a\phi}{2} - \sqrt{3}\sin\dfrac{\sqrt{3}a\phi}{2}\right)\right]/3a^2$

9. $\sin a(\phi-b)\,H(\phi-b)/a$

11. $\phi^{-2/3}/\Gamma(1/3)$

13. $(e^{b\phi} - e^{a\phi})/\phi$

15. $(1-\cos 2a\phi)/2\phi$

17. $e^{-a^2/4\phi}/\sqrt{\pi\phi}$

19. $U(\phi) = \phi^2 + \phi^4/12$

21. $\{[a + c(a^2+1)]\cos\phi + \sin\phi - ae^{-a\phi}\}/(a^2+1)$

23. $\frac{1}{2}(e^\phi + e^{-\phi}) = \cosh\phi$

25. Treating x as a parameter yields

$$\frac{\partial^2 u}{\partial^2 x} - \frac{s^2}{a^2}u = -\frac{s}{a^2}f(x)$$

Using the variation of parameters
method of differential equations
we obtain

$$u(x,\ s) = c_1 e^{sx/a} + c_2 e^{-sx/a} - \frac{1}{a}\int_0^x f(y)\sinh\frac{s(x-y)}{a}\,dy$$

with boundary conditions
$u(0,s) = 0$,

$\lim\limits_{x \to \infty} u(x, s) = 0.$

Hence, $c_1 = c_2 = 0$, so that

$U(x, t) =$

$\mathcal{L}^{-1}\left\{-\dfrac{1}{a}\int_0^x f(y) \sinh \dfrac{s(x - y)}{a}\, dy\right\}.$

27. $\dfrac{\partial^2 u}{\partial x^2} + \dfrac{\mu}{\delta}\dfrac{\partial u}{\partial x} - \dfrac{s}{\delta} u = 0$ has the solution.

$u = c_1\, e^{(-\mu + \sqrt{\mu^2 + 4s\delta}\, x)/2\delta} + c_2\, e^{(-\mu - \sqrt{\mu^2 + 4s\delta}\, x)/2\delta}$

with the boundary conditions

$u(0, s) = \dfrac{c}{s},$

$\lim\limits_{x \to \infty} u(x, s) = 0.$

Hence, $c_1 = 0$, $c_2 = -c/s$ and

$u = \dfrac{ce^{-\mu x/2\delta}}{s}\, e^{(-x/\sqrt{\delta})\sqrt{s + \mu^2/4\delta}}$

so that by the convolution theorem and the first shifting theorem

$u =$

$ce^{-\mu x/2\delta}\, \mathcal{L}\{1\}\mathcal{L}\left\{\dfrac{x}{2\sqrt{\pi^3\delta}}\cdot e^{-(\mu^2 t + x^2/t)/4\delta}\right\}$

Thus,

$u(x, t) = \dfrac{cxe^{-\mu x/2\delta}}{2\sqrt{\pi\delta}}$

$\displaystyle\int_0^t t^{-3/2}\, e^{-(\mu^2 t + x^2/t)/4\delta}\, dt$

APPENDIX A.3

1. $2\sqrt{2}$

3. $\displaystyle\int_0^1 y\sqrt{1 + \tfrac{9}{4}y}\, dy =$

$\dfrac{4}{81}\left[\dfrac{(13)^{5/2} - 4}{20} - \dfrac{(13)^{3/2} - 1}{3}\right]$

5. $\dfrac{-4}{3}$

7. 148.75

9. 0

11. By Green's theorem

$\tfrac{1}{2}\displaystyle\int_\gamma x\, dy - y\, dx =$

$\tfrac{1}{2}\displaystyle\iint_G (1 - (-1))\, dx\, dy = A$

13. By Green's theorem

$\displaystyle\int_\gamma p\, dy + q\, dx =$

$\displaystyle\iint_G (p_x - q_y)\, dx\, dy = 0.$

15. $\displaystyle\int_\gamma xy\, dy = \displaystyle\iint_G y\, dx\, dy =$

$-\tfrac{1}{2}\displaystyle\int_\gamma y^2\, dx = A\bar{y}$

$\displaystyle\int_\gamma xy\, dx = -\displaystyle\iint_G x\, dx\, dy =$

$-\tfrac{1}{2}\displaystyle\int_\gamma x^2\, dy = -A\bar{x}$

INDEX

Page numbers indicate where entry is defined.